63 Topics in Current Chemistry
Fortschritte der chemischen Forschung

Bonding and Structure

Springer-Verlag
Berlin Heidelberg New York 1976

This series presents critical reviews of the present position and future trends in modern chemical research. It is addressed to all research and industrial chemists who wish to keep abreast of advances in their subject.

As a rule, contributions are specially commissioned. The editors and publishers will, however, always be pleased to receive suggestions and supplementary information. Papers are accepted for "Topics in Current Chemistry" in either German or English.

ISBN 3-540-07605-0 Springer-Verlag Berlin Heidelberg New York
ISBN 0-387-07605-0 Springer-Verlag New York Heidelberg Berlin

Library of Congress Cataloging in Publication Data. Main entry under title: Bonding and structure. (Topics in current chemistry; 63). Bibliography: p.Includes index. CONTENTS: Craig, D. P. and Mellor, D. P. Discriminating interactions between chiral molecules.– Gleiter, R. and Gygax, R. No-bond-resonance compounds, structure, bonding, and properties.–Sutter, D. H. and Flygare, W. H. The molecular Zeeman effect. 1. Chemical bonds–Adresses, essays, lectures. 2. Molecular theory–Adresses, essays, lectures. 3. Zeeman effect–Adresses, essays, lectures. I. Craig, David Parker, 1919–Discriminating interactions between chiral molecules. 1976. II. Gleiter, Rolf, 1936–No-bond-resonance compounds, structure, bonding, and properties. 1976. III. Sutter, Dieter Hermann, 1934–The molecular Zeeman effect. 1976 IV. Series. QD1.F58 vol. 63 [QD461] 540'.8s [541'.224] 76-823

Typesetting and printing: Hans Meister KG, Kassel. Bookbinding: Konrad Triltsch, Graphischer Betrieb, Würzburg

Contents

Springer-Verlag	D-6900 Heidelberg 1 · Postfach 105280 Telephone (06221) 487-1 · Telex 04-61723 D-1000 Berlin 33 · Heidelberger Platz 3 Telephone (030) 822001 · Telex 01-83319
Springer-Verlag New York Inc.	New York, NY 10010 · 175, Fifth Avenue Telephone 673-2660

Discriminating Interactions Between Chiral Molecules

David P. Craig

Research School of Chemistry, Australian National University, Box 4, P.O., Canberra, A.C.T. 2600, Australia

David P. Mellor

Department of Chemistry, University of New South Wales, Box 1, P.O., Kensington, N.S.W. 2033, Australia

Contents

I. The Experimental Background

1. Scope of the Article

The fact that molecules now usually referred to as *chiral* were in the past mainly called *optically active,* is a reminder that the optical properties of chiral molecules were practically the only properties that were taken to be characteristic, and certainly the only ones that were at all easily measurable. Optical rotation, which is the longest known of the optical properties depends on the difference in refractive index for left and right-handed circularly polarised light. Once discovered, it proved an easy quantity to measure because it depends on the refractive index difference and does not have to be found by subtracting one very large quantity from another slightly different from it. Other optical measurements, such as *optical rotatory dispersion* (ORD), *circular dichroism* (CD) and even the second-order properties of *induced circular dichroism* (ICD) and *magnetic circular dichroism* (MCD) all have this same feature. The possibilities with isolated molecules probed by fields applied externally are limited. On the other hand, because a chiral molecule is the source of an external field which is inherently chiral, it must be coupled to a second chiral molecule with a strength that depends on the relative handedness of the pair. The field of a chiral molecule is here conceived in very general terms, including the virtual field associated with dispersive interactions and that for short range contacts as well as the more familiar electric field of permanent molecular charge distributions. The possibilities of investigation of *chiral discrimination* are at once much wider when such pairs or clusters are considered instead of isolated molecules. It is thus the condition that the probing field as well as the molecule probed should be chiral that gives a special importance to intermolecular interactions in such studies

Perhaps the most striking applications are in the interactions of biological molecules, where 'biological recognition' of one biomolecule by another, or set of others, is highly specific as to chirality as well as to composition. Discrimination here appears in extreme form. Examples are discussed in I.12, after a review of the results in more conventional systems.

A molecule is chiral or 'handed' if it is not superposable on its mirror image. The general criterion for chirality is that a molecule must not possess an improper axis of rotation In particular it must not possess either a centre of inversion (improper rotation axis with zero angle) or a plane of symmetry (improper rotation by π).

The simplest case of chiral molecules are those of the substituted methanes in which the lack of the mirror plane implies lack of all symmetry. Such molecules are chiral and at the same time asymmetric. Among inorganic molecules there are many examples of octahedral systems which are chiral but do not lack all elements of symmetry. For example the tris bidentate chelate system illustrated in Fig. 1 has one three-fold axis through a triangular face of the octahedron and belongs to the symmetry group $\boldsymbol{D}_3$. It has no improper axis of rotation and is chiral. In such a case symmetry operations of the group transform one enantiomer into itself but never into the other; both belong formally to the same covering symmetry.

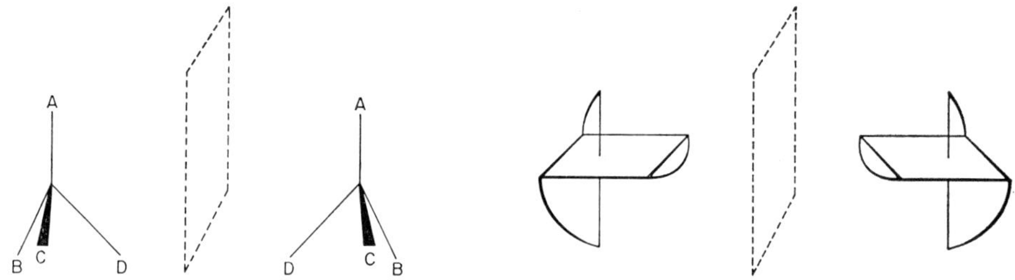

Fig. 1. Schematic diagrams of enantiomeric forms for an asymmetric tetrahedral system lefthand diagram) and for a D_3 octahedral system (right-hand diagram)

Discriminatory interactions of chiral molecules, which are the subject of this review, manifest themselves under many different circumstances; among assemblies of molecules of different chirality such interactions are general. The first section of this review will be devoted mainly to a survey of the phenomena arising from chiral discrimination; the second to a detailed theoretical treatment of the origin and nature of the discriminating interactions.

2. Discriminatory Interactions of Chiral Molecules

(*Chirodiastaltic Interactions*). That there exists a difference in the interaction between enantiomeric molecules (*d* and *l*)[a] and a second chiral molecule which may be

(a) the same species (*d* or *l*) or

(b) another species D or L

has been known for a long time, and certainly since Pasteur's discovery in 1858 of the discriminatory attack by *penicillium* on ammonium tartrate (Section I.12). It is convenient to have a term to describe the part of the interaction between two chiral molecules which discriminates between like and unlike pairs. We have chosen *chirodiastaltic* (diastaltic = 'serving to distinguish'); an alternative is *diastereotopic*, introduced by Bosnich and Watts (1975).

---

[a] *Note on the use of symbols.* In the earlier literature, it was customary to use the symbol *d* (or *l*) to indicate the direction (*dextro* or *levo*) of rotation of the plane of polarisation of plane polarised light by the molecule before the name of which it was placed. In this review, the symbols are used in a general way, as indicated above, to represent the dextro and levo-rotatory molecules themselves; following current practice, the directions of rotation for the sodium lines are shown by the signs (+) and (−) placed before the name of the compound to which they refer. The use of the capital letters D and L to represent a second dextro and levo rotatory chiral molecular species should not be confused with their use to indicate absolute configurations of sugars and amino acids where for example a sugar or an amino acid with a D configuration may be levorotatory. When the species under discussion are ions, this is shown in the usual way d^+, D^+, l^-, L^-. The line joining any two symbols indicates the existence of an interaction of whatever kind between the two chiral molecules; with this device no attempt is made to differentiate between the various possible kinds of interaction forces involved.

We have suggested the former because of its explicit reference to chirality.

The possible chirodiastaltic interactions classified under (a) and (b) may be represented schematically as follows:

(a) The difference between the interactions

d-d and *d-l*	or	*l-l* and *d-l*	where *d-l* is a racemate

(b) The difference between the interactions

d-D and *d*-L	*l*-D and *l*-L	(diastereoisomeric pairs)

In this article a number of manifestations of these differential energy terms will be described. In the theoretical sections several mechanisms will be analysed through which discrimination may occur. The theory has not yet been developed far enough to permit adequate calculations of magnitudes in individual cases, but it already gives some insight, and may suggest experimental approaches helpful in isolating the various theoretically possible modes of chiral discrimination.

Phenomena Dependent on Chirodiastaltic Interactions

3. Melting Points

In two crystals, one formed exclusively from *d* (or *l*) and the other from *d* and *l* species equally the molecules adopt different modes of packing. They interact differently and there are consequent differences in lattice energies which are reflected in the melting points of the two crystals. Differences have been observed between the melting points of

(a) *d-d* and *d-l* species

(b) the species

d-D and *d*-L	and	*l*-D and *l*-L	diastereoisomeric[b] pairs

---

b) It is sometimes convenient to distinguish between two types of diastereoisomers (1) those in which the chiral moieties are linked by covalent bonds as in (b) above; (2) those in which the chiral moieties are charged species held together by ionic bonds.

Table 1. Melting points of crystals of active and racemic compounds[1])

Type (a)		Type (b)	
Substance	M.P. (°C)		M.P.
(+) Tartaric acid	170	(−) Menthyl(+)mandelate	97.2
(±) Tartaric acid (anhydrous)	204—6	(−) Menthyl(−)mandelate	77.6
(+) Usnic acid	203		
(±) Usnic acid	193		
(+) Camphoric acid	187		
(±) Camphoric acid	202		
(+) Lupanine	44		
(±) Lupanine	99		

[1]) From Handbook of Chemistry and Physics, 44th ed. 1963, Chemical Rubber Publishing Co. Cleveland, Ohio.

Examples of (a) are quite common and of (b) much less common. A few are shown in Table 1. It should be observed that there are numerous *d-d* and *d-l* pairs reported as melting as the *same* temperature, as for example the diethyl esters of *d-* and *dl-*tartaric acid (M.P. 17 °C). Where the chiral centre is sequestered within the molecule, and has little or no influence on the packing shape, differences of packing energy may well be too small to be measured except under the most refined conditions (see also Section I.12). The effects of chemical contamination must in any case put in doubt the interpretation of small differences in the search for evidence of discrimination.

4. Discrimination at the Solid/Vapour Interface

A mixture of *dextro* and *levo* crystals of the same substance may react selectively with a chiral vapour. Lin, Curtin and Paul (1974) report that crystals of the (+) and (—) forms of 2,2-diphenyl-cyclopropane-l-carboxylic acid placed on a microscope slide near a few drops of L-phenylalanine and exposed to its vapour, undergo a solid-gas reaction. The top surface of the (—) acid crystal became opaque after 5 min; the (+) acid crystal was essentially unchanged. A similar experiment with D-phenylalanine showed selectivity toward the (+)-acid crystal.

5. Phenomena at the Solid/Liquid Interface

Chirodiastaltic interactions occurring at solid/liquid interfaces involve differences in the lattice energies of the solids and other factors including differences in solvation energy of the chiral molecules in the presence of other dissolved chiral species and, in electrolytes, differences in interionic forces. Interfaces with soluble solids will be considered here and those with insoluble solids in Section I.6.

Some or all of the differences referred to give rise to differences in solubility of related chiral species as listed in Table 2.

Table 2. Solubilities in water of pairs of *d* (or *l*) and *d-l* species

Organic[1)]		Complex ionic[2)]	
Substance	Solubility (g/100g; °C)	Substance	Solubility (10^{-5}M; 25°C)
(+) Glutamic acid	0.89^{25}	+ $[Ru(phen)_3](ClO_4)_2$	169
(±) Glutamic acid	2.64^{25}	± $[Ru(phen)_3](ClO_4)_2$	65.3
(−) Aspartic acid	2.71^{75}	+ $[Ru(bipy)_3](ClO_4)_2$	425
(±) Aspartic acid	4.75^{75}	± $[Ru(bipy)_3](ClO_4)_2$	187
(−) Isoleucine	6.08^{75}		
(±) Isoleucine	4.83^{75}		
(+) Tartaric acid	343[3)] ^{100}		
(±) Tartaric acid	185[3)] ^{100}		

1) Data from source quoted for Table 1.
2) Mizumachi (1973).
3) g/100 ml.

A second and historically the most significant example under this heading is the solubility difference between diastereoisomeric pairs, *e.g.*,

(a)		(b)		(c)
d-D		*l*-D		d^+-D^-
and	or	and	or	and
d-L		*l*-L		d^+-L^-

The difference in the solubility of pairs of diastereoisomers, discovered by Pasteur, is important for two reasons. It was the first chirodiastaltic interaction to be discovered. Secondly Pasteur used it as the basis of a method for resolving (separating) racemates into their enantiomers and ever since it has been the technique most widely used for this purpose. Pasteur treated racemic acid ((±) tartaric acid) with the equivalent quantity of (—) cinchonine so forming a mixture of

(—) cinchonine (+) tartrate

(—) cinchonine (—) tartrate

The diastereoisomers are not mirror images of one another and differ in solubility sufficiently to enable this separation. There appear to be few, if any, quantitative data on the solubilities of diastereoisomeric pairs.

Heats of solution also provide useful evidence. From the compilation by Greenstein and Winitz (1961) we quote $\Delta H_{DL} - \Delta H_L = 370$ J mol^{-1} for the difference in heats of solution of the DL and L forms of alanine. The value for leucine is 1240 J mol^{-1}.

The implications of such facts were not followed up in any great detail and theoretical understanding of intermolecular interactions was not ample enough at the time to allow much to be done. The subject was enlarged by the work of F.P. Dwyer and collaborators beginning in the early 1950's. Two new aspects were the use of conditions in which explanations in terms of diastereoisomers were much less plausible, and the study of optically stable as well as labile systems.

The new work was based on the difference in solubility of (+) and (—) forms of metal complexes in the presence in solution of a second chiral (D or L) ionic species. There appear to be no similar data on molecular species. For ions, the only case so far studied, we have:

$$\left.\begin{matrix} d^{+}\text{-L}^{-} \\ \text{and} \\ l^{+}\text{-L}^{-} \end{matrix}\right\} \qquad \left.\begin{matrix} d^{+}\text{-D}^{+} \\ \text{and} \\ l^{+}\text{-D}^{+} \end{matrix}\right\}$$

In their study Dwyer, Gyarfas and O'Dwyer (1951, 1956) measured the solubilities of the perchlorates of optically active *tris*-o-phenanthroline ruthenium(II) in solutions of optically active substances finding, for example, that in aqueous 1% (+)bromocamphorsulphonate the solubilities of the (+) and (—) complexes are 0.232 and 0.235 grams per 100 ml solution. In 2% potassium *d*-tartrate the solubilities are 0.215 and 0.220 grams per 100 ml. The enantiomers (+) and (—) $[Ru(phen)_3]\ (ClO_4)_2$ are equally soluble in water. It follows that the activity products are equal and hence $a_{d+} = a_{l+}$. Addition of sodium chloride affects the solubility of the enantiomers to the same extent. This is not so in the presence of a second chiral anionic species. From the existence of solubility differences Dwyer and collaborators concluded that, for activity a and activity coefficient (γ),

$$a_{d+} \neq a_{l+} \text{ and } \gamma_{d+}\,\mathrm{C} \neq \gamma_{l+}\,\mathrm{C}$$

The activity coefficients of a chiral ion in the presence of the chiral ion of a second species thus contain a factor dependent on configuration of the ion. This was called "configurational activity"; it is not altogether a happy term because of the dual use of the word activity (optical activity and activity coefficient). Nevertheless the concept is an important one and must be taken into account in the solution chemistry of chiral molecules and ions. It is worth noting that the effect of the presence of a second optically active cationic species on the solubility of $[Ru(phen)_3](ClO_4)_2$ has not been investigated. In view of the effects produced by chiral ions of the same charge in other phenomena (racemisation and oxidation-reduction potentials) it would probably be worth while studying their effects on solubility.

A closely related phenomenon is the difference in solubility of d and l species in a chiral solvent (D or L). Again the only case that appears to have been studied is that involving ionic species (d^{+} and l^{+}) [Mizumachi (1973), Bosnich and Watts (1975)].

Thus we have:

$$\left.\begin{array}{c} d^{+}\text{-L} \\ \text{and} \\ l^{+}\text{-L} \\ \text{and} \\ d^{+}l^{+}\text{-L} \end{array}\right\}$$

Typical results are given in Table 3.

Table 3. Solubilities in (—) 2-methyl-l-butanol[1]) and (—) 2,3 butanediol[2])

In (—) 2-methylbutanol		In (—) 2,3 butanediol	
Complex salt	Solubility (10^{-5}M; 25 °C)	Complex salt	Solubility (mol l^{-1}; 30 °C)
(+) $[Ru(phen)_3](ClO_4)_2$	4.00	(+)-cis-$[Co(en)_2Cl_2]ClO_4$	2.6×10^{-3}
(—) $[Ru(phen)_3](ClO_4)_2$	4.70	(—)-cis-$[Co(en)_2Cl_2]ClO_4$	1.25×10^{-3}
(±) $[Ru(phen)_3](ClO_4)_2$	0.584	(±)-cis-$[Co(en)_2Cl_2]ClO_4$	0.6×10^{-3}
(+) $[Ru(bipy)_3](ClO_4)_2$	2.03		
(—) $[Ru(bipy)_3](ClO_4)_2$	1.85		
(±) $[Ru(bipy)_3](ClO_4)_2$	0.164		

1) Mizumachi (1973).
2) Bosnich and Watts (1968).

It was, and still is, possible to assign the whole of the configurational activity to the influence of diastereoisomers in solution, and therefore to assign it to inter-ionic contacts, but it now appeared in conditions where ion pairs were not of dominant importance in accounting for other aspects of solution behaviour (Sections I.8 and I.9).

6. The Solid/Liquid Interface. Adsorption on Insoluble Solids

Two types of solid will be discussed.

(a) a crystal the structural units of which are chiral molecules *e.g.* lactose

(b) a chiral crystal the structural units of which are achiral *e.g.* quartz. $(SiO_2)_n$.

When a solution of a racemate (*dl*) is brought into contact with a crystal built from chiral molecules, D, *e.g.* lactose, preferential adsorption takes place; *d* may be more readily adsorbed than *l* or vice-versa again pointing to a difference between *d*-D and *l*-D. A number of racemates have been resolved on columns of lactose. Moeller and Gulyfas (1958) for example, resolved $[Co(aca)_3]^\circ$ and $[Cr(acac)_3]^\circ$ on lactose hydrate by passing a solution of the complexes in benzene-petroleum through the column. [See also Henderson and Rule (1939), Lecoq (1943), Prelog and Wieland (1944)].

Partial resolution of racemates has been achieved by the use of finely powdered quartz (derived from crystals of identical chirality). One method is to shake an aqueous solution of the racemate with the powder and then filter the solution before measuring the rotational change. Columns of powdered quartz have also been employed. It would seem that lactose is the more effective of the two in bringing about resolutions: for further details see Dwyer and Mellor (1964).

7. Diffusion

The possibilities include diffusion of *d*, *l* or *dl* molecules into either a chiral solvent or the solution of a second chiral molecule. The only case so far studied appears to be one involving ions (d^+ and l^+) and an uncharged chiral molecule. Carassiti (1958) has observed different rates of diffusion of (+) $[Co(en)_3]^{3+}$ and (—) $[Co(en)_3]^{3+}$ in sucrose solution, reflecting different degrees of association of the complex cations with the molecules of sucrose. Studies of the diffusion of these and similar complex ions in chiral solvents like (—)-2,3-butanediol and (—)-2-methyl-butanol would no doubt reveal similar differences. Studies of the diffusion of coloured ions like (+) and (—) $[Co(en)_3]^{3+}$ in a solution of a colourless chiral cation seem potentially useful lines of enquiry.

8. Effects in Oxidation/Reduction Systems

The redox potentials of oxidation-reduction systems involving enantiomeric complex cations are influenced by the presence in solution of a second species of chiral ion of the same or opposite charge. An example involving optically stable species [Barnes, Backhouse, Dwyer and Gyarfas (1956)] is the redox potentials of the systems:

$$(+)\text{-}[Os(dipy)_3]^{2+} / (+)\text{-}[Os(dipy)_3]^{3+}$$

and

$$(-)\text{-}[Os(dipy)_3]^{2+} / (-)\text{-}[Os(dipy)_3]^{3+}$$

which in water and solutions of sodium chloride are identical within the limits of experimental error of ± 0.2 mv. In solutions of ammonium (+) bromocamphorsulphonate (D^-) at ionic strength 0.001 the differences in potential are 1.2 and 2.5 mv respectively. The ions d^{2+}, d^{3+}, l^{2+} and l^{3+} have different activity coefficients in the presence of D^-. At first one is inclined to attribute this to the influence of the attractive interionic forces leading to diastereoisomeric ion pairs. However similar differences of potential have been produced by using as the second species the cations (+) and (—) $[Co(en)_3]^{3+}$. This suggests the influence of long range forces. The work of Pfeiffer and Quehl (1931, 1932) [see also Schipper (1974)] had already shown that chiral cations as well as anions could cause asymmetric transformation in trischelated complexes of Zn^{2+} and Cd^{2+}. The cations were those of strychnine and cinchonine; they are large, with positive charge widely spread, and their cationic character is of less significance than that of the smaller and triply charged Co^{III} cations earlier referred to.

9. Racemisation and 'Enantiomerization'

The rates of racemisation of d^+ and l^+ ions differ considerably when racemisation takes place in the presence of a second chiral species D^- (or D^+). This was first observed by Rây and Dutt (1941, 1943) who studied (+) and (—)-tris biguanidinium cobalt^{III} chloride in the presence of (+) tartrate ion. Later Dwyer and Davies (1954) found that the rates of racemisation of (+) and (—) $[Ni(o\text{-}phen)_3]Cl_2$ differed in the presence of (+) bromocamphorsulphonate ion (D^-) and in that of the cationic species (+) cinchoninium ions (D^+).

The racemisations terminated in equilibria in which one component was in excess over the other with equilibrium constants different for the two chiral media. From the temperature dependence of the equilibrium constants they found the differences of heat contents ($\Delta H = 1.6$ and 1.8 kJ mol^{-1}) and Gibbs standard free energies ($\Delta G° = 0.30$ and 0.34 kJ mol^{-1}) respectively for bromocamphorsulphonate and cinchomium ions as the added chiral species.

The same equilibria are attained if to a solution of racemate in an initially achiral medium (equal concentrations of *d* and *l* species) there is added another chiral species D or L. The equilibrium is then displaced in favour of one or other of the constituents of the racemic mixture. This process has recently been termed enantiomerization, although examples of optically labile systems in equilibria sensitive to the presence of other chiral molecules or ions have long been recognized. A typical example of what was earlier termed an asymmetric transformation of the *first kind* (no second-phase involved) is that of Read and McMath (1925) in which solutions in dry acetone of (—) or (±) chlorobromomethanesulphonic acid (d^-l^-) together with (—)-hydroxyhydrindamine (L^+) showed a change of optical rotation interpreted in terms of an equilibrium

$$\mathrm{L}^+l^- \rightleftharpoons \mathrm{L}^+d^-$$

strongly favouring the left-hand side.

In this as in other examples analysed by Jamison and Turner (1942) an essential condition appeared to be the existence of ion pairs in solution, *i.e.* of diastereoisomers as close-coupled entities. Salts seemed not to show this displacement of optical equilibrium under conditions favouring ionic dissociation. Subsequently the force of this distinction has largely been lost, although the effect is generally much smaller in the latter cases. We shall return in later sections to the problem of distinguishing the importance of very short-range interactions, as in ion pairs, from that of influences propagated at longer range, beyond that of van der Waals contacts. In another set of findings, on asymmetric transformations of the second kind, the equilibrated product separated as a solid diastereoisomeric salt, and the effect could then be seen as due to lattice forces, as in Pasteur's separations, and again as a property of short range contacts.

Later work on enantiomerization by Bosnich and Watts (1975) depends upon equilibrium constant measurements of $[Ni(phen)_3]Cl_2$ in (—)-2,3-butanediol in the temperature range 277—373 K. The equilibrium constants as a function of temperature give standard enthalpies and entropies $\Delta H° = —495$ J mol^{-1} and $\Delta S° = —1.17$ J K^{-1} mol^{-1}, with the (—) form of the cation the more stable in the (—) solvent. In kinetic terms the (+) ion inverts faster than the (—) ion. In earlier work [Bosnich and Watts (1968)], (—) 2,3-butanediol had been show to be remarkably effective as an enantiomerizing solvent towards $[Co(en)_2Cl_2]^+$, giving a free energy difference $\Delta G° = —3.8$ kJ mol^{-1}. So large a value might be explained as a specific solvent effect, suggested by Bosnich and Watts to be hydrogen bonding to the solvent by the N—H bonds of ethylenediamine. The enthalpy difference of about —500 J mol^{-1} found for the enantiomers of $[Ni(phen)_3]^{2+}$ can be taken as more representative of examples of a general rather than a specific solvent effect.

10. Boiling Points of Active and Racemic Compounds

In principle one would expect differences of boiling point arising from chirodiastaltic interactions. The letter must be far smaller than in solids and the evidence so far as it exists is barely significant when account is taken of the uncertainties in the measurement of the boiling points of compounds difficult to purify in both the chemical and chiral senses. Moreover the enhanced rate of racemisation at the boiling point must be reckoned with. Examples which appear to show chirodiastaltic effects are given in Table 4. They have been assembled by Dr. E. V. Lassak from Guenther's (1949) data. A much more thorough investigation of boiling points is needed before any decision can be reached about the existence or non-existence of differences between *d* and *dl* species.

Table 4. Boiling points (°C) of active and racemic compounds

Compound	B.P.	Refractive index[1)]
d terpinene 4-ol	209—12	1.4785^{19}
dl terpinene 4-ol	212—14	1.4803^{20}
l menthone	209—10	1.4481^{25}
dl menthone	206—7	1.4492^{25}
l piperitone	232.5—234.7	1.4845^{20}
dl piperitone	235—237	1.4845^{20}

[1)] At temperature (°C) shown as superscript.

11. Discrimination in Metal Complexes of Chiral Ligands

Two chiral molecules joined together in a complex give a discriminating term to the binding energy distinguishing *d-d* and *d-l* pairs. This is an intramolecular discrimination, in contrast to intermolecular examples hitherto. It is the analogue of joining chiral fragments to form an organic molecule in either active or meso forms, as will be discussed in a particular case in Section V.2.

Bennett (1959) reported that bis (L) asparaginato copper II[c)] (MLL) is more stable than (D) (L) asparaginato copper II (MDL). From a proton magnetic reso-

[c)] As in the original papers absolute configurations are used throughout this and the following section.

nance study of the octahedral bis histidine complexes of cobalt II McDonald and Phillips (1963) found that the MDL form is stabilised by about 1.34 kJ mol^{-1} over the MLL or MDD forms. Ritsma *et al.* (1969) confirmed these differences by potentiometric studies. The stability constants of the cobalt II and nickel II complexes are given in Table 5.

Table 5. Stability constants of metal-histidine complexes. [Ritsma (1969)]

Compound	log k_1	log k_2
CoII DL(hist)	6.865	5.517
CoII LL(hist)	6.864	5.390
NiII DL(hist)	8.645	7.058
NiII LL(hist)	8.656	6.841

The excess Gibbs energy of $\frac{1}{2}$(MDD + MLL) relative to MDL was found to be 1.29 kJ mol^{-1} for the CoII compound and 2.04 kJ mol^{-1} for NiII, the former value being in very good agreement with that of McDonald and Phillips.

More recently, Barnes and Pettit (1970), on the basis of calorimetric measurements of the bis-histidine complexes of ZnII and NiII, have shown that the MDL complexes are more stable than the MDD or MLL. For example:

$$\text{ZnDL(hist)} \quad \Delta H_f^\circ = -49.2 \text{ kJ/mole}$$

$$\text{ZnLL(hist)} \quad \Delta H_f^\circ = -47.7 \text{ kJ/mole}$$

Histidine and asparagine function as tridentate chelates. In the asparagine complexes, the mixed (DL) form is the more stable while the reverse is true of the histidine complexes. Potentiometric studies of the complexes of various metals (NiII, CuII, CoII, ZnII) with a wide range of amino acids all of which function as bidentate chelates have failed to reveal any differences in the stability of MDL and MLL forms. (Ritsma *et al.* 1965; Gillard *et al.* 1966). The reason why the chirodiastaltic interactions should be so much more marked with tridentate than bidentate chelates calls for investigation in terms of the intramolecular force system.

X-ray crystal analysis (Harding and Long 1968, Candlin and Harding 1970) has revealed the structural consequences of chirodiastaltic interactions in the octahedral cobalt II complexes of D and L histidine. In both bis-(L-histidino) cobalt II monohydrate and D-histidino-L-histidino cobalt II dihydrate each histidine is bound to the cobalt atom by an amino nitrogen, an imidazole nitrogen and an oxygen atom. In the LL form the imidazole nitrogen atoms occupy trans positions. If one considers a cobalt atom with one L histidine molecule attached to it, an approaching D histidine has two options as far as the imidazole nitrogens are concerned. It may coordinate in such a way that these nitrogens occupy either *cis* or *trans* positions. Candlin and Harding find that D ligand nitrogen atoms occupy cis positions as shown in the illustration taken from their paper. At first sight it might appear that the DL form is the analogue of mesotartaric acid,

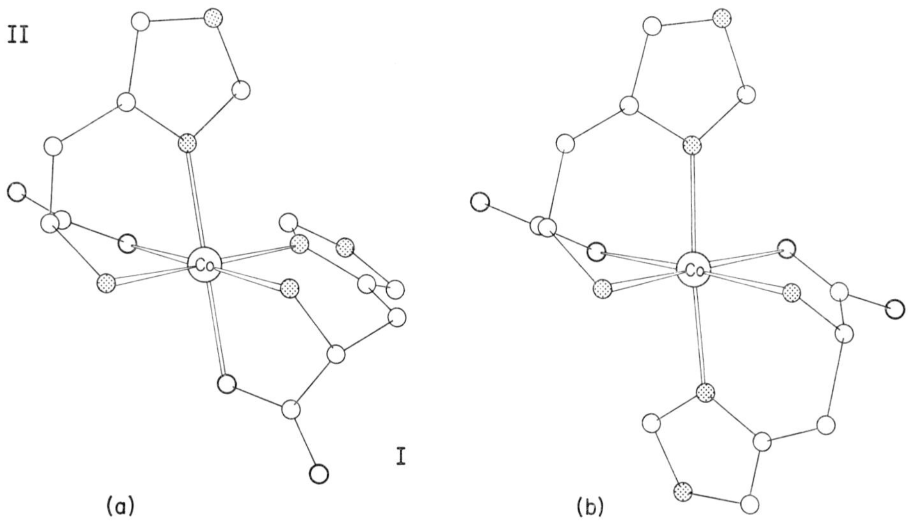

Molecular structures of cobalt II histidine complexes, reproduced with permission from Candlin and Harding (1970). Left-hand: MDL, with molecule I in the D form and molecule II in the L. Right-hand: MLL. Nitrogen atoms are shaded; oxygen atoms are shown heavier

the D ligand being the mirror image of the L, but in fact the ligands are so coordinated with cobalt that the molecule as a whole has no symmetry. It must therefore be optically active. The result of complex formation with a racemic mixture of ligand molecules is thus a racemic mixture of chiral complexes, rather than inactive complexes with internal compensation.

12. Chirodiastaltic Interactions in Biological Systems

Most life processes involve chiral molecules and discrimination can be expected to be a common feature of the interactions. We refer here first to two special aspects, in the physiological responses of taste and odor. More than a century ago Pasteur noted that the (+) and (—) forms of asparagine tasted differently — the former sweet, the latter insipid or almost tasteless. Since then, the often widely different physiological effects of (+) and (—) forms of various natural and synthetic compounds have been brought to light.

Under the heading of taste two classes of compounds only will be discussed, namely amino acids and sugars. The difference between the taste of (+) and (—) forms of amino acids, first noted by Pasteur, has proved to be a general one. This is illustrated in Table 6 where taste is correlated with absolute configuration rather than the sign of the optical rotation of enantiomeric pairs. Amino acids

Table 6. Taste and chirality of amino acids[2]

Amino acid[1]	D[3]	L[3]
Alanine	Sweet	Sweet
asparagine	Sweet	Tasteless
Histidine	Sweet	Tasteless
Isoleucine	Sweet	Bitter
Leucine	Sweet	Bitter
Tryptophane	Sweet	Tasteless
Tyrosine	Sweet	Bitter

[1]) Glycine, the first member of the α amino acid series, is achiral.
[2]) Data from Shallenberger (1971).
[3]) Absolute configurations.

derived from proteins are always L; D amino acids which are much rarer in nature are generally sweet. That it is possible to distinguish the taste of enantiomeric pairs implies that the taste bud receptor site has a chiral structure.

Of related synthetic compounds, one of the most interesting is (+)-6-chloro-tryptophane, which according to Kornfeld [see Chedd, (1974)] is 1000 times sweeter than sucrose; "all the sweetness resides in the unnatural (+) isomer".

Contrary to some earlier findings, it now seems fairly certain that there are no differences in the taste of the enantiomeric forms of sugars [Shallenberger (1969)]. In a test in which seven pairs of D and L sugars were submitted to a panel of tasters, Schallenberger found no statistically significant difference between the taste of each pair. D-glucose was just about as sweet as L-glucose.

As to odor, there has for long been uncertainty mainly because of doubt concerning the chemical purity of the enantiomeric isomers, the completeness of the separation of the isomers and the techniques for testing the odors. These difficulties now appear to have been overcome by two teams of investigators, Russell and Hills (1971), and Friedman and Miller (1971) working independently — interestingly enough in some instances, on the same or closely related compounds. Both teams studied R-(—) and S-(+) carvone which had been carefully purified by gas liquid chromatography. Not all individuals can detect the odor of carvone[d] but by those who can the R isomer was unanimously described as having a spearmint odor; the S, the odor of caraway. Enantiomeric pairs of other closely related compounds were similarly distinguishable.

Friedman and Miller also established that synthetic R-(+) and S-(—) limonene have the odor of oranges and lemons respectively. They also demonstrated the differences between other enantiomeric pairs such at R-(+) and S-(—) amphetamine. There remains no doubt that some enantiomeric pairs do have different odors. Whatever the nature of the sensory detector involved, it must, in order to be able to discriminate between enantiomeric pairs, itself be locally chiral. Hence

[d]) The present day convention is to use capital letters D and L to indicate the absolute configurations of sugars and amino acids; the absolute configurations of all other molecules are indicated by the symbols R and S. For the convention see Cahn *et al.* (1966).

we may conclude that odor discrimination of the kind discussed here is an example of chirodiastaltic interaction. The same reasoning applies to other differences in physiological action of enantiomers. Not all enantiomeric pairs have different odors, perhaps because the chiral centres are sequestered within the molecule. Thus (+) and (—) camphor are said to be indistinguishable in odor as are also (+) and (—)-2-octanol. This raises an interesting problem for any theory of odor perception; another is that 8 per cent of the individuals studied by Friedman and Miller were unable to detect the odor of carvone (either R- or S-). The authors who speak of "carvone odor-blindness" describe this as an example of "specific chiral anosmia".

Other examples of chiral discrimination in physiological reactions may be cited. They have been listed by Albert (1965).

1. (—) isopropyl noradrenaline (isoprenaline) is 800 times a more effective bronchodilator than its (+) isomer [Luduera *et al.* (1957)].

2. Natural (—) adrenaline has twenty times greater activity, on various test objects, than its (+) isomer.

3. (+) Acetyl β-methylcholine is about 250 times more active on the gut than the (—) form [Blaschko (1950)].

Other instances where the evidence points to the existence of either very feeble discrimination or of achiral receptors are:

1. The (+) and (—) forms of cocaine are equally powerful local anaesthetics [Gottlieb (1923)].

2. The (+) and (—) forms of chloroquinine are equally effective anti-malarials [Riegel *et al.* (1949)].

Evidence for chirodiastaltic interactions in biological membranes is to be found in experiments with the (+) and (—) forms of $[Ru^{106}(phen)_3]^{++}$ [Koch *et al.* (1957)]. When administered intraperitoneally in equivalent doses to rats and mice, the (+) form reaches the blood stream twice as rapidly as the (—) form. Whether the chirodiastaltic interactions occur at the surface or within the biological membrane (or at both locations) it is not possible to decide on the basis of the presently available evidence.

We refer finally to the broad area of chirodiastaltic metabolic action of living systems, a field of enquiry beginning with Pasteur's (1858) discovery that on adding the mold *penicillium glaucum* (*P. expansum*) to dilute ammonium racemate it grew at the expense of the *dextro* acid leaving the *levo* acid unaffected. This discovery marked the opening up, slowly at first, of a wide field covering similar reactions initiated not only by molds but by yeasts, bacteria and higher animals. A few examples only, involving the actions of molds and yeasts on the racemates of amino acids and sugars will be quoted.

A racemate of the sugar galactose, obtained by hydrolysis of agar agar or synthesis, when fermented with "galactose adapted yeasts", yields L-galactose free from the D isomer (Anderson 1933). Schulze and Bosshard (1886) used *penicillium glaucum* to isolate the D isomers of leucine and glutamic acid from their respective racemates. Early this century, Ehrlich (1906) treated a wide variety of racemic amino acids with a yeast which, like *penicillium glaucum*, metabolised only the L isomer. In this way he obtained yields of 60—70 per cent of the D isomers of each of nine amino acids. The reaction in which the enzyme decarboxylase is involved may be represented:

$$\left.\begin{array}{l}\text{L-Amino acid} \\ \text{D-Amino acid}\end{array}\right] \begin{array}{ll}+\ \text{yeast} & \rightarrow \text{amine} + CO_2 \\ \text{(decarboxylase)} & \rightarrow \text{D-Amino acid}\end{array}$$

From these and many similar examples it became evident that discrimination between enantiomers is often a matter of degree. Absolute discrimination, however, is shown by specific oxidases like D-amino acid oxidase of mammalian kidney and L-amino acid oxidase of snake venom. "No one [member] of this class of biological catalysts has yet been known to attack measurably an amino acid antipodal to its normally susceptible category of substracts"[e] [Greenstein and Winitz (1961)] [Zellor and Maritz (1945)]. Equally selective is the phosphorylation of mevalonic acid by the enzyme mevalonic kinase; the R- form is phosphorylated, the S- form is unaffected (Tchen 1958).

While noting that reactions between biochemical systems provide perhaps the most striking examples of chiral discrimination and selectivity, we do not wish to leave the impression that they may be treated theoretically in the same way as interactions between small molecules by the methods outlined in following sections. A large biomolecule cannot be treated as a chiral entity participating as a whole as one component in a pairwise intermolecular coupling. Particularly, where the second system is relatively small, its approach may be toward one functional group of the large system and depends on the *local* molecular structure near that group. It is then the local chiral character and not the chiral character of the whole which counts. An illustration is provided by the X-ray and NMR studies of Quoicho *et al.* (1971) on the differential inhibition by D- and L-phenylalanine of the activity of manganese-carboxypeptidase A (Mn CPA). The effects are demonstrably associated with the stereochemical differences between D- and L-phenylalanine in relation to the metal ion and its immediate molecular environment in Mn CPA.

---

[e] Equally discriminating are the dehydrogenases a long list of which, together with their chiral substrates, has been prepared by Popjak (1970).

II. Electrostatic Discrimination

1. Introduction

The central theoretical problem is to account for chiral discrimination in terms of the forces acting between the molecules in the situation that the discriminating interaction energy is small compared with the total intermolecular energy. For small molecules the calculation of energy as a function of separation and orientation by *ab initio* methods is now feasible, but the discrimination could hardly be found accurately enough as the difference of the energy totals for *d-d* and *d-l* pairs to be helpful in understanding the phenomenon. For molecules of the types for which experiments have been made (Section I) there is no alternative yet to empirical methods, in which the calculated energies can be scaled with the help of measured quantities such as intermolecular spacings and energies, and molecular electric and magnetic moments.

Because observations of discrimination are often made in the liquid phase, knowledge of pairwise interactions in free space needs to be supplemented by consideration of the effects of solvent. A highly charged chiral ion will be enclosed in a solvent sheath of which the surface shape may imitate that of the ion, and so have chiral character. One thus sees that the influence of a chiral system might be propagated over several molecular diameters simply by contacts, *i.e.* short range interactions, transmitted through one or more layers of solvent. Such processes of 'relayed' chiral influence are not considered in this article, though many chemists believe them to be significant.

There may also be specific chemical sources of discriminating interactions such as differential hydrogen bonding, as mentioned in Section I.8 in connection with the work of Bosnich and Watts (1968). Another interesting suggestion is that of differential covalent hydration as a possible explanation of discrimination in certain complexes of o-phenanthroline [Gillard (1973, 1974)].

Among other approaches not based directly on pairwise interactions there is the statistical mechanics of solutions in which the constituents are chiral molecules, treated by extension of the hard-sphere or other approximations. This is also more or less unexplored, and there are only the briefest of references in Section V.

Within the scope of pairwise interactions we begin the analysis in Section II.2 with the complete Hamiltonian, but divide the discussion according to different mechanisms of coupling such as 'contact' terms, electrostatic terms etcetera. These act together in the real situation, but individually are dominant under particular conditions of molecular separation and molecular constitution. An additional reason for proceeding in that way is that the magnitudes of the separate terms can be roughly assigned from a knowledge of measurable molecular quantities. In this approach we classify the forces and associated energies in terms of their ranges, namely their dependence on the distance R separating the interacting bodies, beginning with long-range electrostatic forces, through dispersion interactions, which are of intermediate range, to the extremely short range repulsions. After a brief description of these three types we treat them in detail in their application to discrimination beginning in Section II.2 with electrostatic terms.

The longest ranged interactions between ground state systems are electrostatic (see Section II.2 *et seq.*). Between ions of charge Z_1 and Z_2 the Coulomb's law interaction Z_1Z_2/R has the greatest range. For uncharged molecules the leading component is the dipole-dipole interaction depending on R^{-3} according to

$$E(R) = \mu_i^{(a)}\mu_j^{(b)} (\delta_{ij} - 3\,\hat{R}_i\hat{R}_j)\,/\,R^3 \tag{1.1}$$

where $\mu_i^{(a)}$ is the i-th component of the dipole moment of molecule (a) in a cartesian axis system, and $\hat{R}_i$ is the component of a unit vector along the intermolecular join $\boldsymbol{R}$. δ_{ij} is the Kronecker symbol, and summation over repeated indices is implied. Dipole forces are non-discriminating but can be important in combination with quadrupole forces. When the species carry no net charge or dipole moment there may be significant energies arising from the interaction of moments of higher order, notably quadrupole moments. The quadrupole-quadrupole interaction has an energy varying as R^{-5} and is strongly dependent on orientation, as will be discussed at length in a later part of this article (Section II.3). Still higher multipole moments, certainly up to octupole, may also be important. The variation with distance of the interaction of multipoles of order n(a) and n(b) is $R^{-n(a)-n(b)-1}$, the order being that of the covering spherical harmonics, namely $n = 0,1,2 \ldots$ for free charges, dipoles, quadrupoles, and higher orders.

Intermolecular forces between neutral and nondipolar molecules are usually dominated by the dispersion interaction (Section III). This is always attractive between ground state systems and, as first described by London, arises (in the leading term) from the coupling of electric dipoles, one being a dipole fluctuation in one molecule and the other the dipole induced by it in the other. The distance dependence is as R^{-6}. The ordinary dispersion interaction is non-discriminating, but as we shall see (Section III) the electric-magnetic analogue is capable of a weak discrimination. In a classical picture this is the coupling *via* electric forces of a dipole in one molecule to a dipole in the other, the first dipole being induced by magnetic coupling of a charge fluctuation in the second system. It varies as R^{-6}. The magnetic-magnetic dispersion force is again non-discriminatory.

Short range 'contact' forces (Section V) come into play when the filled electron shells of different molecules begin to interpenetrate, as in the simplest possible case of two helium atoms at distances in the neighbourhood of 0.1 nm. The interaction energy between two bodies mutually acted on by these forces is given in an approximate way by cR^{-12}, c being an empirical constant. Since however the electron density of an atom falls off at distances from about the van der Waals radius and greater according to an exponential law, one expects that a better representation of the interaction energy will be by an exponential $ae^{-\alpha R}$ where α may be calculated from the atomic wavefunctions, but is usually fitted empirically. The repulsion energy from this cause varies so quickly with distance that it can be considered to arise from the interference of atoms with 'hard' surfaces and is essentially a contact interaction. No more than a very small interpenetration of closed electron shells can be tolerated, but outside the contact distance the energy is extremely small.

Closed shell repulsions are known to be the main determinants of the packing patterns of aromatic hydrocarbons in crystals, and play a part also in fixing the optimum intramolecular configuration where some freedom is permitted by the valence forces, as for example in the hindered rotation of ethanes.

The contact terms are largely responsible for the orientations adopted by nonspherical ions and molecules in crystal lattices, and must underlie the lattice energy discriminations in the packing of diastereoisomers, as in Pasteur's separations. For a long time they were thought to be the only interactions involved in asymmetric transformations and crystallizations of chiral systems.

There is one further source of discrimination, in a different category from the others. If the *d-d* and *d-l* pairs are of chemically identical molecules, the members of the pairs have identical sets of *energy levels*. One molecule of such a pair which has been excited to an upper level can resonate with the other member by exchanging excitation energy with it. The energy shifts caused by the resonance are different for *d-d* and *d-l* pairs. There is thus a discriminating resonance energy (Section IV), the leading term of which varies with distance as R^{-2}. No experimental method has yet been devised to exploit this interesting coupling, but the magnitudes involved suggest that it should not be impossible to do so.

We note finally that the calculation of discrimination can be made only with assumptions on the orientation of the coupled molecules. The discriminating parts of the total energy are usually too small to influence orientation, and therefore have to be found for orientations which are at energy minima for the *total* interaction. The cases are

(a) The locked-up limit of molecules in fixed relative orientations, apart from librational oscillations, determined by short-range repulsions. This applies to crystal lattices, and perhaps to ion-pairs in solution.

(b) Intermediate cases in which the dependence of total energy on orientation is comparable to the thermal energy kT. There may be partial locking, as when one axis is fixed relative to the intermolecular join $\boldsymbol{R}$ and rotation about that axis is more or less free.

(c) The limit of free relative motion, in which the total interaction is very much less than kT, as for widely separated molecules.

2. The Interaction Hamiltonian

In order to include all types of discrimination we first give a complete hamiltonian (II.1), expressed as the sum of the isolated molecule hamiltonians H_{a} and H_{b}, their electrostatic interaction H_{E}, the coupling by radiation, and the hamiltonian for the radiation field,

$$H = H_{\mathrm{a}} + H_{\mathrm{b}} + H_{\mathrm{E}} + \sum_i \boldsymbol{p}_i \cdot \boldsymbol{A}_i - \tfrac{1}{2} \sum_i \boldsymbol{A}_i^2 + H_{\mathrm{rad}} \qquad \text{(II.1)}$$

$$H_{\mathrm{E}} = -\sum_{iq} \frac{Z_q}{r_{iq}} - \sum_{jp} \frac{Z_p}{r_{jp}} + \sum_{pq} \frac{Z_p Z_q}{r_{pq}} + \sum_{ij} \frac{1}{r_{ij}} \qquad \text{(II.2)}$$

$$H_{\mathrm{rad}} = (8\,\pi)^{-1} \int (\boldsymbol{E}^{\perp 2} + \boldsymbol{B}^2)\, dV \qquad \text{(II.3)}$$

The sums in expression (II.2) run over the electrons i and the nuclei p (charge Z_p) of molecule a, and over j and q in b. $\boldsymbol{p}_i$ is the particle momentum operator,

$\boldsymbol{A}_i$ the vector potential at the site of particle i, $\boldsymbol{E}^{\perp}$ the transverse electric field and $\boldsymbol{B}$ the magnetic induction field. H_{rad} is an integral over the volume occupied by the system specified by the total hamiltonian. In a perturbation theory approach, the radiation dependent terms give corrections to the unperturbed ground state energies belonging to $H_{\mathrm{a}}+H_{\mathrm{b}}$ only in second and higher orders, representing the coupling of the molecules through photons emitted and absorbed in virtual transitions. H_{E} gives the first order correction to the energy of molecules in their ground states and includes all the electrostatic terms to be considered in this section. To first order we need use only the first three terms of (II.1), namely $H_{\mathrm{a}}+H_{\mathrm{b}}+H_{\mathrm{E}}$. At distances R greater than that for interpenetration of the electron shells H_{E} is expanded in a multipole series given in full by Hirschfelder, Curtiss and Bird (Molecular Theory of Gases and Liquids 1954) and quoted in (II.4) in the notation of Craig and Schipper (1974) on which much of this section is based,

$$H_{\mathrm{E}} = \sum_{(s)} i^{l-|l|-l'+|l'|} (-1)^{n'+m} F(|s|)\, \overline{Q}^*(nl)\, \overline{Q}(n'l')\, D^*(n:lm)\, D(n':l'm) r_{\mathrm{ab}}^{-n-n'-1} \tag{II.4}$$

where

$$F(|s|) = \frac{[(n-|l|)!\,(n'-|l'|)!]^{\frac{1}{2}}(n+n')!}{[(n+|m|)!\,(n-|m|)!\,(n+|l|)!\,(n'+|l'|)!\,(n'-|m|)!\,(n'+|m|)!]^{\frac{1}{2}}}$$

The expansion gives the interaction operator as a series of multipole-multipole operators, which can be taken term by term. The quantities $\overline{Q}(nl)$ are the components of multipole moments referred to axes fixed in the molecules, primes distinguishing quantities for centre b. The moments transform as spherical harmonics, and are defined in (II.5),

$$Q(nl) = \sum_i e_i r_i^n P_n^l (\cos\theta_i) \exp(il\phi_i) \tag{II.5}$$

e_i being the charge at distance r_i from the origin. The real components will be denoted by $Q(nl^+)$, $Q(nl^-)$.

For example the components of the dipole are given in (II.6),

$$\left.\begin{aligned}
Q(1,\pm 1) &= \sum_i e_i r_i \sin\theta_i \exp(\pm i\phi_i)\\
Q(1,0) &= \sum_i e_i r_i \cos\theta_i = \sum_i e_i z_i\\
Q(1,1^+) &= \sum_i e_i r_i \sin\theta_i \cos\phi_i = \sum_i e_i x_i\\
Q(1,1^-) &= \sum_i e_i r_i \sin\theta_i \sin\phi_i = \sum_i e_i y_i
\end{aligned}\right\} \tag{II.6}$$

the quantities x_i, y_i and z_i being cartesian displacements. Here and elsewhere the Q's may appear as operators or as expectation values according to context. The representation coefficients D are functions of the angles of rotation which take the molecular axes into axes fixed to R, and the numbers F are independent of the signs of the azimuthal and magnetic quantum numbers l,l' and m, included in the set S.

When the expectation value of (II.4) is taken over the product of the ground state wave functions for molecules a and b, the result is the sum of coupling energies for the permanent electric moments in the two molecules. If the moments are known, or taken as parameters, the electrostatic interaction energy is known for a chosen orientation, and can be compared for *d* and *l* species. By forming averages over angles we can make the calculation for molecules in relative rotational motion. First it is useful to examine the symmetry restrictions imposed by chiral character, and to see how the moments in one chiral enantiomer are related to those in the other.

3. Symmetry Properties

Chiral molecules can belong only to one of the symmetry groups which lack all improper rotations. The possible groups are $\boldsymbol{C}_n$, $\boldsymbol{D}_n$, $\boldsymbol{T}$, $\boldsymbol{O}$ and $\boldsymbol{I}$. Only the first two are known in examples including, trivially, the group $\boldsymbol{C}_1$, containing only the identity, which covers the simplest chiral molecules shown on the left of Fig. 1. If the external electric effects of a chiral molecule are to be represented by a field of which point multipoles are the sources, the multipoles must display the same covering symmetry as the molecule. Now each simple multipole component $Q(nl)$ possesses elements of symmetry, and is usually more symmetric than the molecule of which it simulates the field. The unwanted symmetry is removed by taking combinations of two or more multipoles. For example a dipole moment is symmetric to reflection in any plane containing the dipole axis, and a quadrupole moment has both planes and a centre of inversion. Neither separately can be chiral; but together they constitute a combined source with no inversion centre and one which is chiral if the symmetry planes of the quadrupole do not contain the dipole axis. We describe the arrangement as the skewed dipole-quadrupole; it is the simplest chiral combination of electric moments, and applies to asymmetric systems (symmetry $\boldsymbol{C}_1$).

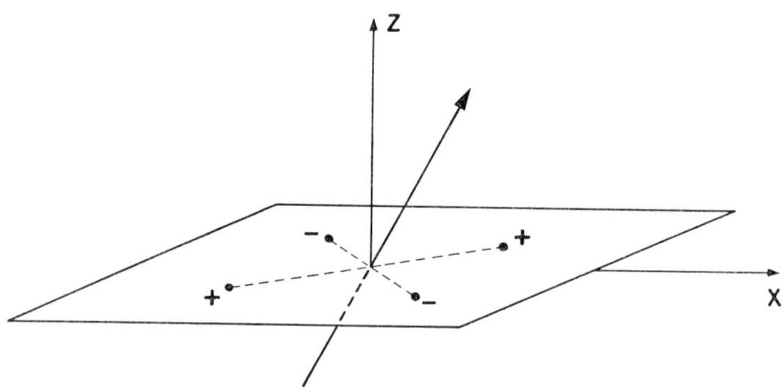

Fig. 2. Schematic chiral dipole-quadrupole combination. The dipole lies in the (xz) plane. The quadrupole component is symmetric to mirror planes containing the z axis and either of the bisectors of the x and y axes. The combination has neither centre nor plane of symmetry

More general results can readily be found. All moments are from now on referred to the same body axis system, the z axis being the common polar axis. The essential basis is in Equation (II.7), which give the transformation properties of the multipole components under the operations of inversion (i), reflection in the xy plane (σ_h) reflection in a plane containing the z axis (σ_v), and improper rotation about the z axis by $2\pi/p$, (iC_p).

$$\left.\begin{aligned} iQ(nl) &= (-1)^n\, Q(nl) \\ \sigma_h Q(nl) &= (-1)^{l+n}\, Q(nl) \\ \sigma_v Q(nl) &= Q(n,-l) \\ iC_p Q(nl) &= e^{2\pi i l/p}\, Q(nl) \end{aligned}\right\} \qquad \text{(II.7)}$$

The third relation in (II.7) can be written in a more general form to apply to reflection in a plane containing the z axis and displaced by an angle ξ from the molecule-fixed x axis. We denote this operation by $\sigma_v(\xi)$; then

$$\sigma_v(\xi) Q(nl) = e^{2il\xi}\, Q(n,-l) \qquad \text{(II.8)}$$

The relations (II.7) and (II.8) show that any one multipole component is symmetric to at least one improper rotation; thus the combination of two components is a minimum requirement for chirality. One component must belong to a multipole of even order and one of odd, in order to remove the inversion symmetry. The simplest combinations of components can be found in the following way. For a molecule belonging to group $\boldsymbol{C}_n$ ($n>2$) the polar axis is taken along the n-fold axis and x,y chosen arbitrarily. The non-zero moment components are restricted to those invariant to rotations by angles ϕ which are multiples of angles $2\pi/n$, namely those with unit characters χ in the expression (II.9),

$$\chi_n(\phi) = \frac{\sin\,(2n+1)\phi/2}{\sin\,\phi/2} \qquad \text{(II.9)}$$

We first find which moment components can be present in any group $\boldsymbol{C}_n$, and then find combinations of them that can be chiral. For example in $\boldsymbol{C}_2$ the allowed components are $Q(1,0)$, $Q(2,0)$, $Q(2,\pm 2)$, $Q(3,\pm 2)$, $Q(4,0)$, $Q(4\pm 2)$, $Q(4\pm 4)$... Inasmuch as the sole dipole component $Q(1,0)$ is symmetric to all reflections $\sigma_v(\xi)$, and all other components are symmetric to at least one such reflection, namely reflection σ_v of $Q(nl^+)$, we see that no binary combination including the dipole component can be chiral in $\boldsymbol{C}_2$. The simplest combination is a quadrupole-octupole $Q(2,2+)$ $Q(3,2-)$ or the equivalent $Q(2,2-)$ $Q(3,2+)$. For group $\boldsymbol{C}_3$ the simplest pair is the octupole-4-pole $Q(3,3+)$ $Q(4,3-)$ or its partner $Q(3,3-)$ $Q(4,3+)$. In the general $\boldsymbol{C}_n$ the necessary components are one each of orders n and $n+1$. In the asymmetric case, formally belonging to $\boldsymbol{C}_1$, the simplest pairs are $Q(1,1+)$ $Q(2,1-)$ and $Q(1,1-)$ $Q(2,1+)$, for general directions of the polar axis. If the polar axis is taken to be along the direction of the dipole moment, so that the dipole belongs to $Q(1,0)$, the chiral dipole-quadrupole cannot be confined to two components but requires at least three.

For molecules belonging to the other set of groups $\boldsymbol{D}_n$ which also include actual examples in particular $\boldsymbol{D}_3$ (see Fig. 1 right hand side), we find by a similar use of the relations (II.7) and (II.8) that the simplest combinations are $Q(n,n+)$ $Q(n+1, n-)$. In $\boldsymbol{D}_3$ this implies one octupole component plus one 4-pole component.

4. Fixed Relative Molecular Orientations

The limit of nearly free relative orientations is not of much physical interest in the case of electrostatic forces. The intermolecular interaction is a Boltzmann-weighted average of the form

$$<H_{\mathrm{E}}>_{\mathrm{av}} = \frac{\iint <H_{\mathrm{E}}> \exp\left(-<H_{\mathrm{E}}>/kT\right) d\omega \, d\omega'}{\iint \exp\left(-<H_{\mathrm{E}}>/kT\right) d\omega \, d\omega'} \tag{II.10}$$

in which the expectation value of the electrostatic intermolecular hamiltonian is evaluated over all orientations ω and ω' of the molecule pair. k is Boltzmann's constant and T the absolute temperature. It is shown elsewhere (Craig and Schipper 1975) that the lowest-order term to give discrimination belongs to $<H_{\mathrm{E}}>^3/(kT)^2$, depending on distance as R^{-17}. This is a much faster variation with distance than the contact interaction (R^{-12}), and is negligible. We conclude that under conditions of nearly free relative motion of the coupled molecules, electrostatic contributions to discrimination are of no importance. This is not true of dispersion terms, as will later be seen.

In the limit of fixed orientations there are larger electrostatic terms. The molecules a and b to be considered are asymmetric, possessing non-zero dipole and quadrupole moments in a chiral arrangement. In an asymmetric molecule there is no natural or preferred origin of coordinates for the multipole expansion and therefore *inter alia* no unique meaning to be attached to the vector $\boldsymbol{R}$ joining a and b. If an infinite expansion were to be used the arbitrary character of the origin choice would be of no importance, insofar as the different relative weightings of the several multipole-multipole contributions are origin dependent, in such a way that the total interaction energy is the same for any choice. However if we cut off the expansion the result is origin dependent, and may be very inaccurate. We must therefore treat the systems in carefully defined ways. For the same reasons care is required in defining the discrimination energy. In many physical situations the interaction energy difference between *d-d* and *d-l* pairs is developed for orientations determined by atom contacts; these orientations need have no symmetry relationship. Again discrimination must be calculated for idealized situations. We give an example in the following paragraphs.

In an asymmetric molecule the origin and coordinate axes are chosen arbitrarily. The molecular origins will in principle be put at the centres of mass, though this will not affect the argument in any way. Then, beginning with dipole and quadrupole moment components specified in an arbitrary (xyz) axis system we transform to a new system through Eulerian angles (α,β,γ). The z' axis is first chosen to be along the axis of the dipole moment, and an axis system ($x'y'z'$) defined by the transformation ($\alpha,\beta,0$). The sole dipole component is $\mu_{z'}$. There are five independent components of the (traceless) quadrupole moment, namely $Q_{x'y'}$, $Q_{x'z'}$, $Q_{y'z'}$, $Q_{x'x'}$, $Q_{y'y'}$, which may be reduced to four by choice of the Euler angle γ to eliminate the (xy) component of the quadrupole. The coordinates will be denoted by $(x_0y_0z_0)$. The new quadrupole moment is given in the symmetric array (II.11). The symbol $\boldsymbol{Q}$ is now employed for the quadrupole, and not the general multipole as before.

$$\boldsymbol{Q} \equiv \begin{Bmatrix} Q_{x^0x^0} & 0 & Q_{x^0z^0} \\ & Q_{y^0y^0} & Q_{y^0z^0} \\ & & Q_{z^0z^0} \end{Bmatrix} \tag{II.11}$$

The trace is zero, leaving four independent components, a number which can only be further reduced by special choice of origin. The dipole-quadrupole has no centre of symmetry. The dipole is symmetric to $\sigma_v(\xi)$ for all ξ; however the quadrupole (II.11) has no symmetry plane in common, as may be seen by applying the third of Eq. (II.7) to $Q_{x^0z^0}$ and $Q_{y^0z^0}$ [viz. to the real forms of $Q(2,1)$ and $Q(2,-1)$]. All σ_v reflections transform the dipole-quadrupole to its enantiomeric form; the simplest choices of reflection plane are those transforming either $Q_{x^0z^0}$ or $Q_{y^0z^0}$ into its negative.

The dependence of the model on the choice of origin has already been stressed, as has also the need to specify the meaning to be attached to the discrimination in terms of the relationslip of one chiral enantiomer to its antipode. To give an indication of magnitudes we take the situation of two molecules in which the dipole moments are large enough to impose a fixed configuration with the dipoles collinear and arranged head-to-tail. The dipole-dipole part of the interaction energy is then independent of the relative rotation of the systems about this fixed axis and the variations are entirely caused by the quadrupole-quadrupole coupling. Realistic values of the quadrupole components are that each is about 1 eÅ^2, or 1 em^{-20}. If the centres are 0.5 nm apart, the *d-d* pair in its most stable orientation is more stable than the most stable *d-l* orientation by $\sim$300 J mol^{-1}, or about the thermal energy at 35 K. We are thus led to believe that in locked or partially locked states the electrostatic discrimination can be of some significance. In this particular case the like entities (both *d*) are more stable than the unlike (*d-l*), but there is no systematic reason why this should be so.

5. Contributions by Permanent Magnetic Moments

Optical activity is almost always related to spectroscopic transitions allowed to both electric and magnetic radiation, consequent on the existence of electric and magnetic dipole transition moments $\boldsymbol{\mu}$ and $\boldsymbol{m}$ joining the ground state to at least one excited state and with the condition that $\boldsymbol{\mu} \cdot \boldsymbol{m} \neq 0$. The symmetry restrictions for chirality, namely the absence of any improper rotation, also ensure that neither of the two types of transition moment is necessarily zero, though they may happen to be very small. One naturally examines whether permanent magnetic moments can be important in discrimination.

In a classical picture the source of a magnetic field is a current distribution, which may be expanded in a multipole series analogous to that for a charge distribution as the source of an electric field. We are concerned only with the magnetic dipole term in this expansion.

Its symmetry properties are not the same as those of the electric dipole, which transforms under pure rotations as the multipole components $Q(1,0)$ and $Q(1,\pm1)$ and is antisymmetric under inversion in a centre of symmetry. The magnetic dipole transforms in the same way under rotations but is symmetric to inversion. The properties are those of an antisymmetric second-rank tensor, or axial vector, and can be rationalized using the model that a magnetic field has as source a plane current loop. The current direction in the loop is invariant to inversion of the spatial coordinates but reversed by twofold rotation about an axis in the plane.

Orbital magnetism in a molecular ground state is possible only if the state belongs to a spatially degenerate representation of the molecular point group, as in a Π state of a linear molecule. If the molecule also had an electric dipole moment the combination of electric and magnetic moments along the molecular axis would appear to give chirality, there being no centre of symmetry, nor any plane, since reflection of the magnetic moment in a plane containing the dipole axis changes its sign. Chirality would thus have appeared in a molecule belonging to a non-chiral symmetry group. However in the absence of an external magnetic field the existence of a molecule with a permanent magnetic moment in a fixed direction with respect to the molecular frame is ruled out by the requirement that a stationary state is symmetric to time inversion. The magnetic moment direction is reversed under time inversion and, in a state of stationary energy, must be considered to be switching rapidly and to average to zero. A molecule thus cannot display chiral character through the existence of a permanent magnetic moment. This is true also if the magnetism arises from electron spin, or both orbit and spin as in the $^2\Pi_{\frac{1}{2}}$ state of NO, upon which there was early discussion along these lines [van Vleck (1931)].

It is of some interest that these arguments do not imply that a *pair* of molecules with permanent electric and magnetic moments cannot show discrimination. We are then concerned with a state given as the product wave function $\psi_a\psi_b$ of the separate free molecule functions, and the property of symmetry to time reversal is required only of the product, so that both factors may change sign. Within the coupled pair each may have a definite direction of magnetic moment in relation to the electric moment and be individually chiral. It becomes possible to analyse the energies of *d-d* and *d-l* pairs as before. In physical terms two linear molecules, both in Π states, could approach end-to-end along the direction of the common dipole axis. Under the influence of the mutual magnetic fields two pair-states would form, one with cancelling magnetic moment (becoming a Σ electronic state at shorter distances) and with reinforcing moment (becoming a Δ electronic state). In a certain sense, the energy difference between the two is the discrimination, though that term could only be used at separation distances beyond distances of electronic overlap, which affects the magnetism in other ways irrelevant to this discussion. In a practical case one might think of the end-on approach of two $^2\Pi_{\frac{1}{2}}$ NO molecules, leading eventually to the formation of a linear dimer in either $^1\Sigma$ or $^3\Delta$ states. The long range splitting of these states is due to electric-magnetic discrimination.

In spite of the restrictions on the existence of ground state magnetic moments it is instructive to work out the magnitudes of interactions. We note that where the covering group is $\boldsymbol{C}_n$ or $\boldsymbol{D}_n$, $n > 3$, the electric and magnetic moments must lie along the rotation axis. In other cases it is only the component of the magnetic moment along the direction of the electric moment which contributes ot the discrimination. A straightforward application of the coupling operator (II.4) for interacting dipoles can be made. If as in the dipole-quadrupole calculation of Section II.4 we suppose that the configuration is locked by the electric dipoles in a head to tail arrangement, the discrimination depends solely on the magnetic interactions. If the component of $\boldsymbol{m}$ along the electric dipole axis is m_z, the discrimination energy is given by

$$4\, m_z^2\, R^{-3} \qquad \text{(II.12)}$$

which is the energy difference between the collinear parallel and antiparallel arrangement of the moments m_z. If m_z is equal to the Bohr magneton, the discrimination energy at $R = 0.5$ nm is 0.17 J mol^{-1}. This is much less, by 2 or 3 orders of magnitude, than the dipole-quadrupole discrimination at the same separation, but since its distance variation is slower, as R^{-3} instead of R^{-5}, it becomes the greater of the two at long enough range, but is then extremely small.

Where the coupled systems are nearly free to rotate independently the calculation proceeds analogously to that in Section II.4, with the replacement in expression (II.10) of the electrostatic coupling operator H_{E} by the sum of operators for electric-electric dipole coupling and magnetic-magnetic dipole coupling $H_{\mathrm{E}} + M_{\mathrm{M}}$. The first term in the expansion of the exponentials in (II.10) is simply the unweighted average over angles of $\langle H_{\mathrm{E}}\rangle + \langle H_{\mathrm{M}}\rangle$. This vanishes: the average interaction energy of multipoles of non-zero order, electric or magnetic, is zero. The second term is proportional to the orientation average of $\{\langle H_{\mathrm{E}}\rangle + \langle H_{\mathrm{M}}\rangle\}^2$, which gives expression (II.13) [Craig and Schipper (1975)],

$$(2/3kT)R^{-6}\,\{\mu_z^2\mu_z'^2 + |\boldsymbol{m}|^2|\boldsymbol{m}'|^2 + 2\,\mu_z m_z \mu_z' m_z'\} \qquad \text{(II.13)}$$

Primed quantities belong to molecule b. The electric-magnetic cross term changes sign when either a or b is replaced by its antipode, because any method of generating the antipode changes the sign either of μ_z or m_z, but not both. The squared terms are unaffected. The discrimination [taking account of the sign in the expansion of (II.10)] becomes

$$\Delta = -\,(8/3kT)\mu_z m_z \mu_z' m_z'\, R^{-6} \qquad \text{(II.14)}$$

and, for moment components of 0.1 enm and 1 Bohr magneton, $\Delta \sim 0.5$ J mol^{-1} at $R = 0.5$ nm and room temperature. This type of permanent moment discrimination, though small, is of a quite different order from the pure electric averaged quantity which, as already seen (Section II.4) has distance dependence in R^{-17}. It is somewhat larger than the dispersive contribution (Section III) but is expected very rarely. The discrimination (II.14) favours the like *d-d* pair over the unlike *d-l*.

6. Chiral Molecules Represented by Separated Dipoles

The viewpoint so far adopted is that the chiral molecule is represented by a superposition of point multipoles, electric or magnetic, at a common origin. This is no more than a convenient postulate. Since, as we recalled earlier there is no natural choice of origin, there is equally no necessity to take the same origin for the several multipolar components. It would be quite possible to locate a dipole in one part of a molecule, and another dipole or a higher moment in another, each being associated with particular molecular substructures. However in treating intermolecular couplings at distances R large compared with the molecular dimensions there is little to be gained by separating the origins. At shorter distances the charge distribution as a source of a chiral field should be treated as spatially extended. This prompts study of a second case in which the dimensions of the source charge distribution are not small compared with the separation of charge from field point. A familiar analogy occurs in the theory of optical activity arising from coupled transition dipole oscillators [see *e.g.* Caldwell and Eyring (1971)] depending on simulation of the chiral molecule by two classical dipole oscillators separated by a distance comparable to the molecular size, the oscillator motions being at right angles to one another and transverse to the separation vector. The two extreme phase relations between the oscillators correspond to the two chiralities, and optical activity arises from the differential response to radiation from the chiral pairs. In the present context of interactions of chiral systems, the simplest model for an extended dipolar system is a pair of permanent point dipoles separated by $\boldsymbol{r}$. Two such dipole pairs can show discrimination if both are chiral, *i.e.* if the dipoles do not lie in the same plane. The arrangement in Fig. 3 can be used to illustrate the magnitudes. The vectors $\boldsymbol{r}$ lie along $\boldsymbol{R}$. Since the longitudinal components do not contribute to discrimination we take transverse moments $\boldsymbol{\mu}_1$ and $\boldsymbol{\mu}_2$, the $\boldsymbol{\mu}_2$ being displaced from $\boldsymbol{\mu}_1$ by an angle δ about the axis. We compare the cases of like and unlike pairs a and b; in the first the systems have the same chirality, b being simply translated from a by $\boldsymbol{R}$. In the unlike case b has opposite chirality, being generated by reflection of a in the plane containing $\boldsymbol{R}$ and $\boldsymbol{\mu}_1$ and then translated along $\boldsymbol{R}$; in the molecule b $\boldsymbol{\mu}_2$ is now rotated by $-\delta$ from $\boldsymbol{\mu}_1$. Proceeding as in Section II.4 for the coupling energies of point dipole-quadrupole combinations we calculate the energies as a function of ψ, the angle by which $\boldsymbol{\mu}_1$ in b is displaced from $\boldsymbol{\mu}_1$ in a about the common axis.

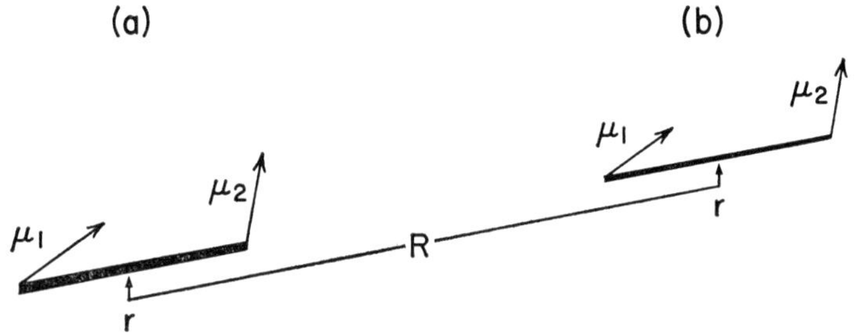

Fig. 3. Pair of chiral electric dipole doublets. Moments are non-coplanar

For the like pair,

$$E_L(\psi) = \frac{\mu_1^2 + \mu_2^2}{R^3} \cos \psi \; \frac{\mu_1 \mu_2}{R_3} \left[\varrho^+ \cos \psi \cos \delta - \varrho^- \sin \psi \sin \delta\right] \tag{II.15}$$

and for the unlike pair

$$E_U(\psi) = R^{-3}[\mu_1^2 \cos \psi + \mu_2^2 \cos(\psi - 2\delta) + \mu_1 \mu_2 \varrho^+ \cos(\psi - \delta)] \tag{II.16}$$

where

$$\varrho^{\pm} = \left(\frac{R}{R+r}\right)^3 \pm \left(\frac{R}{R-r}\right)^3.$$

The functions E_L and E_U are plotted in Fig. 4 for $\mu_1 = \mu_2$, $r/R = 0.5$ enm and $\delta = \pi/2$. The discrimination is the difference between the minima, in this example 0.053 in units of $2\mu_1^2/R^3$, and favours the unlike over the like interaction. The discrimination is easily found directly in special cases. For example if we take $\mu_1 = \mu_2$, $\delta = \pi/2$ the discrimination is given by expression (II.17),

$$\Delta = -\frac{\mu^2}{3}\left[\{4 + (\varrho^-)^2\}^{\frac{1}{2}} - \varrho^+\right] \tag{II.17}$$

as a difference $\Delta = E_L$ (min)—E_U (min). The dependence of the discrimination upon the parameter r/R is shown in Fig. 5, again for $\sigma = \pi/2$, and with $\mu_1 = \mu_2$. At $r/R = 0$ the point dipoles of each system coincide, and must be achiral, with zero discrimination. At $r/R = 0.2$, $R = 0.5$ nm, and $\mu_1 = \mu_2 = \frac{1}{2}$ enm, which represents the close approach of two chiral molecules each with dipoles of about 2.5 D separated by 0.1 nm, the discrimination is near 5.5 kJ mol^{-1}, or more than twice the thermal energy at room temperature.

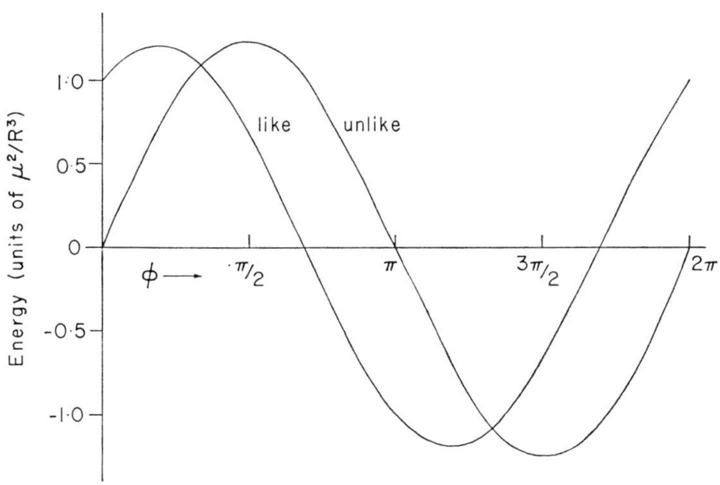

Fig. 4. Interaction energies for pairs of chiral dipole doublets as a function of dihedral angle

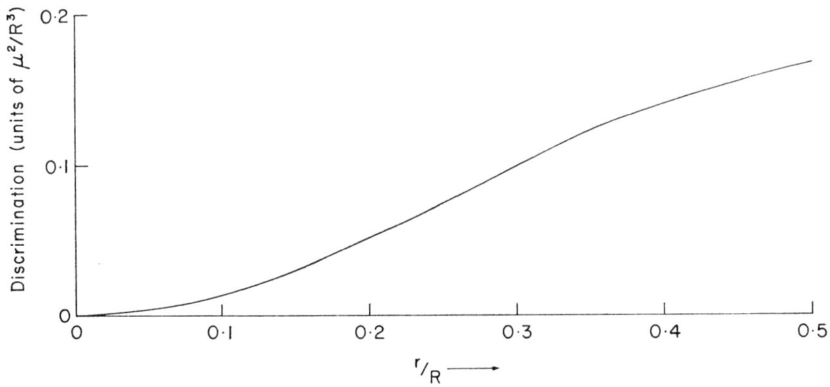

Fig. 5. Discrimination energy for dipole doublets as a function of ratio of doublet separation (r) to pair separation (R)

7. Multipoles on a Lattice

It is apparent that symmetry conditions for chirality impose much more severe restrictions on electric sources at a single site than on distributed sources, as in the separated dipoles of Section II.6. One can see that the expansion of the charge distribution of a chiral molecule about an arbitrary origin will be good only at distances very large compared with molecular size. At shorter distances the structure of the charge distribution becomes of concern. Thus even if the multipolar expansion method is acceptable for the total electric interactions it may not be for the discrimination. We next treat an extreme case, namely that of a long polymer molecule producing a chiral electric field from sources extended along the molecular frame of a lattice of such molecules. The nature of these sources allows the analysis to be taken in two parts. The individual sources may be chiral, as for example if each is a skewed dipole-quadrupole of the type in Section II.4. An array of such sources along a polymeric molecule, or on a lattice, will produce a chiral field at most points, the field being calculable by summing contributions by the several sources. The second possibility is the extension of the dipole-pair of Section II.6 to larger arrays. It is obvious that dipoles arranged regularly along a line and related to one another by translation are not chiral. However if dipoles are arranged helically as in Fig. 6 the field is chiral by virtue of their spatial arrangement, even though the sources at individual sites are achiral. The possible practical application is to helical polymers carrying dipolar groups, such as $—NH_2$ or —OH, regularly placed along the polymer chain. The polymers might be treated singly, or condensed into a lattice. Discriminating interactions might be looked for between pairs of polymer chains and between polymer chains and small chiral molecules.

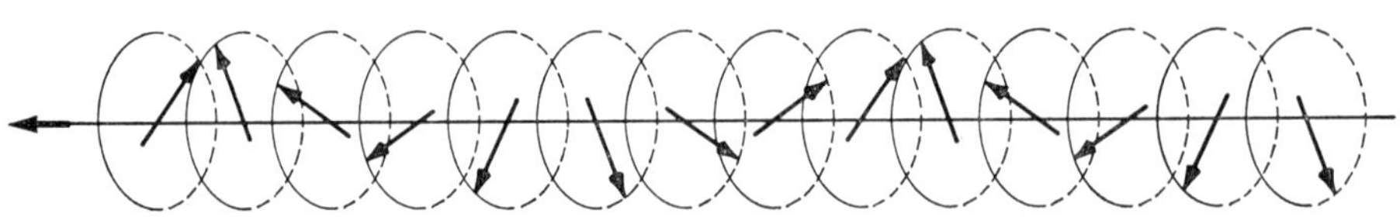

Fig. 6. Helical arrangement of dipoles along transverse axis

We calculate the interaction through the electric field and its derivatives at the site at which a test chiral system is placed. If ϕ is the electric potential, and ϕ_0 its value at the test site the interaction energy with a chiral system specified by electric charge q and dipole, quadrupole and higher moments p_i, Q_{ij}... is given in expression (II.18)

$$V = q\phi_0 - p_i F_i - (1/6)\, Q_{ij}\, F'_{ij} \qquad \text{(II.18)}$$

where

$$F_i = -(\partial\phi/\partial r_i)_0;\ F_{ij} = -(\partial^2\phi/\partial r_i\, \partial r_j)_0 \qquad \text{(II.19)}$$

in (II.18) the sum is taken over repeated indices. The field sources are fixed in position but the test system is not constrained. In the weak-coupling limit the test system is able to rotate nearly freely, and the interaction energy is a Boltzmann average as in expression (II.10). It is again found that the lowest order term showing discrimination is that in $(kT)^{-2}$ and under conditions of realistic fields and electric moments is extremely small in magnitude. In stronger fields we again have partial or full locking, in the sense that the test system is left with one degree of rotational freedom or none at all. In the latter case the only motion is that of libration (rotatory oscillation) about an equilibrium orientation. It has not yet been possible to treat this situation except by calculations on special cases, because the forces responsible for locking the system in its equilibrium orientation may or may not be those which contribute the discrimination energy. For example the orientation may be fixed by dispersion forces and the discrimination by electrostatic forces.

We now quote representative calculations. Where the sources are chiral, here taken to be skewed dipole-quadrupoles, we consider sources at lattice points of an orthorhombic lattice. The dipole component is directed along the c orthorhombic axis with magnitude 1 em^{-10} ($\sim$4.8 Debye) and the quadrupole moment, in the a,b,c axis system is

$$\boldsymbol{Q} = \begin{pmatrix} 1 & 0 & -1 \\ 0 & -1 & 1 \\ 1 & 1 & 0 \end{pmatrix}$$

with each nonzero component equal to 1 em^{-20} in magnitude. The test system, which is placed at a lattice point in place of a source, has the same moments, but now referred to axes fixed in the system, which is free to set itself in the field at its energy minimum. The enantiomeric system (opposite sign for A_{xz}) is similarly treated, and the energy minima compared to give the discrimination energy. Table 6 gives results for three lattices.

The thermal energy at 273 K is 2268 J mol^{-1}. The discriminations come from the quadrupole-quadrupole part of the interaction, and the values given closely follow the expected R^{-5} variation with lattice spacing. The main contributions come from close neighbours of the test site.

The possibility that real polymeric systems may have helical configurations, and so produce chiral electric fields, is of perhaps greater interest. Biopolymers

Table 6. Discrimination in lattices of dipole-quadrupoles

Orthorhombic lattice spacings (nm)			Discrimination energy (J mol^{-1})
$a = 0.3$	$b = 0.5$	$c = 0.7$	−2760
$a = 0.5$	$b = 0.83$	$c = 1.16$	− 215
$a = 0.7$	$b = 1.16$	$c = 1.63$	− 42

with dipolar groups regularly spaced on a helical molecular framework are a case in point. The model system is now a 3-dimensional lattice with a dipole placed at each lattice point. Beginning with a reference dipole placed at an origin, other dipoles are generated from it by pure translations in two of the three crystal directions. In the third direction dipoles are generated by translation and a rotation through an angle ϕ for each unit cell. The result can be visualised as the packing of helices of the type shown in Fig. 6 to make a three-dimensional structure. The angle ϕ is defined by rotations of the dipole from its initial orientation at the origin by angles ϕ_l ϕ_m and ϕ_n about orthogonal axes fixed in the crystal, the rotations being made always in the same order. The calculations in Table 7 refer to a cubic lattice of cell dimensions 0.5 nm. The moments at the test site are as before, constituting a skewed dipole-quadrupole. The source dipole at the origin has magnitude 0.1 enm and is directed along a body diagonal.

Table 7. Discrimination (Δ) in a helical dipole lattice (J mol^{-1})

Modulation (ϕ_l,	ϕ_m,	ϕ_n)	Δ
30,	30,	30	−1430
60,	60,	60	−1760
90,	90,	90	−2175

The discriminations are again comparable to the thermal energy at room temperature. It must be remembered however that the source dipole moments are larger (~4.8 D) than would usually be found in dipolar substituent groups. Values of one half or one third would lead to a proportional reduction in the values in Table 7.

It will be noticed that this discussion of the chirality of an extended molecule in terms of contributions by sources which are individually achiral bears on another broad question concerning discriminating interactions in general. In discussing the interaction of one chiral molecule with a chiral solid, two cases are clearly distinguished. In one the individual sources in the solid are chiral and each pairwise interaction of the external molecule with the sources contributes a discriminating term to the total energy. In the second case the sources in the solid are achiral and individual pairwise terms nondiscriminating; discrimination only appears as a result of interference between pairwise terms, the interferences being different according to the optical form of the external molecule and the chirality of the assembly of sources in the solid. In terms of space groups the distinction is between chiral molecules placed on a lattice belonging to one of the optically inactive space groups and achiral molecules on the lattice points of one of the optically active space groups. There are many examples, such as quartz. Such materials are of obvious importance and wide occurrence in connection with optical and other phenomena arising from chirality; selective absorption of optically active materials on quartz surfaces is a case in point (see Section I.5). It is important to recognise that the pairwise approach to discriminating interactions which we have used in earlier sections is not applicable to this situation of achiral sources in an optically active crystalline arrangement. It is however possible to proceed in the same way as we have done in dealing with the discriminating interactions of extended helical polymers. There is of course the possibility of dealing with optically active solids of this class as if they were giant molecules.

III. Discrimination in the Dispersion Interaction

1. General Description

Mavroyannis and Stephen (1962) considered the dispersion interaction in a general framework including radiative corrections. They noted that when magnetic dipole terms were included there was a difference in the dispersion energy between optically active molecules in *d-d* and *d-l* pairs. It was assumed that the molecules were rotating freely, the interaction being averaged over all orientations. This was the first calculation of a discriminating interaction, though Dwyer and colleagues (*loc. cit.*) had noted the possibility of long range discrimination in the interpretation of their experimental results. Mavroyannis and Stephen's result was recovered in the course of a wider study by Craig, Power and Thirunamachandran (1971), and we use their more elementary method in what follows. Returning to the Hamiltonian (II.1) we recall that magnetic interactions appear in it in a fundamentally different way from electric interactions, even when magnetostatic in type. The first three terms in (II.1) including only the static interactions (II.2), already give the electrostatic coupling between permanent electric moments, and in second order, account for the pure electric parts of the unretarded dispersion interaction. The magnetic interactions depend on electric currents coupled to each other via magnetic fields, as described by the fourth term in (II.1) coupling the particle momentum $\boldsymbol{p}_i$ to the vector potential $\boldsymbol{A}_i$ of the field at the position of the particle. Where the field $\boldsymbol{A}$ varies with time, the electric interactions appear both in the $\boldsymbol{A}$ dependent terms and in the static term H_{E}, whereas the magnetic terms appear only through $\boldsymbol{A}$. The symmetry between electric and magnetic fields can be restored by expressing H in a different way, as will be discussed in Section IV. However where the fields are static a simpler procedure suffices, involving only the first four terms of expression (II.1). $\boldsymbol{A}$ is purely magnetic and is the potential for the magnetostatic field, $\boldsymbol{B} = \text{curl } \boldsymbol{A}$. If the magnetic field is taken to be uniform over each of the molecules (dipole approximation), the components of $\boldsymbol{A}$ are $A_i = \frac{1}{2}(B_j x_k - B_k x_j)$ and its cyclic permutations, x_k and x_j being displacements along the respective axes. It is then readily shown that the magnetic energy operator is, expression (III.1)

$$-\tfrac{1}{2}(\boldsymbol{B}^{(a)} \cdot \boldsymbol{m}^{(a)} + \boldsymbol{B}^{(b)} \cdot \boldsymbol{m}^{(b)}) \qquad \text{(III.1)}$$

$\boldsymbol{B}^{(a)}$ being the magnetic field at molecule a produced by the moment $\boldsymbol{m}^{(b)}$. With the usual expressions for the dipole fields the operator becomes

$$H_{\mathrm{M}} = \frac{\boldsymbol{m}^{(a)} \cdot \boldsymbol{m}^{(b)} - 3(\boldsymbol{m}^{(a)} \cdot \boldsymbol{R})\,(\boldsymbol{m}^{(b)} \cdot \boldsymbol{R})}{R^3} \qquad \text{(III.2)}$$

$$= R^{-3}\, m_i^{(a)}\, m_j^{(b)} \{\delta_{ij} - 3\, \hat{R}_i \hat{R}_j\} \qquad \text{(III.3)}$$

where the indices i and j refer to a rectangular coordinate system (ijk), $m_i^{(a)}$ and $m_j^{(b)}$ are components of the magnetic moment, δ_{ij} is the Kronecker delta and $\hat{R}_i$ the i-th component of a unit vector along $\boldsymbol{R}$. The Hamiltonian in this approximation is

$$H = H_{\mathrm{a}} + H_{\mathrm{b}} + H_{\mathrm{E}} + H_{\mathrm{M}} \qquad \text{(III.4)}$$

in which H_{E} is the dipole-dipole part of the full electrostatic interaction (II.4), namely expression (III.5),

$$H_{\mathrm{E}} = R^{-3} \mu_i^{(a)} \mu_j^{(b)} \{\delta_{ij} - 3 \hat{R}_i \hat{R}_j\} \tag{III.5}$$

$$= R^{-3} \mu_i^{(a)} \mu_j^{(b)} \beta_{ij} \tag{III.6}$$

These expressions, like (III.3), are written with the summation convention.

2. Non-Discriminating Dispersion Terms

In the ordinary theory of the dispersion interaction the electric perturbation term H_{E} alone is included. The dispersion energy is then given by

$$\Delta E_{\mathrm{E}} = - \sum_{n\mathrm{a}, n\mathrm{b}} \frac{<0^{\mathrm{a}}0^{\mathrm{b}}|H_{\mathrm{E}}|n^{\mathrm{a}}n^{\mathrm{b}}> <n^{\mathrm{a}}n^{\mathrm{b}}|H_{\mathrm{E}}|0^{\mathrm{a}}0^{\mathrm{b}}>}{E(n^{\mathrm{a}}) - E(0^{\mathrm{a}}) + E(n^{\mathrm{b}}) - E(0^{\mathrm{b}})} \tag{III.7}$$

where $|n^{\mathrm{a}}>$ indicates the nth excited state of molecule a, $E(n^{\mathrm{a}})$ its energy, and $|0^{\mathrm{a}}>$ the ground state. We suppose that the ground state has no permanent electric or magnetic moment, the effects of which were discussed in Section II. Reference to the form of H_{E} in (III.5) and to the fact that it appears twice in each term of the sum (III.7) shows that the dependence on distance is as R^{-6}, and that the magnitudes are determined by products of four transition electric moments. Explicitly we have in expression (III.8) the dispersion energy ΔE_{E}

$$\Delta E_{\mathrm{E}} = - R^{-6} \beta_{ij} \beta_{kl} \sum_{n\mathrm{a}, n\mathrm{b}}{}' \frac{<0|\mu_i|n^{\mathrm{a}}> <0|\mu_j|n^{\mathrm{b}}> <n^{\mathrm{b}}|\mu_k|0> <n^{\mathrm{a}}|\mu_l|0>}{E(n^{\mathrm{a}}) + E(n^{\mathrm{b}})} \tag{III.8}$$

as a function of the moment components, the energy intervals (ground state energies set to zero), and the dyadics β_{ij} as defined by expressions (III.5) and (III.6). The summation convention applies. If one of the molecules a and b is replaced by its enantiomer, obtained by inversion of the origin of coordinates, the new ΔE is the same as the old, because the signs of *two* of the four moments μ_i are changed in each numerator of the double sum in (III.8). Thus the pure electric dispersion energy is the same for *d-d* and *d-l* pairs.

The interaction operator in (III.4) is the sum of electric and magnetic terms $H_{\mathrm{E}} + H_{\mathrm{M}}$, so that in the complete dispersion energy (III.9)

$$\Delta E = - \sum_{n\mathrm{a}, n\mathrm{b}}{}' \frac{<0^{\mathrm{a}}0^{\mathrm{b}}|H_{\mathrm{E}} + H_{\mathrm{M}}|n^{\mathrm{a}}n^{\mathrm{b}}> <n^{\mathrm{a}}n^{\mathrm{b}}|H_{\mathrm{E}} + H_{\mathrm{M}}|0^{\mathrm{a}}0^{\mathrm{b}}>}{E(n^{\mathrm{a}}) + E(n^{\mathrm{b}})} \qquad \text{III(.9)}$$

there is a purely magnetic term ΔE_{M} involving the magnetic dipole operator (III.2) twice and a cross term involving both H_{E} and H_{M}. The pure magnetic contribution is different from the pure electric (III.8) only in the appearance of the m_i in place of the μ_i, and it is non-discriminating and extremely small.

The ratio of magnetic to electric dipole coupling strength for moments of 1 Bohr magneton and 1 Debye is $10^{16}(4.8\mathrm{h}/4\pi mc)^2 \approx 10^{-4}$, where h is Planck's constant, m the mass of the electron and c the velocity of light. In practice the electric transition dipole moment for at least one of the molecular transitions, even in a small molecule, is $\sim$10D while the magnetic moment is no more than a few Bohr magnetons, and a ratio of 10^{-3} is more realistic. Thus ΔE_{M}, the magnetic analogue of ΔE_{E} in (III.8) is less by six or perhaps eight orders of magnitude. In a typical molecular example the electric dispersion term might be dominated by a single transition with $\mu = 0.2$ enm, $E(n) = 3$ eV. At a separation of 0.5 nm with $\Delta E_{\mathrm{E}} \approx 6.7$ kJ mol^{-1} (560 cm^{-1}) a magnetic term ΔE_{M} less by six or eight orders of magnitude is of little interest.

3. Discriminating Terms

The electric-magnetic cross term in the total dispersion energy (III.9), namely

$$\Delta E_{\mathrm{E-M}} = -2\,\mathrm{Re} \sum_{n^{\mathrm{a}} n^{\mathrm{b}}}{}' \frac{\langle 0^{\mathrm{a}}0^{\mathrm{b}}|H_{\mathrm{E}}|n^{\mathrm{a}}n^{\mathrm{b}}\rangle \langle n^{\mathrm{a}}n^{\mathrm{b}}|H_{\mathrm{M}}|0^{\mathrm{a}}0^{\mathrm{b}}\rangle}{E(n^{\mathrm{a}}) + E(n^{\mathrm{b}})} \tag{III.10}$$

includes only the real part of the sum over states. The corresponding sums in ΔE_{E} and ΔE_{M} are purely real for any choice of basis states n^{a} and n^{b}. Written out as in (II.8), we have

$$\Delta E_{\mathrm{E-M}} = -2\,R^{-6}\,\beta_{ij}\,\beta_{kl}\,\mathrm{Re} \sum_{n^{\mathrm{a}},n^{\mathrm{b}}} \frac{\langle 0|\mu_i|n^{\mathrm{a}}\rangle \langle n^{\mathrm{a}}|m_k|0\rangle \langle 0|\mu_j|n^{\mathrm{b}}\rangle \langle n^{\mathrm{b}}|m_l|0\rangle}{E(n^{\mathrm{a}}) + E(n^{\mathrm{b}})} \tag{III.11}$$

If molecule b is replaced by its enantiomer generated by inversion in the origin, the new $\Delta E_{\mathrm{E-M}}$ differs from the old by sign change of $\langle 0|\mu_j|n^{\mathrm{b}}\rangle$, but not of $\langle m^{\mathrm{b}}|m_l|0\rangle$ because m_l as the component of an axial vector is invariant to inversion. There is thus an overall sign change of $\Delta E_{\mathrm{E-M}}$, which discriminates between *d-d* and *d-l* interactions. A simplification of (III.11) is to restrict it to a single pair of upper levels n^{a} and n^{b}, by assuming that the transition moments to this pair dominate. The simpler expression is given in (III.12)

$$\Delta E_{\mathrm{E-M}} \simeq -2\,R^{-6}\,\beta_{ij}\,\beta_{kl}\,\frac{(R_{ik})_{0n^{\mathrm{a}}}(R_{jl})_{0n^{\mathrm{b}}}}{E(n^{\mathrm{a}}) + E(n^{\mathrm{b}})} \tag{III.12}$$

where

$$(R_{ik})_{0n^{\mathrm{a}}} = \langle 0|\mu_i|n^{\mathrm{a}}\rangle \langle n^{\mathrm{a}}|m_k|0\rangle \tag{III.13}$$

The quantities defined in expression (III.13) are the components of the optical-rotatory pseudotensor. $\boldsymbol{R}_{0n^{\mathrm{a}}}$ transforms under rotations like the second rank tensor for the quadrupole moment ex_ix_j, but is antisymmetric to inversion in the origin. Its contraction $\sum_i \mu_i m_i = \boldsymbol{\mu} \cdot \boldsymbol{m}$ is a pseudo-scalar of which the imaginary part determines the optical rotatory power of the molecule.

Expression (III.11) depends on the coupling of two electric moments for the upward transitions $n^a \leftarrow 0$ and $n^b \leftarrow 0$, and the coupling of the magnetic moments for the corresponding downward transitions. There is no term for the coupling of an electric to a magnetic moment in this static limit, though we shall see that there is an important term of this type in the retarded interaction (Section IV). The time ordered diagrams for ΔE_E and ΔE_{E-M} are shown in Fig. 7.

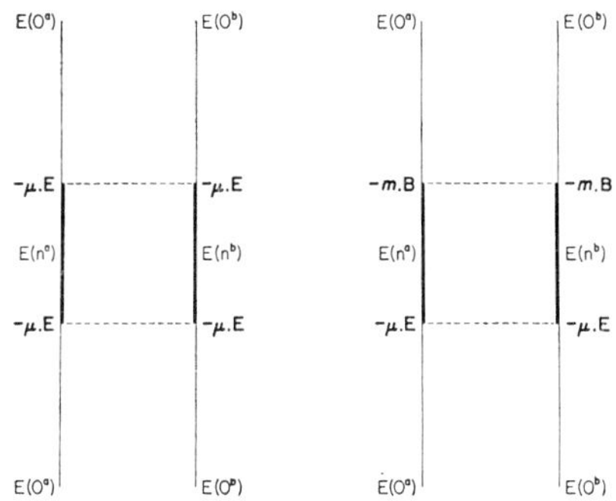

Fig. 7. Time ordered graphs for static dispersion interaction. Left-hand diagram, electric-electric terms. Right-hand diagram, electric-magnetic terms

The magnitude of the discriminating term differs from that of the pure electric dispersion energy through the appearance of the magnetic dipole coupling in place of one of the electric dipole couplings in expression (III.8). By the argument given earlier it should be less by about three orders of magnitude. Thus a total dispersion energy of 6.7 kJ mol^{-1} might be accompanied by a discrimination of 5—10 J mol^{-1} in locked configurations.

4. Orientation Averages of the Discrimination

The question how to apply these results to actual chemical situations is not straightforward. If the interacting molecules are held fixed in a lattice the calculation can be made with the help of expression (III.11) but the meaning of discrimination energy is then the difference between *d-d* and *d-l* energies each calculated for the optimum *d* and *l* orientations. Since these orientations are determined to some degree by all the forces acting, and certainly not only by the dispersion forces, they cannot be treated in a systematic way. In other situations than a crystal lattice it is perhaps realistic to suppose that at distances greater than those for contacts between closed shells the molecules have one axis in each fixed relatively to the other, but are free to rotate about that axis. For example the electric transition dipole direction might be fixed and the magnetic dipole free to rotate about it. As a special case the electric dipoles might be lined up along the intermolecular axis, as would be expected if the electric-electric dispersion term were dominant.

In that case the minimum energy configuration is that with the electric dispersion term (III.8) minimized (representing maximum attraction between a and b). Since the direction of these dipoles is undetermined, there are two distinct relative positions of the molecules satisfying this condition. A second averaging is over all configurations, applicable when the motions of the two molecules, at any distance, are uncorrelated, a situation which might arise in solution, where the effect of solvent molecules is to produce force fluctuations destroying pairwise correlation between the motions of the chiral molecules.

The treatment of these averages [Craig, Power and Thirunamachandran (1971)] is simplified by a decomposition of the magnetic moment into components parallel and perpendicular to the electric moment $\boldsymbol{\mu}$ according to $\boldsymbol{m}_{\mathrm{b}} = \boldsymbol{m}_{\mathrm{b}}^{\parallel} + \boldsymbol{m}_{\mathrm{b}}^{\perp}$. If b is replaced by its enantiomer we must specify the conditions of the replacement: the enantiomer being generated by inversion in the origin, the new moments are $-\boldsymbol{\mu}_{\mathrm{b}}$, $\boldsymbol{m}_{\mathrm{b}}$.

Two corresponding auxiliary expressions derived from the optical-rotatory tensor $\boldsymbol{R}$, are the following:

$$\left.\begin{aligned} R^{\parallel} &= \boldsymbol{\mu} \cdot \boldsymbol{m} = \mu m^{\parallel} \\ R^{\perp} &= |\boldsymbol{\mu} \times \boldsymbol{m}| = \mu m^{\perp} \end{aligned}\right\} \qquad \text{(III.14)}$$

As already remarked the imaginary part of the pseudo-scalar $R^{\parallel}$ determines the optical rotatory power. $\boldsymbol{\mu} \times \boldsymbol{m}$ is a polar vector perpendicular to the electric dipole direction and $R^{\perp}$ its (scalar) magnitude. In both the averages to be discussed the contribution by $R^{\perp}$ averages to zero. Where the molecules are locked with electric dipole directions fixed, but are each free to rotate independently about these axes, the averaged discriminating term is given by expression (III.15)

$$\Delta E_{\mathrm{E-M}}^{\mathrm{Av(I)}} = -2\, R^{-6} \frac{(\cos\gamma - 3\cos\theta_{\mathrm{a}}\cos\theta_{\mathrm{b}})^2\, \mathrm{Re}\{R^{\parallel}(n^{\mathrm{a}}) R^{\parallel}(n^{\mathrm{b}})\}}{E(n^{\mathrm{a}}) + E(n^{\mathrm{b}})} \qquad \text{(III.15)}$$

This average depends only on the contracted $\boldsymbol{R}$ tensor $\boldsymbol{\mu} \cdot \boldsymbol{m}$, apart from the angle parameters specifying the directions of the electric moments $\boldsymbol{\mu}^{\mathrm{a}}$ and $\boldsymbol{\mu}^{\mathrm{b}}$ relative to the separation $\boldsymbol{R}$ as polar direction. θ_{a}, ϕ_{a} are the polar angles of $\boldsymbol{\mu}^{\mathrm{a}}$ and γ is the angle between $\boldsymbol{\mu}^{\mathrm{a}}$ and $\boldsymbol{\mu}^{\mathrm{b}}$, according to (III.16),

$$\cos\gamma = \cos\theta_{\mathrm{a}} \cos\theta_{\mathrm{b}} + \sin\theta_{\mathrm{a}} \sin\theta_{\mathrm{b}} \cos(\phi_{\mathrm{a}} - \phi_{\mathrm{b}}) \qquad \text{(III.16)}$$

where the electric dipoles are aligned along $\boldsymbol{R}$ we have the two special cases of parallel and antiparallel arrangement.

The parallel arrangement is formed when molecule b is related by a translation to molecule a, and then allowed rotational freedom about $\boldsymbol{R}$. The antiparallel arrangement involves translation and a rotation by π of molecule b about an axis perpendicular to $\boldsymbol{R}$, so that the electric dipole sense is reversed from that of a. In the first (parallel) arrangement $\theta_{\mathrm{a}} = \theta_{\mathrm{b}} = \gamma = 0$ and the interaction between molecules of the same chirality is

$$\Delta E_{\mathrm{E-M}}(d\text{-}d) = -8\, R^{-6} \frac{\mathrm{Re}\{R^{\parallel}(n^{\mathrm{a}}) R^{\parallel}(n^{\mathrm{b}})\}}{E(n^{\mathrm{a}}) + E(n^{\mathrm{b}})} \qquad \text{(III.17)}$$

In the antiparallel arrangement the sign is changed. To deal with molecules of opposite chirality we change the sign of $R^{\|}(n^{\mathrm{b}})$, and note that this again changes the sign of the interaction as found for molecules of like chirality. We thus have for the discrimination energy, namely $\Delta E(d\text{-}d) - \Delta E(d\text{-}l)$,

$$\Delta E_{\mathrm{disc}} = -16\, R^{-6} \frac{\mathrm{Re}\{R^{\|}(n^{\mathrm{a}})R^{\|}(n^{\mathrm{b}})\}}{E(n^{\mathrm{a}}) + E(n^{\mathrm{b}})} \qquad \text{(III.18)}$$

for parallel molecules, and for antiparallel the same quantity with opposite sign.

The average over all orientations of both molecules is readily found from (III.15) by averaging over θ_{a}, θ_{b} and $\phi_{\mathrm{a}} - \phi_{\mathrm{b}}$; the required average is

$$\Delta \,{}^{\mathrm{Av}\,(2)}_{\mathrm{E-M}} = \frac{1}{8\pi} \int_0^{2\pi}\int_0^{\pi}\int_0^{\pi} \Delta \,{}^{\mathrm{Av}\,(1)}_{\mathrm{E-M}} \sin\theta_{\mathrm{a}} \sin\theta_{\mathrm{b}}\, d\theta_{\mathrm{a}}\, d\theta_{\mathrm{b}}\, d(\phi_{\mathrm{a}} - \phi_{\mathrm{b}})$$

$$= -\,(4/3)\, R^{-6} \frac{\mathrm{Re}\{R^{\|}(n^{\mathrm{a}})R^{\|}(n^{\mathrm{b}})\}}{E(n^{\mathrm{a}}) + E(n^{\mathrm{b}})} \qquad \text{(III.19)}$$

Expression (III.19) was first found by Mavroyannis and Stephen (1962); the discrimination energy is twice this quantity.

We have still to discuss the absolute sign of (III.17) and (III.19), taking account of the sign of the real part of the product of the contracted R tensors. We can usually choose the basis molecular wave functions to be real. The expectation value of the magnetic moment, and therefore of $\boldsymbol{\mu} \cdot \boldsymbol{m}$, is then purely imaginary. The product in braces in (III.17) and (III.19) is real and negative, and the overall sign of both expressions is positive, showing that the discrimination favours the unlike (*d-l*) over the like (*d-d*) interaction. The like species repel and unlike attract, so far as the contribution by the electric-magnetic term is concerned. The overall dispersion interaction is strongly attractive. The ratio of the averaged discriminating energy to the averaged total is again 10^{-3}—10^{-4}.

IV. Resonance Discrimination

1. Physical Ideas

The topic in this section is one for which no practical application has yet been suggested, though various possibilities exist. Its inclusion here is not simply a *jeu d'esprit*. The magnitudes are small but comparable to those in the dispersion calculation and the physical ideas are novel. They come from a starting point fundamentally different from that of the permanent moment coupling in Section II, and that of the static or instantaneous coupling of transition moments in Section III.

We first recall the essential features of resonance coupling in a context in which there is no discrimination. A pair of identical molecules a and b possesses identical energy levels $|n^a>$ and $|n^b>$ with energies $E(n^a)$ and $E(n^b)$ as before. If a and b are well separated any state function of the molecule pair is approximately the product $\psi_a(n^a)\ \psi_b(n^b)$ of states of the isolated pair. Pair states of the type $\psi_a(0)\ \psi_b(n^b)$ and $\psi_a(n^a)\ \psi_b(0)$ in which one molecule is in its ground state and one in the n-th excited state have identical energies and are in resonance. If we take as Hamiltonian the first three terms of (II.1), namely

$$H = H_a + H_b + H_E \tag{IV.1}$$

and expand H_E to the dipole-dipole term we find

$$H = H_a + H_b + R^{-3}\,\mu_i^{(a)}\,\mu_j^{(b)}\,\beta_{ij} \tag{IV.2}$$

the μ_i's being as before operators for the components of the dipole moment. The effect of the third term in (IV.2) is to couple the states denoted by $(0;n^b)$ and $(n^b;0)$ so that the excitation is exchanged between molecules a and b, and over a sufficiently long time the stationary states are given by the wave functions (IV.3),

$$\psi^{\pm} = \frac{1}{\sqrt{2}}\,\{\psi_a(0)\ \psi_b(n^b)\ \pm\psi_a(n^a)\ \psi_b(0)\} \tag{IV.3}$$

separated by an energy

$$2\,\varepsilon = 2\,R^{-3}\,\beta_{ij}\,<0\{\mu_i\,|n^a>\ <n^b|\,\mu_j\,|\,0> \tag{IV.4}$$

the quantity ε being the resonance energy, depending on the inverse third power of the separation distance R, and appearing as the first order perturbation correction by the dipole-dipole interaction. The electric-electric dispersion energy in expression (III.7) is a second-order correction due to the same operator. The transfer time of excitation energy between molecules is $\tau \sim h/\varepsilon$ and, for resonance coupling of dipole-allowed excitation at $R \simeq 0.5$ nm, is about 10^{-14}s. The matrix element for this static dipole-dipole coupling corresponds to the time ordered diagram on the left hand side of Fig. 8.

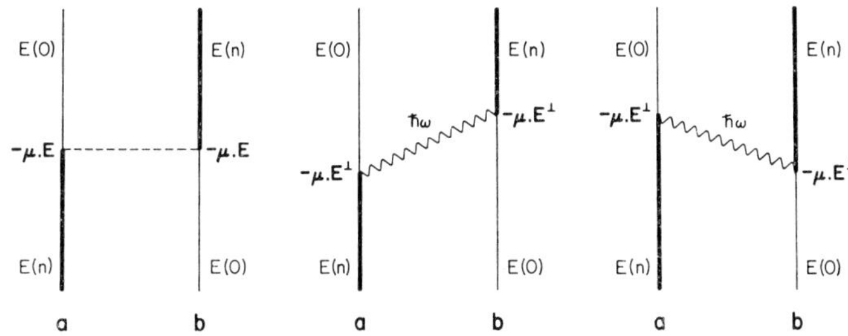

Fig. 8. Time ordered graphs for static and fully retarded electric-electric resonance interaction

Resonance excitation transfer and its associated resonance energy ε do not arise only by the static mechanism in the Hamiltonian (IV.2), and a more general approach is of special importance for resonance discrimination, as will be seen. This is the coupling of the excited and unexcited molecules by the emission and absorption of a photon. This radiative interaction is contained in the terms $\boldsymbol{p}.\boldsymbol{A}$ and $-\frac{1}{2}\boldsymbol{A}^2$ of the Hamiltonian (II.1), appearing there as an addition to the static H_{E}. The structure of the complete Hamiltonian is that at long distances the static interaction is cancelled by a part of the radiative terms, and an alternative form of the Hamiltonian given by Power and Zienau (1959) provides better insight as well as greater convenience in calculation. The Hamiltonian is given in (IV.5) and (IV.6)

$$H = H_{\mathrm{a}} + H_{\mathrm{b}} + H_{\mathrm{int}} + H_{\mathrm{rad}} \tag{IV.5}$$

$$H_{\mathrm{int}} = -\boldsymbol{\mu}^{(\mathrm{a})} \cdot \boldsymbol{E}\perp(\mathrm{a}) - \boldsymbol{\mu}^{(\mathrm{b})} \cdot \boldsymbol{E}\perp(\mathrm{b}) - \boldsymbol{m}^{(\mathrm{a})} \cdot \boldsymbol{B}(\mathrm{a}) - \boldsymbol{m}^{(\mathrm{b})} \cdot \boldsymbol{B}(\mathrm{b}) \tag{IV.6}$$

this form of H_{int} being confined to the dipole approximation. The appropriate graphs are given in the centre and right-hand side of Fig. 8. H_{rad} has the form given in (II.3). Features of (IV.6) are that the static interaction does not appear separated out from the complete electric interaction, but is included with the radiative coupling in the first two terms involving the transverse electric field. The analogous term in Section III for static coupling depended on the complete electric field vector $\boldsymbol{E}$.

The full result for the electric dipole-electric dipole resonance interaction arising from the complete Hamiltonian (IV.5) was given by McLone and Power (1964). This generalisation of the static result in (IV.4) is given in expression (IV.7),

$$\varepsilon_{\mathrm{E}}^{(\mathrm{R})} = \mu_i\, \mu_j \left\{ \beta_{ij} \left(\frac{\sin(R/\lambda\!\!\!^{-})}{bR^2} + \frac{\cos(R/\lambda\!\!\!^{-})}{R^3} \right) - \alpha_{ij} \frac{\cos(R/\lambda\!\!\!^{-})}{\lambda\!\!\!^{-2}R} \right\} \tag{IV.7}$$

where μ_i and μ_j are the expectation values of the transition moments in the two molecules, $\lambda\!\!\!^{-} = \lambda/2\pi = \hbar c/E(n)$ is the reduced characteristic wavelength for the resonance transition, and α_{ij}, defined in (IV.8),

$$\alpha_{ij} = \delta_{ij} - \hat{R}_i\, \hat{R}_j \tag{IV.8}$$

is the dyadic for coupling of the transverse parts of the dipole moments, whereas β_{ij} as earlier introduced in (III.6) is for full dipolar coupling, including the longitudinal parts. The matrix elements for the calculation of the result in (IV.7) are associated with the right hand graphs in Fig. 8. It is readily seen that at short distances $\varepsilon_E(R)$ tends toward (IV.4), as the static interaction becomes dominant, and at long distances goes into the pure transverse coupling given by the final (radiation) term, varying with distance as $\cos(R/\lambda\!\!\!^{-})\lambda\!\!\!^{-2}R$. At intermediate distances the first ('induction') term has some importance.

2. The Electric-Magnetic Resonance

The coupling now to be described is the analogue of the electric-electric dipole coupling (IV.7) with graphs as in Fig. 9.

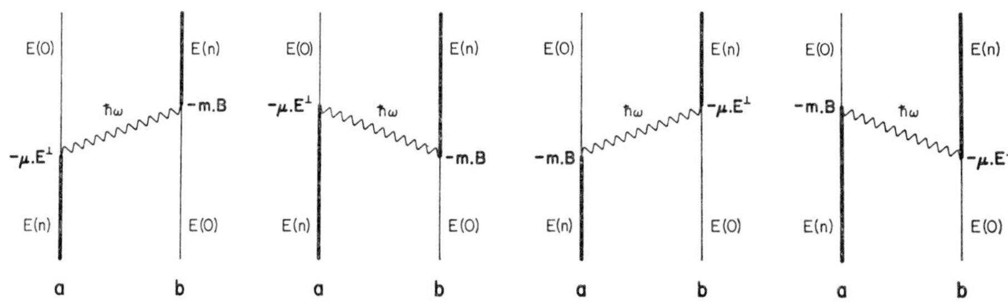

Fig. 9. Time ordered graphs for electric-magnetic resonance interaction

This coupling is that of the electric dipole in one molecule to the magnetic dipole in the other. We know that a permanent electric dipole moment is not acted on by a magnetic field, but time-dependent electric and magnetic moments both interact with an electromagnetic field. A photon emitted by one molecule through its electric transition dipole acted upon by the electric vector of the field can be absorbed by the other through the magnetic dipole and magnetic field vector, and *vice versa*.

Resonance discrimination is evidently confined to pairs of molecules with identical energy levels, but in addition the level or levels concerned must possess both electric and magnetic moments joining them to the ground state. The molecules must therefore be chiral as well as chemically identical, and the meaning of discrimination is that the resonance energy is unequal for *d-d* and *d-l* resonance. The total resonance energy, in close analogy with the dispersion energy, has pure electric contributions ε_E, and pure magnetic ε_M. ε_M differs from ε_E in (IV.7) only by the substitution of the magnetic moment components m_i and m_j for μ_i and μ_j. The cross term ε_{E-M} [Craig, Power and Thirunamachandran (1971)] is given in expression (IV.10)

$$\varepsilon_{E-M} = \varepsilon_{ijk}\,\hat{R}_k\left(\frac{\cos(R/\lambda\!\!\!^{-})}{\lambda\!\!\!^{-}R^2} + \frac{\sin(R/\lambda\!\!\!^{-})}{R^2}\right)\Big\{ <n^a|\mu_i|0> <0|im_j|n^b> + <n^b|\mu_i|0> <0|im_j|n^a> \Big\} \qquad \text{(IV.10)}$$

Repeated indices are summed, and the three index symbol ε_{ijk} is $+1$ and -1 for cyclic and non-cyclic sequences of the indices. For real basis wave functions

$\varepsilon_{\mathrm{L-M}}$ is itself real. It contains no static term (*i.e.* no term which is non-zero as $\bar{\lambda} \to \infty$) and at short distances goes to a limiting form (IV.11) with a dependence on distance as R^{-2},

$$\varepsilon_{\mathrm{E-M}}\,(R \to 0) = \varepsilon_{ijk}\hat{R}_k \frac{1}{\bar{\lambda} R^2} \Big\{ \langle n^{\mathrm{a}} | \mu_i | 0 \rangle \langle 0 | \mathrm{i} m_j | n^{\mathrm{b}} \rangle + \langle n^{\mathrm{b}} | \mu_i | 0 \rangle \langle 0 | \mathrm{i} m_j | n^{\mathrm{a}} \rangle \Big\} \tag{IV.11}$$

This quantity is small at all distances. Comparing it with the static electric dipole-dipole resonance interaction we see that instead of the electric moment appearing quadratically, (IV.9) contains the magnetic moment and the electric moment; also the denominator $\bar{\lambda} R^2$ is greater than R^3 by 2 or 3 orders of magnitude in the important range of R. Thus the resonance discrimination is perhaps 10^{-5} times the total resonance interaction and probably not over 0.1 cm^{-1} in spectroscopic units. Its directional properties are those of $(\boldsymbol{\mu} \times \boldsymbol{m}) \cdot \boldsymbol{R}$, namely a transverse dipole coupling maximized for the electric moment of one molecule at right angles to the magnetic moment in the other, and both at right angles to $\boldsymbol{R}$ These are the conditions for optimum interaction through an electromagnetic field joining molecules one through the electric and the other through the magnetic vector.

3. Possible Application

The discussion of magnitudes in Section IV.2 was on the term $\varepsilon_{\mathrm{E-M}}$ only, which one expects to be the only contributor to discrimination in any actual case. Replacement of molecule b by its enantiomer according to the prescription in Section III changes the value calculated from expression (IV.10). However it also changes the sign of the pure electric ε_{E} in (IV.7) leaving ε_{M} unchanged. Alternatively by changing the prescription for generating the enantiomer one could change the sign of ε_{M} and not of ε_{E}. There is a second difficulty in the fact that each of ε_{E}, ε_{M} and $\varepsilon_{\mathrm{E-M}}$, when averaged over all orientations of a and b, vanishes, so that there is no way of calculating the resonance discrimination that is independent of orientation, or indeed independent of the choice of molecular origins as discussed in Section II. This contrasts with the dispersion calculations in Section III, where averaging over all orientations leaves non-zero contributions, which are well defined apart from sensitivity of the various moments to the choice of origin. The most useful approach appears to be to define the enantiomer in a way which leaves the electric moment invariant. Then ε_{E} is the same for *d-d* and *d-l* interactions, and ε_{M}, which changes sign, is extremely small. The discrimination is then given by $2\varepsilon_{\mathrm{M}}$ together with the difference of $\varepsilon_{\mathrm{E-M}}$.

In trying to devise ways in which the resonance discrimination might produce detectable effects the suggestion was made [Craig, Power and Thirunamachandran (1971)] that the replacement of isolated *d* molecules in a crystal by the *l* isomers might produce spectral or other changes of a sufficient size. The full exploration of this crystal system by Dissado (1974) suggests that a discriminating term of about 10^{-4} times the total resonance interaction is expected, amounting in favourable cases to a spectroscopic interval of ~ 1 cm^{-1}.

V. Discrimination by Short Range Forces

1. Nature of Short-Range Discrimination

We pointed out in Section I that the origin of the discrimination in crystal packing, underlying the Pasteur method of separating diastereoisomers by fractional crystallization, is in the 'contact' interactions or very short range repulsions corresponding physically to the exchange force opposing the interpenetration of closed electron shells. It is a consequence of the difference in range of the dispersive attractions ($\sim R^{-6}$) and repulsions ($\sim R^{-12}$) that the former may be responsible for the stability of a crystal while the latter dominates the detailed crystal packing and structure. Where the units are ions or are strongly dipolar and the binding forces ionic rather than dispersive, repulsions again determine local structure, although where alternative local structures are not too different in energy there is evidence that permanent electric moments have an effect.

Some consequences were discussed in Section I. If a pair of chemically identical (*d-d*) or nonidentical molecules (*d*-D) are packed together, and this packing compared with the corresponding *d-l* or *d*-L pair, each for minimum total energy, there is no relation between the pair structures to enable them to be treated in a systematic way. Each case depends on the particular molecular composition and the particular atomic non-bonded radii which collectively are responsible for the surface 'shape' of the molecule conceived as bounded by a hard surface. For each case it is of course possible to make calculations of packing energy and of optimum structure in the way now commonly followed for the packing in molecular crystals. Such calculations have not been reported so far as we are aware.

2. The Intramolecular Analogue of Discrimination by Short-Range Forces

If we consider a substituted ethane in which each methyl moiety is chiral (considered as rigid) and with the same set of substituents, the problem of optimum packing of the methyls is that of finding the optimum dihedral angles in the two possible situations, in which the methyl moieties are of the same of opposite chirality. There is one degree of freedom, the dihedral angle, in place of the six degrees in a typical unconstrained intermolecular contact. This primitive form of intramolecular discrimination is perhaps the simplest model of the effect of short range forces. We now refer to a recent calculation of it [Craig, Radom and Stiles (1975)]. The substituted ethane is 2,3-dicyanobutane, each methyl moiety carrying CN, CH_3, and H attached to the central carbon. The moieties are asymmetric, and may be joined in two ways, to give a *meso* form and two enantiomeric chiral forms, as in Fig. 10.

The active forms belong to the dissymmetric group $\boldsymbol{C}_2$ and are chiral in all configurations. The meso form belongs to $\boldsymbol{C}_{1h}$, with mirror plane normal to the C—C bond for one eclipsed configuration but is otherwise asymmetric: the two enantiomers however interconvert by rotation about the C—C bond and the molecule is inactive for the expected barrier heights.

The calculation by *ab initio* methods is of the total molecular energy as a function of the dihedral angles. The potential has three maxima coinciding with the eclipsed configurations, and three minima at the staggered configurations. The

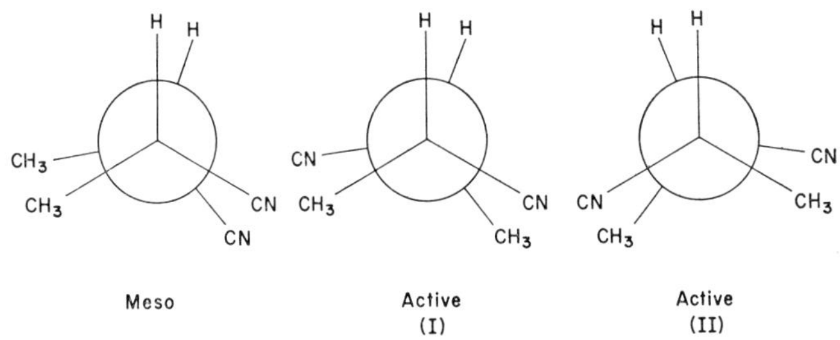

Fig. 10. Projections of substituted ethane molecules

depths of these minima and their populations weighted by appropriate Boltzmann factors determine, as differences between meso and active forms, the intramolecular discrimination. The value found is 3.0 kJ mol^{-1}, a value several orders of magnitude greater than any likely dispersion term, and some ten times larger than permanent electric moment discrimination. How far such a result can be transferred to intermolecular cases is unclear. In the ethane calculation the chiral fragments are held together by a covalent bond and cannot relax and reduce the repulsions so readily as intermolecular pairs held only by weaker constraints. The found 3.0 kJ mol^{-1} is probably on the high side of likely intermolecular energies of discrimination.

3. The Statistical Approach

The differences in thermodynamic properties between chirally pure systems and racemates have their microscopic origins in the discriminating interactions between *d-d* and *d-l* pairs, and certainly in many cases to a dominating degree in short-range interactions which can in the limit be analysed in terms of contacts between hard surfaces. We have already noted that the treatment of microscopic properties of contact interactions is difficult, mainly because the optimum orientations depend sensitively on molecular shape, and general microscopic approaches have not yet been developed. The possibility of a statistical treatment in terms of the theory of a classical fluid of non-spherical molecules has been investigated in a preliminary way by Sawford (1975). Discrimination is looked for in the second virial coefficient. The molecule is modelled as an aggregate of spherical atoms with hard surfaces, and the intermolecular interactions are treated as sums of atom-atom contributions. The second virial coefficient must be evaluated by numerical integration. Sawford's preliminary calculation is confined to a two-dimensional 'flatland', in which chiral molecules appear as projections on a plane. For example the pair (a) and (b) shown in Fig. 11 is chiral within symmetry operations permitted in flatland, *i.e.* excluding motions out of the plane. The second virial coefficient is the excluded area on the projection plane averaged over all orientations. The numerical integration is done using a simple Monte-Carlo method following Rigby (1970). In a given number of trial configurations the second virial coefficient is proportional to the number of configurations in which the two molecules overlap. The calculation of the discrimination, allowing for error associated with the number of configurations sampled, gave 0.2—0.3% or less of the total excluded volume, the excluded volume being smaller for interactions of like-like type. Similar results were found for other simple projected 'molecules'.

Again in flatland, exact calculations are possible for the volumes excluded by simple re-entrant polygons [Fig. 11 (c) and mirror image], which while bearing little relation to projected actual molecules, help to confirm the reality of discrimination arising from space-filling differences in pairwise contacts. The excluded volumes for like-like and like-unlike shapes (c) are 206.619966 a and 206.852126 a respectively, showing discrimination of about 0.1% with like-like lower.

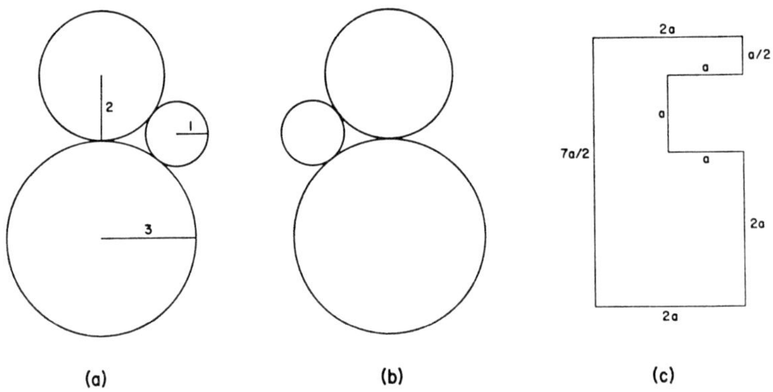

Fig. 11. 'Chiral' systems in two dimensions

In general molecular shape is of less importance than size and interaction energy in determining the properties of mixtures and solutions (Rowlinson 1970) and the small second virial discriminations are not surprising. Calculations on more complicated systems, though extremely difficult at the moment, seem likely to provide a rather direct route to some quantitative understanding of chiral effects in solutions and pure liquids.

Acknowledgements. We are glad to thank a number of colleagues for advice and critical reading of the manuscript, particularly Drs E. V. Lassak, P. J. Stiles, A. M. Sargeson and T. Thirunamachandran.

VI. References

Albert, A.: Selective toxicity, 5th ed., 1973, Chapman.
Anderson, E.: J. Biol. Chem. *100*, 249 (1933).

Barnes, D. S., Pettit, L. D.: Chem. Commun. *1970*, 1000.
Barnes, G. T., Backhouse, J. R., Dwyer, F. P., Gyarfas, E. C.: Proc. Roy. Soc. N.S.W. *89*, 151 (1956).
Bennet, W. E.: J. Am. Chem. Soc. *81*, 246 (1959).
Blashko, J.: Proc. Roy. Soc. (London), Ser. B. *137*, 307 (1950).
Bosnich, B., Watts, D. W.: J. Am. Chem. Soc. *90*, 5744 (1968).
Bosnich, B., Watts, D. W.: Dissymmetric solvent interactions: Thermodynamic parameters for the enantiomerization of tris(o-phenanthroline) nickel (II) ions in (—)-2,3-butanediol. 1975, to be published.

Cahn, R. S., Ingold, C. K., Prelog, V.: Angew. Chem. Intern. Ed. Engl. *5*, 385 (1966).
Caldwell, D. J., Eyring, H.: The theory of optical activity, p. 15. New York: Interscience 1971.
Candlin, R., Harding, M. M.: J. Chem. Soc. A. *1970*, 384.
Carassiti, V.: J. Inorg. Nucl. Chem. *8*, 227 (1958).
Chedd, G.: New Scientist *62*, 299 (1974).
Craig, D. P., Power, E. A., Thirunamachandran, T.: Proc. Roy. Soc. (London), Ser. A *322*, 165 (1971).
Craig, D. P., Radom, L., Stiles, P. J.: Proz. Roy. Soz. (London), Ser. A *343*, II (1971).
Craig, D. P., Schipper, E.: Chem. Phys. Letters *25*, 476 (1974).
Craig, D. P., Schipper, E.: Proc. Roy. Soc. (London), Ser. A *342*, 19 (1975).

Dissado, L. A.: J. Phys. Chem. *7*, 463 (1974).
Dwyer, F. P., Davies, N. R.: Trans. Faraday Soc. *50*, 24 (1954).
Dwyer, F. P., Mellor, D. P.: Chelating agents and metal chelates. London–New York: Academic Press 1964.
Dwyer, F. P., Gyarfas, N., O'Dwyer, M. F.: Nature *168*, 29 (1951).
Dwyer, F. P., Gyarfas, N., O'Dwyer, M. F.: Proc. Roy. Soc. N. S. W. *86*, 146 (1956).

Ehrlich, F.: Biochem. Z. *1*, 8 (1906); Ber. *41*, 1453 (1908), Biochem. Z. *182*, 245 (1927).
Friedman, L., Miller, J. G.: Science *172*, 1044 (1971).

Gillard, R. D.: Chem. Commun. *1973*, 585.
Gillard, R. D.: Inorg. Chim. Acta *11*, L21 (1974).
Gillard, R. D., Irving, H. M., Parkins, R. M., Payne, N. C., Pettit, L. D.: J. Chem. Soc. A *1966*, 1159.
Gottlieb, R.: Arch. Exp. Path. Pharmakol. *97*, 113 (1923).
Greenstein, J. P., Winetz, M.: Chemistry of the amino acids, Vol. I, p. 545. New York: John Wiley 1961.
Guenther, E.: The essential oils, Vol. II. New York: D. van Nostrand 1949.

Harding, M. M., Long, H. A.: J. Chem. Soc. A *1968*, 2554.
Henderson, G. M., Rule, H. G.: J. Chem. Soc. *1939*, 1568.
Hirshfelder, J. O., Curtiss, C. F., Bird, R. B.: Molecular theory of gases and liquids, p. 843. New York: Wiley 1954.

Jamison, M. M., Turner, E. E.: J. Chem. Soc. *1942*, 437.

Koch, J. H., Rogers, W. P., Dwyer, F. P., Gyarfas, E. C.: Australian J. Biol. Sci. *10*, 342 (1957).

Lecoq, H.: Bull. Soc. Roy. Sci. Liege *12*, 316 (1943).
Lin, C. T., Curtin, D. Y., Paul, I. C.: J. Am. Chem. Soc. *96*, 6200 (1974).
Luduera, F., von Euler, L., Tullar, B., Landa, A.: Arch. Intern. Pharmacodyn. *11*, 392 (1957).

McDonald, C. C., Phillips, W. D.: J. Am. Chem. Soc. *85*, 3736 (1963).
McLone, R. R., Power, E. A.: Mathematika *11*, 91 (1964).
Mavroyannis, C., Stephen, M. J.: Mol. Phys. *5*, 629 (1962).
Mizumachi, K.: J. Coord. Chem. *3*, 191 (1973).
Moeller, T., Gulyas, E.: J. Inorg. Nucl. Chem. *5*, 245 (1958).

Pasteur, L.: Compt. Rend. *46*, 615 (1858).
Pfeiffer, P., Quehl, K.: Ber. *64*, 2667 (1931); *65*, 560 (1932).
Popjak, G.: Stereospecificity of enzymic reactions, Chap. 3. In: The enzymes (ed. P. D. Boyer). New York: Academic Press 1970.
Power, E. A., Zienau, S.: Phil. Trans. Roy. Soc. (London), Ser. A *251*, 427 (1959).
Prelog, V., Wieland, P.: Helv. Chim. Acta *27*, 1127 (1944).

Quiocho, F. A., Bethge, P. H., Lipscomb, W. N., Studebaker, J. F., Brown, R. D., Koenig, S. H.: Symp. Quantit. Biol. *36*, 561 (1971).

Rây, P., Dutt, N. K.: J. Indian Chem. Soc. *18*, 289 (1941); *20*, 81 (1943).
Read, J., McMath, A. M.: J. Chem. Soc. *127*, 1572 (1925).
Riegel, B., Sherwood, L.: J. Am. Chem. So. *71*, 1129 (1949).
Rigby, M.: J. Chem. Phys. *53*, 1021 (1970).
Ritsma, J. H., Van de Grampel, J. C., Jellinek, F.: Rec. Trav. Chim. *88*, 411 (1969).
Ritsma, J. H., Wiegers, G. A., Jellinek, F.: Rec. Trav. Chim. *84*, 1577 (1965).
Rowlinson, J. S.: Discussions Faraday Soc. *49*, 30 (1970).
Russell, G. F., Hills, J. I.: Science *172*, 1043 (1971).

Sawford, B. L.: 1975, unpublished results.
Schipper, P. E.: Inorg. Chim. Acta *14*, 161 (1975).
Schulze, E., Bosshard, E. Z.: Physiol. Chem. *10*, 134 (1886).
Shallenberger, R. S.: Nature *221*, 555 (1969).
Shallenberger, R. S., Acree, T. E.: In: Handbook of sensory physiology, Vol. IV (ed. Beidler). Berlin–Heidelberg–New York: Springer 1971.

Tchen, T. T.: J. Biol. Chem. *233*, 1100 (1958).

Van Vleck, J. H.: Electric and magnetic susceptibilities, p. 280. Oxford 1931.

Zeller, E. A., Maritz, A.: Helv. Chim. Acta *28*, 365 (1945).

Received May 6, 1975

No-Bond-Resonance Compounds, Structure, Bonding and Properties

Prof. Dr. Rolf Gleiter

Institut für Organische Chemie der Technischen Hochschule D-6100 Darmstadt

Dipl.-Chem. Ruedi Gygax

Physikalisch-Chemisches Institut der Universität CH-4056 Basel

Dedicated to Professor Dr. H. Behringer on the occasion of his 65th birthday

Contents

I. Introduction and Scope

About 33 years after its first synthesis [1] it was discovered [2,3] that the reaction product between diacetylacetone and phosphorouspentasulfide does not have the seven membered ring structure *A*. It has the rather unusual structure *B*. X-ray analysis [2] of *B* as well as spectroscopic investigations [3] of *C* showed that unusual

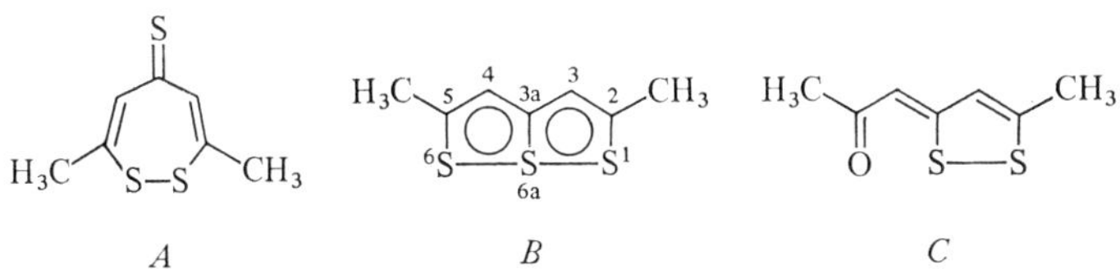

bond lengths are present in these compounds. After these discoveries these compounds aroused special interest due to their peculiar type of bonding, sometimes referred to as "no-bond-resonance" [2,4,5]. This term is meant to express that the σ-bond between the sulfur centers is delocalized as exemplified by the two resonance formulas shown below analogous to the well known π delocalization

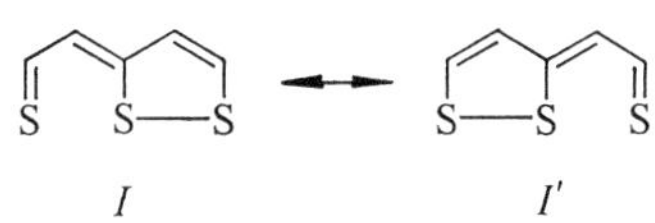

in benzene.

This kind of description accounts for the uncommonly long distances between the sulfur atoms found by X-ray analysis (see next chapter) as well as for the fact that it has not been possible to isolate two different isomers [6,7] which would arise from the following equilibrium.

(1)

Even at low temperatures none of the expected isomers could be detected by NMR spectroscopy (see Chapter II.3.).

During the last years many experiments have been carried out that promoted a wider understanding of the bonding and the structure in these molecules. Preparative chemists have developed new methods of introducing all kinds of substituents and to replace the sulfur atoms by other heteroatoms. Spectroscopists used new methods designed to elucidate the structure, and together with model calculations, to understand the bonding in these species.

This review is devoted to a critical appreciation of those efforts whose aim was the clarification of the structure and the comprehension of the bonding and properties of no bond resonance compounds.

In Scheme 1 a short summary of the most important ways of synthesizing trithiapentalenes is presented.

1) From triketones

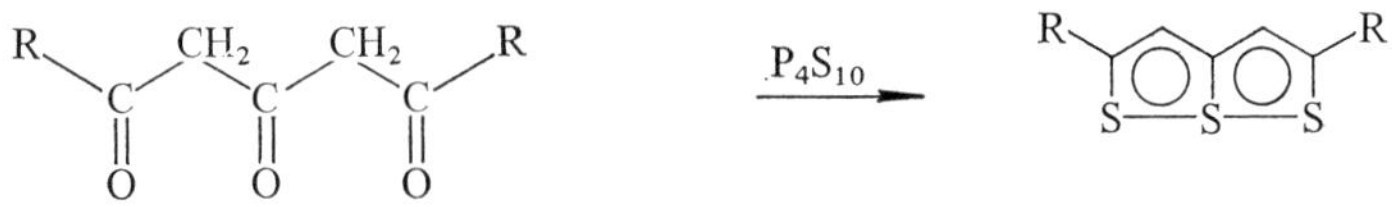

(2) [1,7,8-12]

2) Condensations

a) $C_6H_5-C\equiv C-CO-CH_2-R$

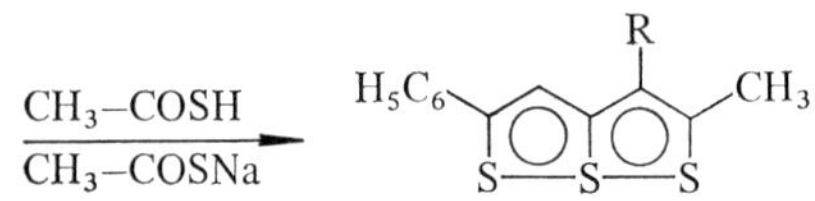

(3) [13]

b)

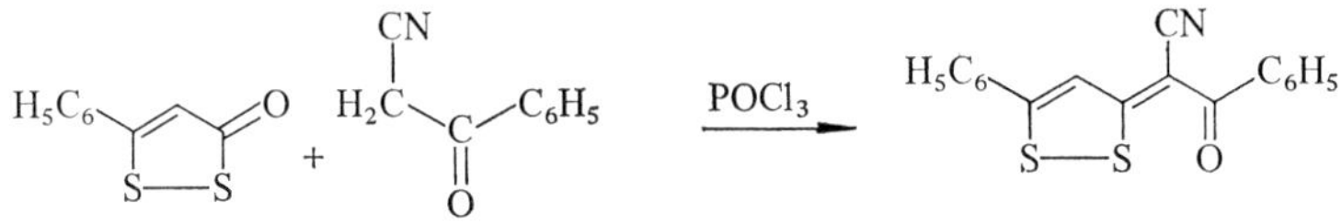

(4) [9,14,15]

c)

(5) [14,15]

d)

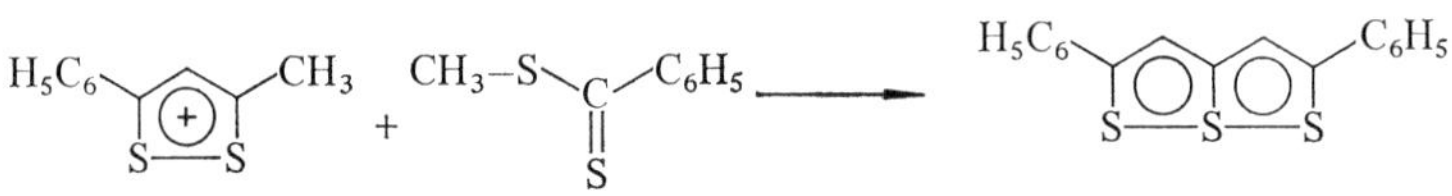

(6) [16]

e)

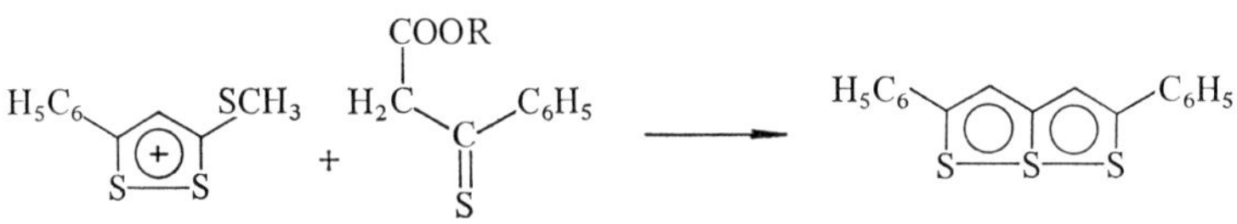

(7) [17]

3) From γ pyrone derivatives

a)

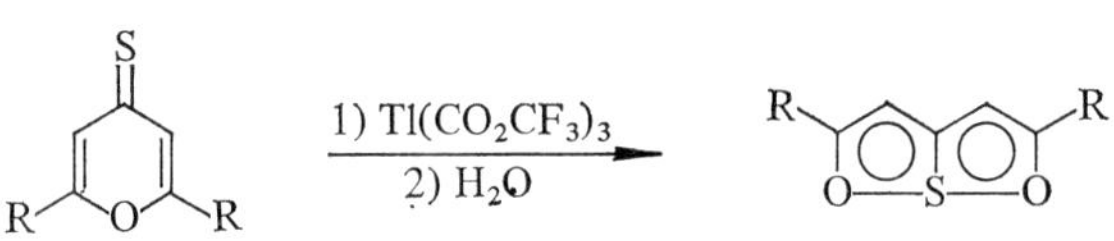

(8) [18]

b)

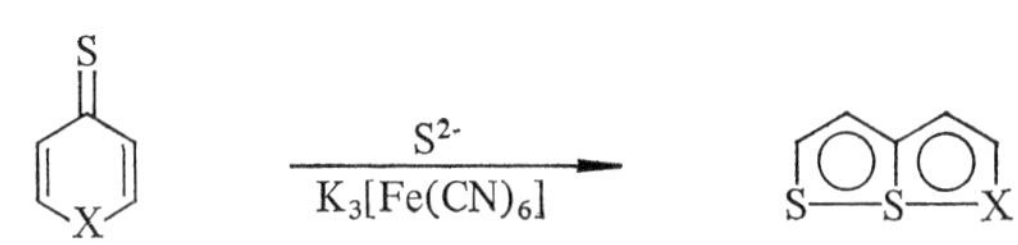

(9) [19,20]

4) From Vilsmeier salts

a)

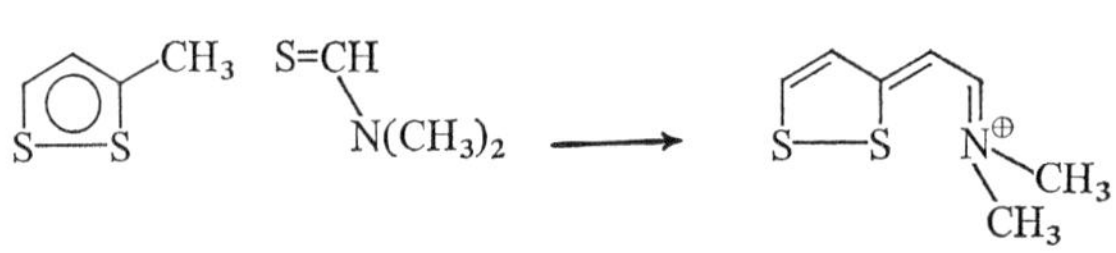

(10) [21-23]

Scheme 1

b) 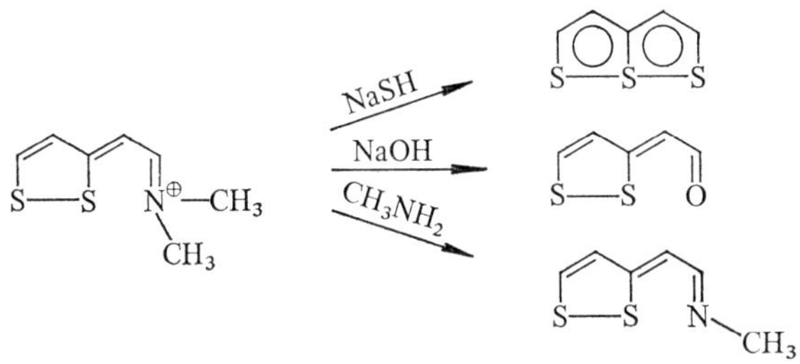 11)[21–23]

5) Interconversions

a) 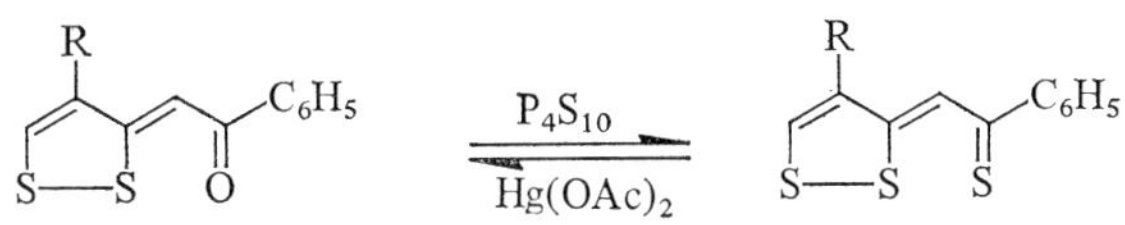 (12)[14]

b) 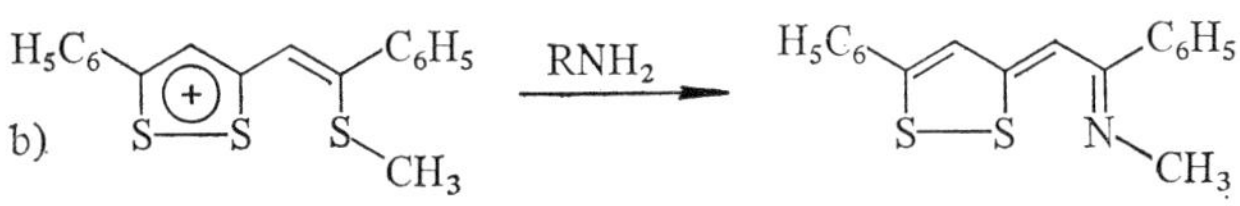 (13)[24]

c) 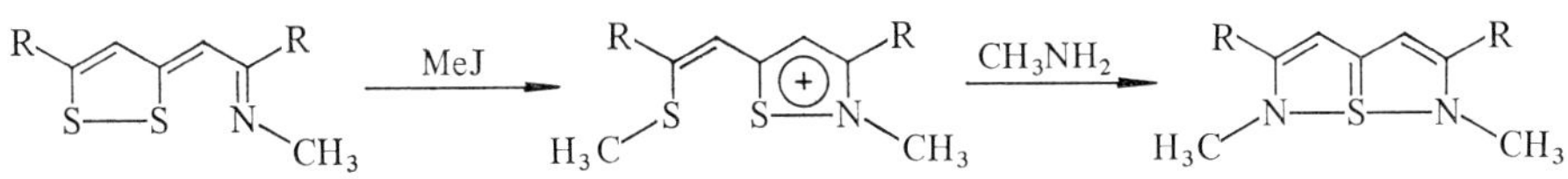 (14)[22]

Scheme 1 (continued)

A more detailed description of synthetic approaches has been given in recent reviews[4,8,25].

II. Molecular Structure of Trithiapentalenes

1.1 X-Ray Measurements on Trithiapentalene and Substitution Products

Fig. 1a shows a collection of information from an X-ray analysis of the unsubstituted trithiapentalene [26]. In Table 1 the distances between the sulfur atoms of selected differently substituted trithiapentalenes are listed.

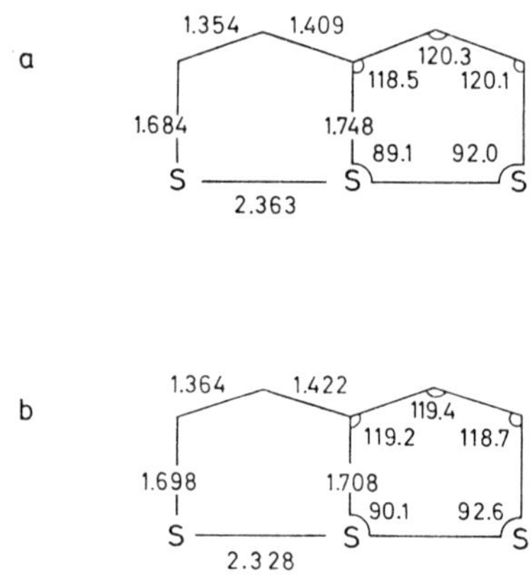

Fig. 1. Comparison between the structural data of trithiapentalene (*Ia*) as obtained by X-ray (a) and electron diffraction (b) analysis

A comparison of the structural data indicates that in some symmetrically substituted compounds (*Ia, Ib*) the two S—S bond lengths are equal, while in others (*Id, Ie, Ih*) this is not the case. This could be due to different angles of rotation of the aryl groups, thus creating an unsymmetrical environment in the crystal lattice.

Another interesting point arises from a comparison between *If* and *Ig*. A substitution of a phenyl group by a p-dimethylaminophenyl group in the 5-position has a large influence on the S—S bond length in *I*.

These results indicate that the S—S bond lengths are easily influenced by substitution in the carbon skeleton. However, the geometry of the carbon skeleton itself remains essentially constant. The distances C_2—C_3 and C_4—C_5 are always shorter than C_3—C_4 and C_{3a}—C_4.

The S—S distances collected in Fig. 1 and Table 1 should be compared with the corresponding sum of the Van-der-Waals radii [R(S···S) = 3.7 Å] [36] on the one hand, and with the covalent radii in cyclic disulfides on the other. Typical values [R(S—S)] for the latter vary from 2.0 to 2.1 Å [37]. This comparison reveals that the S—S bonds in *I* are significantly longer than the expected S—S single bond [36] but considerably shorter than the sum of the two sulfur Van-der-Waals radii.

Table 1. Sulfur-sulfur bond lengths in some substituted trithiapentalenes. For phenylgroups the angles of rotation out of the molecular plane are given in parenthesis

Topology	Notation	S—S bond length [Å] left	S—S bond length [Å] right	Ref.
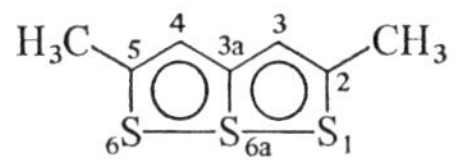	*Ib*	2.358	2.358	[27]
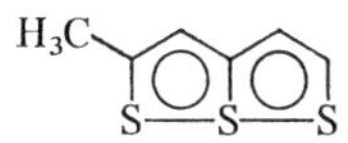	*Ic*	2.431	2.308	[28]
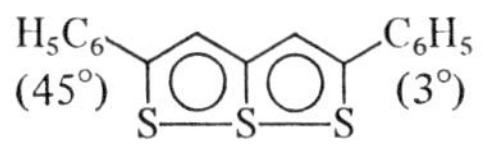	*Id*	2.362	2.304	[29]
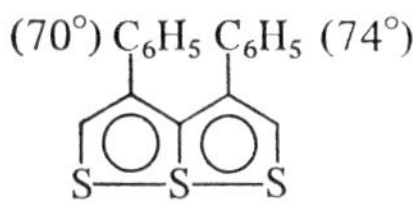	*Ie*	2.232	2.434	[30]
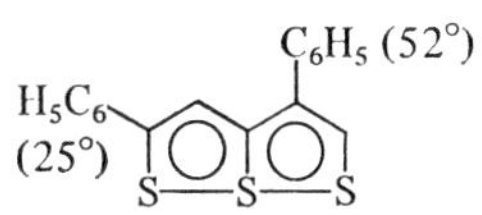	*If*	2.504	2.222	[31]
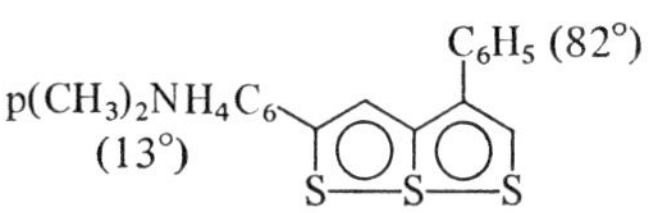	*Ig*	2.348	2.350	[32]
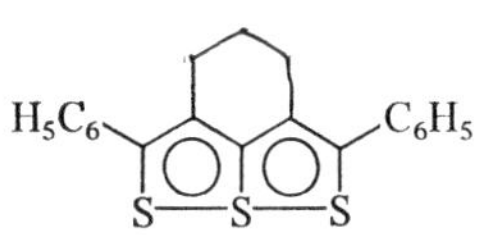	*Ih*	2.329	2.288	[33]
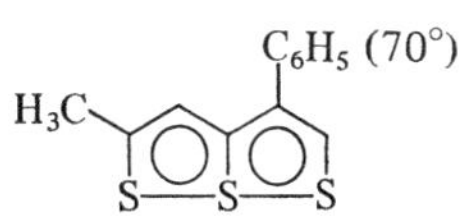	*Ii*	2.481	2.242	[34]
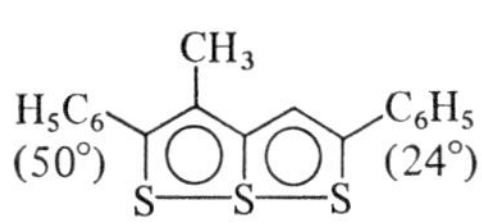	*Ik*	2.255	2.398	[35]

1.2. X-Ray Data of Aza-, Oxa-, and Selenaderivatives of Trithiapentalenes

Table 2 displays data for molecules derived from *I* by replacing one, two or three of the sulfur atoms by N—R, O and Se respectively. As in the case of *I* these molecules exhibit a similarly elongated single bond between the heteroatoms. The corresponding sums of the Van-der-Waals radii are [36]: S···O = 3.25 Å, S···N = 3.35 Å, Se···Se = 4.00 Å, and S···Se = 3.88 Å.

Table 2. Bond lengths between the heteroatoms in some analogues of trithiapentalene. For phenylgroups the angles of rotation out of the molecular plane are given in parenthesis

Topology	Notation	Bond length left	Bond length right	Ref.
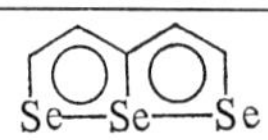	*IIa*	2.586	2.579	38)
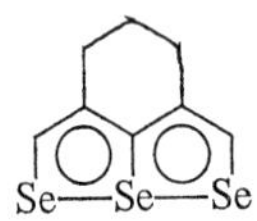	*IIb*	2.563	2.548	39)
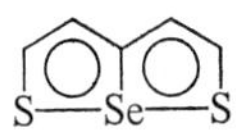	*IIIa*	2.446	2.446	40)
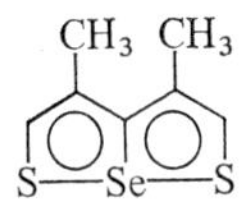	*IIIb*	2.414	2.414	41)
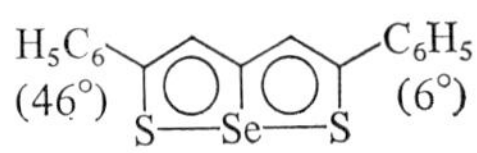	*IIIc*	2.433	2.419	41)
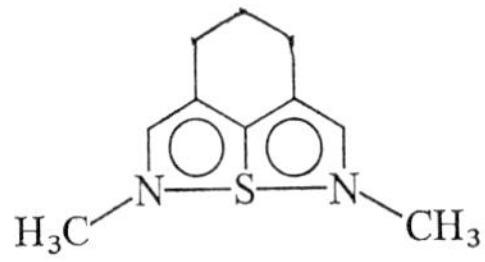	*IVa*	1.901	1.948	42)
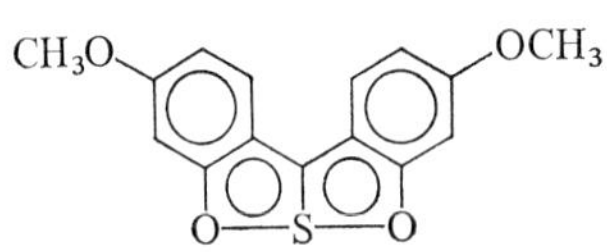	*Vc*	1.878	1.879	43)
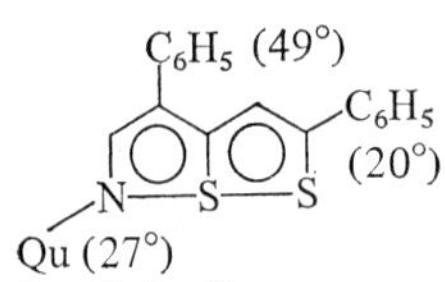 Qu=Quinoline	*VIa*	1.887	2.364	44)
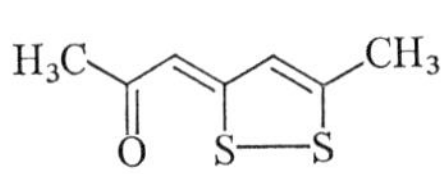	*VIIb*	2.41	2.12	45)
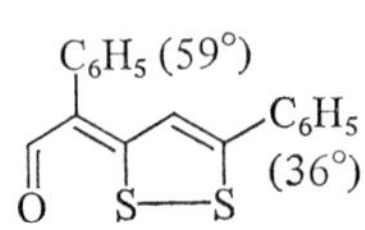	*VIIc*	2.382	2.106	46)
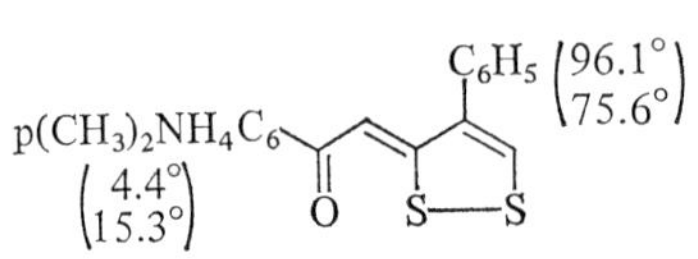	*VIId*	2.443 2.284	2.101 2.111	48)

Table 2 (continued)

Topology	Notation	Bond Length		Ref.
		left	right	
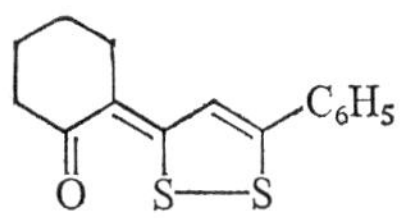	*VIIe*	2.255	2.126	47)

Table 2 is far from being complete. Numerous other related compounds have been studied.

Especially Se compounds similar to *V* have been investigated [49]. The measured S—O and Se—O distances are effectively the same as those found in tetracoordinated sulfur and selenium compounds [50].

2. Electron Diffraction Analysis of Trithiapentalene

An electron diffraction analysis of *Ia* has been published [51] as a valuable supplement to the X-ray data. The best accord with experiment is obtained with a model assuming C_{2V} symmetry as shown in Fig. 1b. Remarkably, the vibrational amplitude of the S_1—S_{6a} bond is found to be considerably larger than the S_1—S_6 amplitude. On the whole these data are consistent with the ones obtained from X-ray analysis and ESCA spectroscopy (see Chapter IV 4.).

3. NMR Spectroscopic Data

NMR data on *I* and *VII* have been collected in the literature [8,21–24,52] and we shall present only a few typical examples shown below. The chemical shifts

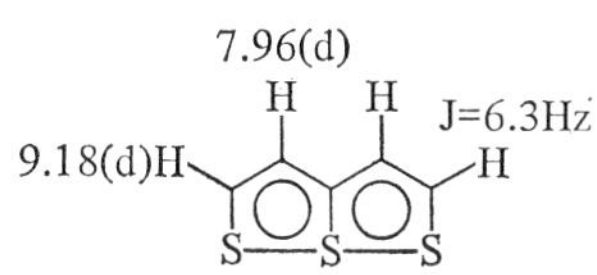

Ia 21)

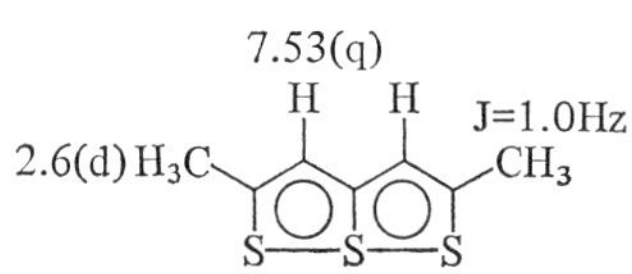

Ib 53)

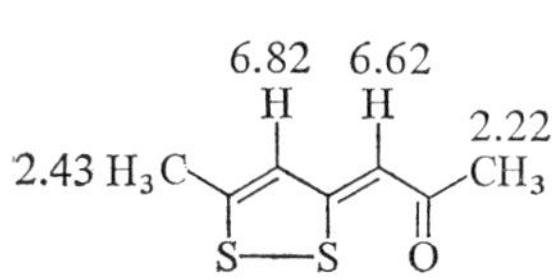

VIIb 54)

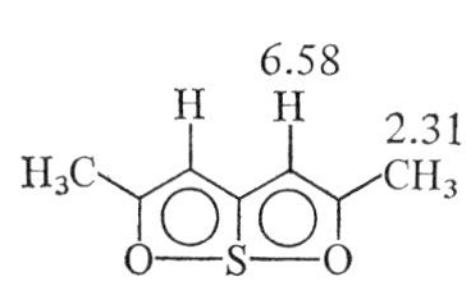

Vb 18)

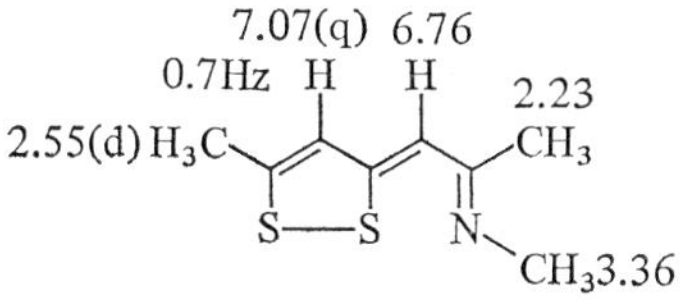

VIb 23)

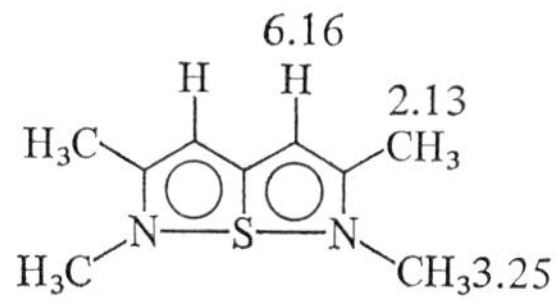

IVb 22)

(ppm) as well as the coupling constants (Hz) and the splitting pattern (d = doublet) are indicated.

Recently the C^{13} NMR spectra of *Ia* and a number of substitution products have been reported [55]. Below the C^{13} chemical shifts of *Ia* and *Ib* are given in ppm downfield from internal TMS.

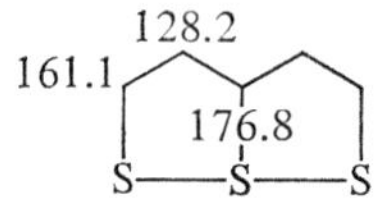

Ia [55]

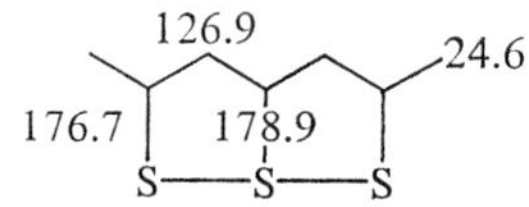

Ib [55]

For our purposes it is important to mention that the H^1 as well as C^{13} NMR spectra of symmetrically substituted species can be interpreted only with a model of C_{2V} symmetry.

From the substituent chemical shift effects in the C^{13} NMR spectra it was concluded[55] that the behaviour of *I* is more akin to that of olefines than of aromatic systems.

The H^1 chemical shifts in *Ia* have been invoked as evidence for a strong ring current [21,56]. However, without a detailed study on model-compounds this conclusion seems rather speculative.

4. Infrared Spectra

The analysis of the IR spectrum of *Ia* together with a normal coordinate analysis should give valuable information of its structure. So far no work on such experiments has been reported.

More revealing on a qualitative basis are the IR spectra of type *VII*. Due to the relative short distance between O and S the expected carbonyl frequency is shifted to lower frequencies (1500—1610 cm^{-1}) [8]. It was possible to assign the band with a large $\tilde{\nu}_{CO}$ character on the basis of O^{18} studies [57]. Supplementary to these studies were the measurements of the carbonyl stretching frequencies of the *trans-VII* compounds [58] [see Eq. (19)]. The observed shifts $\Delta\tilde{\nu}$ have been in the order of 40—80 cm^{-1} towards higher wave numbers in the *trans* configuration. This research could lead to a correlation between $\tilde{\nu}$ or $\Delta\tilde{\nu}$ and the S··O bond length.

5. Dipole Moments

Dipole moments of types *I* and *VII* have been measured and reviewed recently [8]. At the moment they are useful for the assignment of structures in this area of chemistry in a qualitative way.

III. Bonding-Models

1. General Remarks on Potential Surfaces

For the absence of the two distinguishable forms *I* and *I'* (as indicated in the introduction and implied by physical data already discussed), there are two possible interpretations: Either the two formulas *I* and *I'* represent resonance forms or two metastable forms exist, described by *I* and *I'*. In the latter case the equilibration of *I* and *I'* must be so fast that they have eluded the experiments reported.

In order to clarify this discussion we shall start with a simple model of a four electron three center bond [59,60].

Let us assume that there are three centers A(1), B, A(2) linearly arranged and that the A(1)··A(2) distance has a constant value. We can discriminate between two cases as follows:

case 1: resonance A(1): B··A(2) $\longleftrightarrow$ A(1)··B :A(2) C_{2V}

case 2: equilibrium[a)] A(1): B··A(2) $\rightleftharpoons$ A(1)··B :A(2) C_S

In the two cases we can describe the potential energy as a function of two coordinates, the distances R[A(1)-B] and R[A(2)-B]. The corresponding potential-curves are shown in Fig. 2.

In a more general case a set of internal coordinates q_i (see *e.g.* Ref. [62)] and [63a)]) define a geometry and we are dealing with a multidimensional energy surface instead of the two dimensional exemplified above. The ground state geometry (or geometries) is (are) then defined as a minimum (minima) on the potential surface with respect to variation of all q_i.

In case 2 two electronically equivalent ground state geometries exist.

This is generally the case whenever the graph which is described by one set of geometry parameters q_i cannot be transformed to the corresponding graph described by the set q_i' (*e.g.* describing the mirror image) by a simple rotation [61b)]. Note that such a formulation does not imply whether an interconvertibility of the equivalent forms is chemically feasible or not.

The most famous controversy whether a molecule belongs to case 1 or 2 is the one of benzene.

It is important to note the difference in the experimental evidence for the two cases. *While a proper evidence for case 2 can be taken as a proof for the double minimum, its absence in case 1 can only be advocated by missing evidence for case 2.*

Typical examples of case 2 with different heights of the activation barrier of the surface include: H-bridge between like molecules, the transition region of a degenerate S_N2 reaction, pyramidal molecules like NH_3 [61d)], symmetrically substituted *cis-trans* isomers [63b)] and — with a very high barrier — a pair of optically isomers.

---

a) The term equilibrium should be used with caution since it is a classical term. If the "activation barrier" between the two structures is low, the system has to be described by quantum mechanics. A stationary state wave function is then obtained by linear combination of two structures say α and β [61)], *e.G.* Ψ stationary $= \frac{1}{\sqrt{2}}\,1\,(\Psi_\alpha + \Psi_\beta)$.

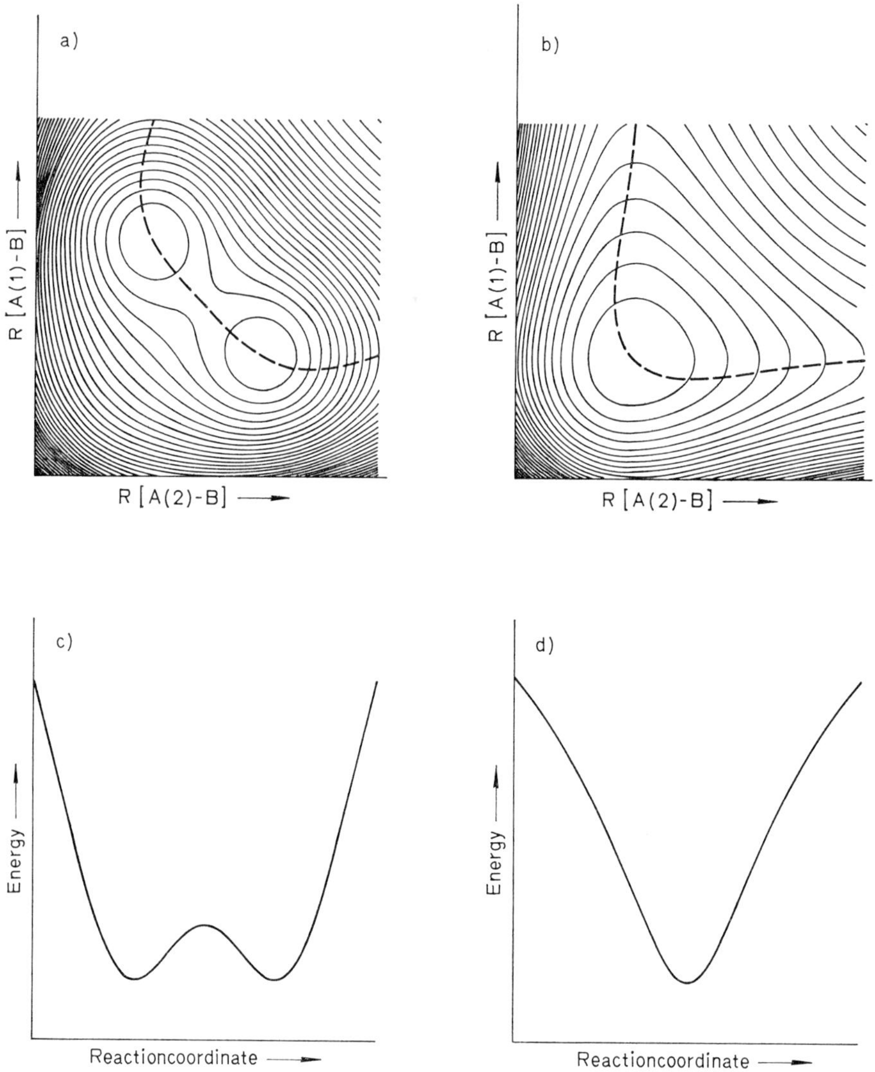

Fig. 2. Two dimensional stretching surfaces for linear A(1)-B-A(2). In *a* a surface corresponding to case 2 (equilibrium), and in *b* a surface corresponding to case 1 (resonance) is shown. In *c* and *d* the energy as a function of the reactioncoordinate along the dotted line is drawn

The symmetrically substituted molecules of type *I*, *IV* and *V* considered in this article may be candidates for case 2, although at best with a low energy barrier.

2. Molecular Orbital Models for Trithiapentalene

As shown in Chapters II.1. and II.2. trithiapentalenes reveal unusually long sulfur-sulfur bonds. To rationalize this, several theoretical studies have been put forward [60,64–66,70–75]. The first models were based on Hückel MO calculations including $p\sigma$ orbitals on sulfur [64], the inclusion of $3d$ orbitals on the central sulfuratom [66] or even simple π-considerations neglecting the σ frame [65].

Before discussing more sophisticated treatments we shall anticipate that symmetrically and unsymmetrically substituted trithiapentalenes exhibit similar electronic and photoelectronic spectra, indicating a similar π-electronic structure.

These results suggest that in a simple model the π-system might be omitted and the carbon skeleton replaced by three hydrogen atoms. In a further step

we also neglect the "lone pair" orbitals on the sulfur atoms which are mainly $3s$ in character and thus energetically different from the $3p$ orbitals. We thus end up with three S—H fragments. On each S—H unit there is a p-orbital directed along the y-axis. The $3p_x$ orbitals are assumed to be completely involved in the S—H σ-bonds.

Interaction between the three $3p_y$ orbitals leads to three linear combinations shown in Fig. 3a. The symbols "*S*" and "*A*" refer to the symmetry properties (**S**ymmetric and **A**ntisymmetric) with respect to the vertical plane containing

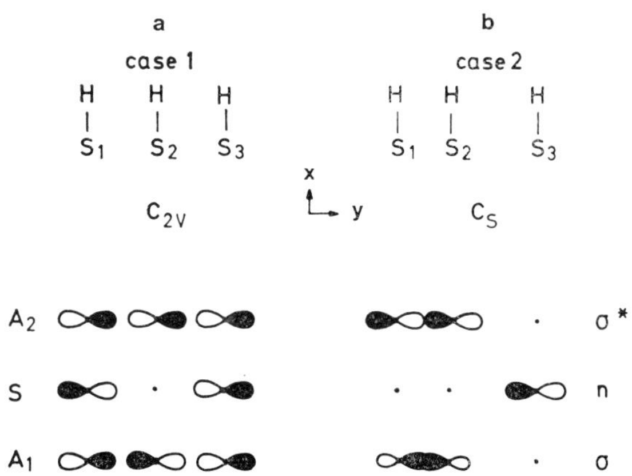

Fig. 3. Schematic representation of the three linear combinations of the p_x orbitals on the centers S_1, S_2 and S_3 for C_{2V}(a) and C_s symmetry (b)

the middle S—H fragment. In a one-electron approach we have to fill in four electrons occupying the lower A and the S orbital. The former is strongly bonding while the latter is only weakly bonding.

Note that the four electron three-center bond is different from two two-center bonds; the latter also involve four electrons but four basis orbitals instead of three.

To come back to our problem, the understanding of the trithiapentalene structure, we have to compare the four electron three-center bond (electron-rich three center bond) with a system of C_S symmetry involving one S—S single bond composed of two $3p_y$ orbitals and a $3p$ lone pair on the third center (see Fig. 3b). Qualitatively it is hard to say which of the two cases should be energetically favoured. Shortening the S—S distances in the first case (C_{2V} symmetry) the two occupied MO's are stabilized. In the second case the shortening of one S—S distance lowers the energy of the σ orbital and leaves the energy of the lone pair constant.

In order to elaborate this comparison and to test our approximations made in the beginning Extended Hückel [67] (EH) calculations have been carried out on the above model systems. In Fig. 4a the energy levels of the orbitals are plotted as a function of the S—S distances for case 1 and for the S_1—S_2 distance for case 2 keeping S_3 at infinity (Fig. 4b).

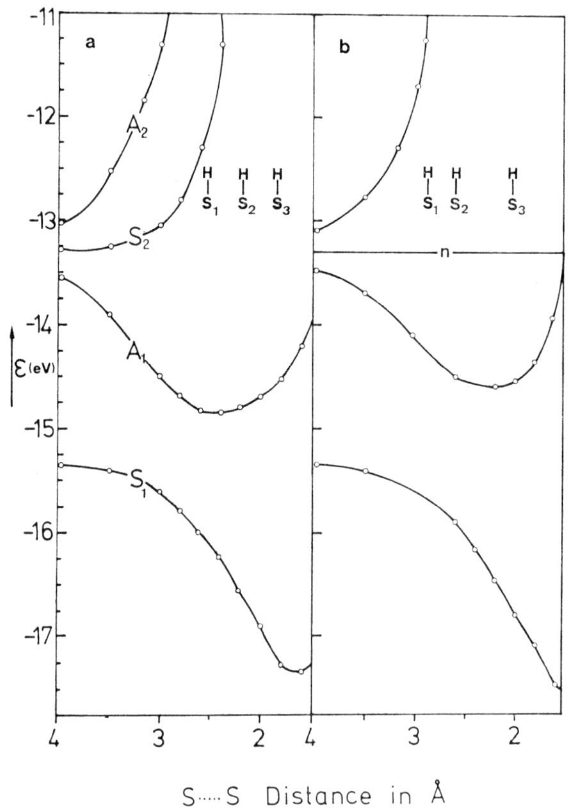

Fig. 4. The behaviour of the individual energy levels between —11 and —17 eV for a three center symmetrical approach of three SH units. (a) and a two center approach of two SH units keeping a third SH at infinity (b)

In contradiction to our expectation the S_2 orbital rises in the symmetrical case as the S—S distance is lowered. This is due to a mixing with a high lying σ orbital which we omitted in our former consideration. This mixing—clarified in

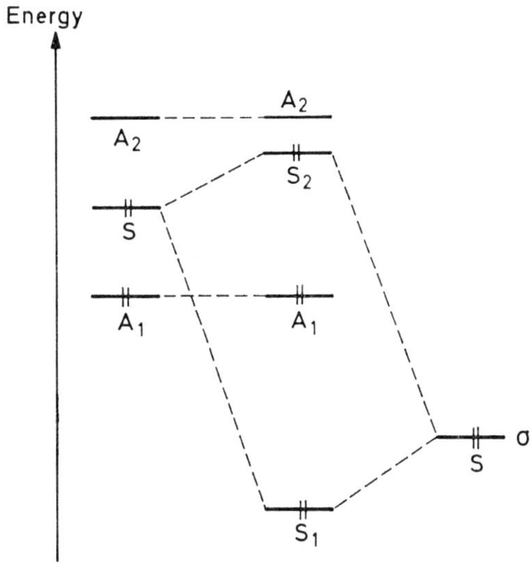

Fig. 5. Interaction diagram between the electron-rich three center bond and a high σ-orbital

Fig. 5 — is important for smaller distances. At longer distances a structure with C_{2V} symmetry is predicted while at smaller distances the unsymmetrical case competes favourably.

If we extrapolate these model calculations we might predict that trithiapentalene prefers a structure with C_S symmetry. However, there are reasons why the structure with C_{2V} symmetry may be stabilized. One of these reasons is the interaction with low lying unoccupied orbitals of S-symmetry, *e.g.* a $3d$ or $4s$ orbital on the central sulfur atom.

We can compare these qualitative arguments with the results of several semiempirical calculations on *Ia* of various degrees of sophistication using the EH [67], CNDO/2 [68] and MINDO/3 [69] method.

Varying the S—S distance and keeping all other distances constant, the EH method predicts the C_S structure to be the most stable [60,70], as anticipated from the model calculations discussed above. The barrier of the valence isomerization reaction is considerably lowered (from 1 eV to 0.5 eV) by inclusion of 3d orbitals on the sulfur atoms.

The CNDO/2 method predicts a structure with C_{2V} symmetry [70–72] if one varies only the S—S distances. A relatively low force constant for the vibration of the central sulfur atom along the S—S—S axis is predicted. Inclusion of $3d$ orbitals lowers this force constant even further.

Varying all geometrical parameters of *Ia* with the MINDO/3 method assuming only C_S symmetry yields a structure with C_{2V} symmetry [73]. The predicted S—S distance of 2.19 Å is considerably shorter than the experimental one.

A detailed analysis of the CNDO/2 and MINDO/3 results shows that the symmetrical structure is due to the dominance of the nuclear-nuclear repulsion. The electronic part of the total energy favours a structure with C_S symmetry for *Ia*.

Recently ab initio calculations on *Ia* have been carried out [74,75] using a fixed geometry taken from X-ray analysis. It was concluded from a comparison between the results with and without the inclusion of 3*d* functions that the influence of the *d* functions in the ground state is a minor one.

All these calculations show shortcomings. Either they do not vary any or only one geometrical parameter or their result indicates an imperfect parametrization.

3. Molecular Orbital Models for Structures Involving Two, Four and Five Sulfur Atoms

The qualitative concept of "no-bond-resonance" has been extended by Klingsberg [4] to other structures (see also Chapter VI) with less and more sulfur atoms than *I*.

Here we shall briefly compare the most important σ MO's for *I*, *VIII* and *IX* assuming a geometry with C_{2V} or C_S symmetry for *I* and *VIII* or equal and

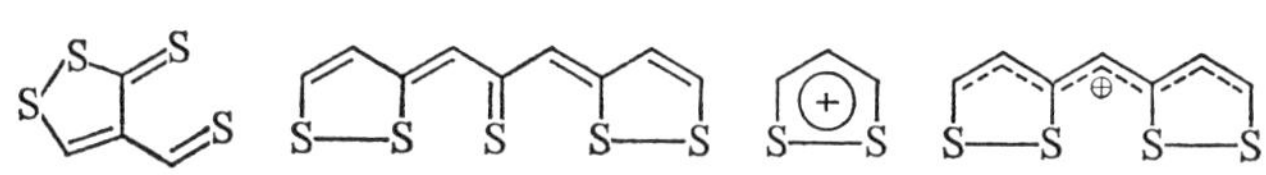

VIII *IX* *X* *XI*

non equal S—S bond lengths for *IX* [76].

In Fig. 6 this comparison is shown for *I*, *VIII* and *IX*.

The calculated activation energies for the structural changes $Ia(C_S)$ to $Ia(C_{2V})$ (for the small negative or positive values see foregoing chapter), $VIII(C_S)$ to $VIII(C_{2V})$ (1.5 eV) [76] and *IX* to *IXb* (2 eV) [83] are reflected by the three correlation diagrams in Fig. 6. In the first case [$Ia(C_S)$ to $Ia(C_{2V})$] the slope of the correlation lines connecting the occupied orbitals is small (HOMO) or negligible. In the two other cases [$VIII(C_S)$ to $VIII(C_{2V})$ and *IX* to *IXb*], however, there is a steep ascent by the correlation line connecting an occupied σ orbital with a lone pair combination which amounts to a considerable activation energy. This result is corroborated by the fact that derivatives of *VIII* and *IX* have been isolated (see Chapter VI) and valence isomerization for *VIII* has been reported [77].

This can be rationalized by assuming that the breaking of *one* S—S single bond (ca. 61 kcal/Mol [36]) is nearly compensated for by the electron-rich three center bond. In other words, the linear arrangement of three centers as in *Ia* favours an electron-rich three-center bond. The energy difference between an S—S σ-bond and a 3*p*-lone pair on the third sulfur atom on one side and an electron-rich three-center bond on the other is small.

In the case of *IXb*, however, the electron-rich five-center bond cannot compensate the breaking of *two* S—S single bonds present in *IX*.

In the case of *VIII* (C_{2V}) the nonlinear arrangement of the four sulfur atoms is not favourable for an electron-rich four-center system. Another difference between *VIII* (C_{2V}) and $Ia(C_{2V})$ or *IXb* is that in case of *VIII* (C_{2V}) the HOMO is predominantly antibonding. In the other two examples the HOMO is nonbonding.

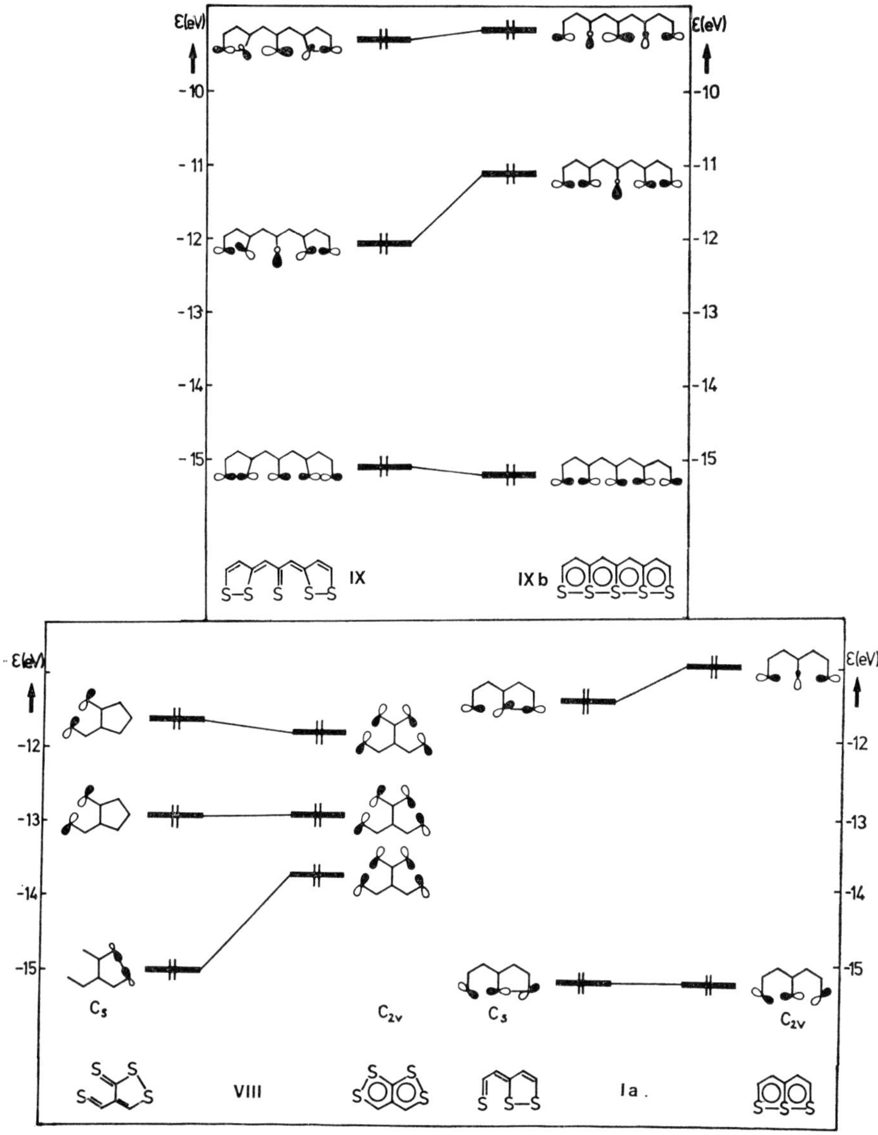

Fig. 6. Correlation diagram between the highest occupied σ orbitals of *I*, *VIII* and *IX* with equal and non equal S—S bond lengths

In *IXa* there is still another possibility indicated to delocalize the σ orbitals between the sulfur centers. EH calculations on *IX*[b]), *IXa*[b]) and *IXb*[b]) predict that *IXa* is *ca.* 0.9 eV less stable than *IX*. The calculated energy difference between structure *IXb* and *IX* is *ca.* 2 eV.

---

b) For *IX* the experimental [78]) S—S bond distances were taken. For *IXa* two S—S bond lengths as indicated in the drawing were made equal (2.38 Å), for *IXb* all S—S bond lengths were made equal (2.38 Å). All other parameters were kept constant.

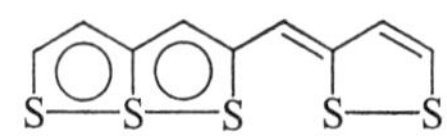

IXa

For *X* and *XI* the possibilities of existing in the valence isomer forms *Xa* and *XIa* has been discussed in the literature [4] but there is no experimental evi-

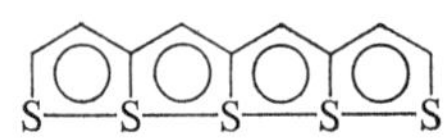

IXb

Xa

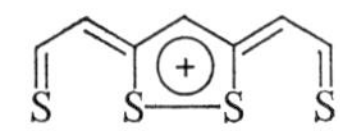

XIa

dence for these forms [79]. To achieve a minimum for *Xa* or *XIa* on the corresponding potential surface an orbital crossing has to occur. The HOMO [$b_2(\pi)$] has to cross the $b_1(\sigma^*_A)$ combination. In other words the 6π-system *X* has to be transformed to the 4π-system *Xa*.

This crossing never occurs between 2 Å and 3 Å (see Fig. 7).

As a corollary of this rationalization two possible ways to stabilize systems like *Xa* and *XIa* result: i) one adds two more electrons to the system *e.g. XII* and *XIII* or ii) a strong donor atom is combined with the system (*XIV* and *XV*). Derivatives of *XIII* have been reported [80].

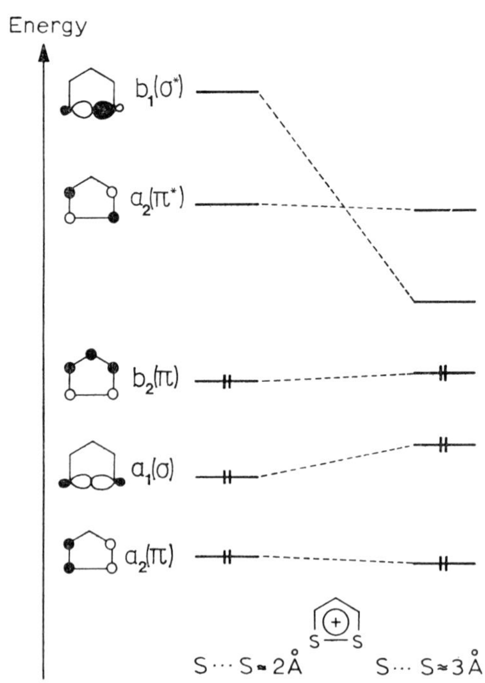

Fig. 7. Qualitative correlation diagram of the lone pair combinations on the sulfur centers and some π-orbitals of the dithiolium cation for an S..S distance of ca. 2 and ca. 3 Å

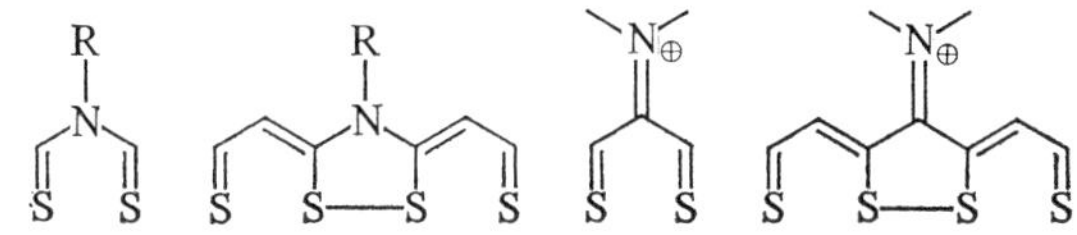

XII *XIII* *XIV* *XV*

As it were a two dimensional extension of the 'no-bond-resonance' concept has been synthesized in *XVI* [81]. The structure of this compound has been described as a sulfur analogue of coronene [82] since it comprises a 24 π-electron system.

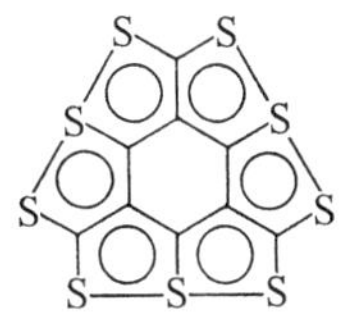

XVI

Extended Hückel calculations suggest [83] that the π-orbitals of *XVI* are of minor importance. Essential, as in *I*, are the linear combinations of the $3p$ orbitals on the sulfur atoms in the molecular plane. In Fig. 8 we have shown the highest occupied MO's of *XVI*. From this its analogy with *I* is evident (see Fig. 6).

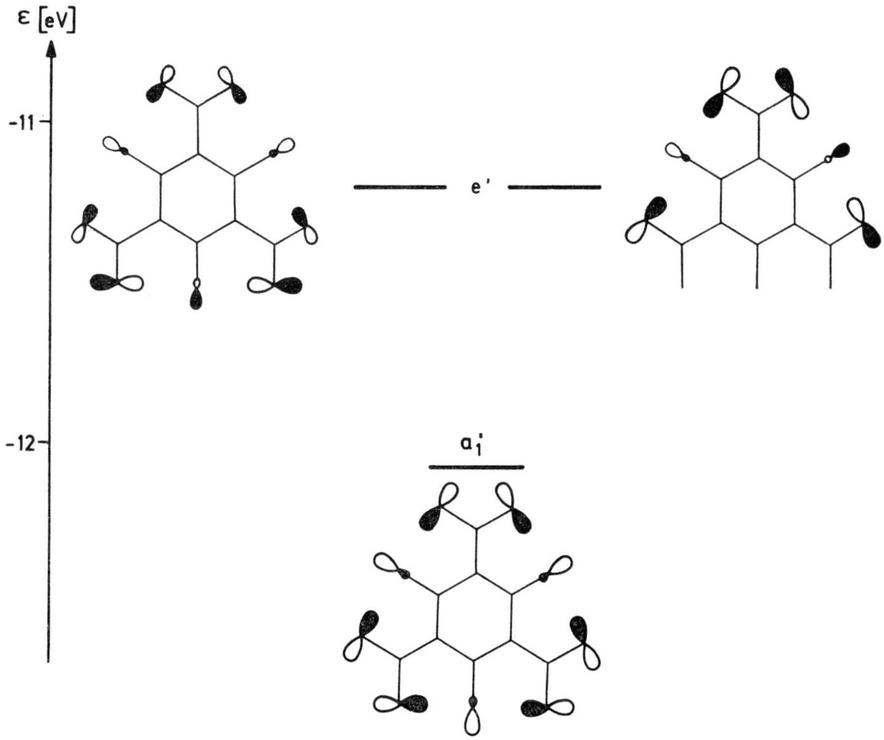

Fig. 8. Schematic representation of the highest occupied MO's of *XVI* as derived from an Extended Hückel calculation assuming D_{3h} symmetry

IV. Spectroscopic Investigations

At the first glimpse the X-ray and NMR-data mentioned above (see Chapters II.1. and II.3.) suggest a structure with C_{2V} symmetry for *I*. In order to understand these findings fully, the limitations of these methods have to be stated.

The X-ray data are gathered from the solid state. If the barrier between the two valence isomers is very low, intermolecular forces could affect the picture. We remind the reader of biphenyl [84] where the dihedral angle between the phenyl groups is 42° in the gas phase and 0° in the solid state.

The NMR experiment is relatively "slow". The presence of two rapidly interconverting species would escape its detection. Let us assume a difference of 0.5 ppm between one H atom in one valence isomer and the same H atom in the other one (*e.g.* *I* and *I'*). Then the two isomers could not be detected by NMR spectroscopy if they were interconverting faster than about 10 msec (100 MHz spectrometer). This corresponds to an activation barrier of about 7 kcal/mol according to transition state theory, assuming 150° K.

From the spectroscopic methods considered so far only the electron diffraction experiment is not affected by one of the shortcomings mentioned above. The result that the amplitude of the central sulfur atom is higher than for the outer ones leaves open the possibility of a very low activation barrier or a broad U-shaped potential.

1. Electronic Spectra of Trithiapentalenes

Absorption spectra of differently substituted trithiapentalenes as well as oxygen and nitrogen analogues are reported in the literature [7–25,85]. Several approaches to analyze these data by using semiempirical calculations [86–90] or by qualitative [7,85] models have been attempted.

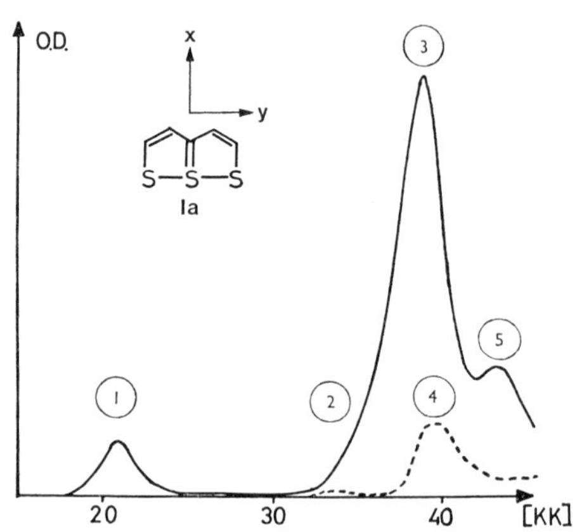

Fig. 9. Electronic spectrum of *Ia* measured in stretched polyethylene film with light parallel (——) and perpendicular (----) to the *y*-axis of the molecule. The spectrum is computer corrected for the non ideal orientation of the molecules in the film. C_{2V} symmetry for *Ia* is assumed

In Fig. 9 the electronic spectrum of *Ia* measured in stretched polyethylene film is shown. In Table 3 electronic spectroscopic data of four trithiapentalenes are collected and the direction of the transition moment is given assuming C_{2V} symmetry.

Table 3. Observed transitions of trithiapentalenes. For the estimation of the direction of the transition moment C_{2V} symmetry was assumed

Compound	Band	$\tilde{\nu}$[kK][1]	Direction of polarization δ[1]	log ε[2]
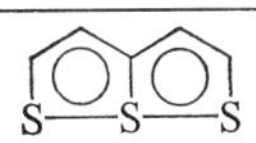	①	21	90°	3.7
	②	33.5	0°	< 3.6[3]
	③	39	90° }	4.7
Ia	④	39.5	0° }	
	⑤	43	90°	4.3
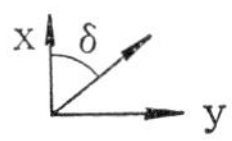				
	①	21	90°	3.8
	②	33	0°	< 3.6[3]
	③	38	90° }	4.8
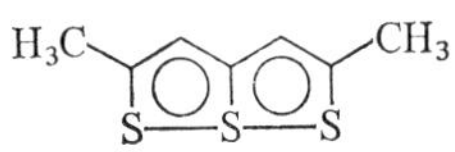	④	39	0° }	
	⑤	42	90°	< 4.3[3]
Ib	①	20.5	90°	3.9
	②	34	0°	< 3.0[3]
	③	37.5	0° }	4.7
	④	38	90° }	
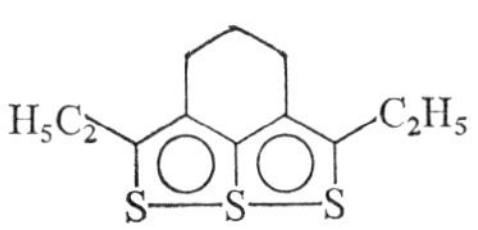	⑤	42.5	90°	4.3
	①	19.5	90°	4.1
Ic	②	28.5	0°	< 3.9[3]
	③	32	90°	4.4
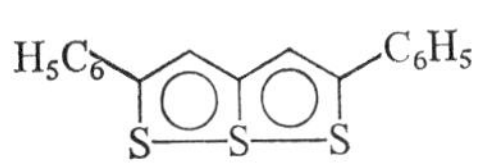	④	34.5	0°	—[3]
	⑤	38.5	90°	4.7
	⑥	42.5	0° }	4.4
Id	⑦	43	90° }	

[1]) In polyethylene.
[2]) In cyclohexane.
[3]) Shoulder.

To interpret these data four models have been discussed in the literature. Three models [90] *D*, *E* and *F* considering the π-electrons only and a fourth one which takes all valence electrons into account [88].

As mentioned before, photoelectron (584 Å) and electron spectra support the conclusion that the electronic structure does not depend very much on the S—S distance in *I*, *i.e.* in both cases one deals with a 10 π-system.

For a 10 π-system, considering only the π orbitals, three models are possible.

For model *F* agreement between a PPP calculation and experiment was obtained using standard parameters.

C_{2V} structure

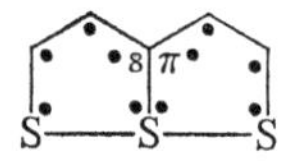

D

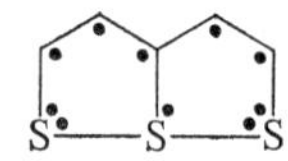

E

C_S structure

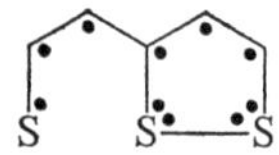

F

For model *E* only a satisfactory agreement between a PPP calculation and experiment could be achieved by the assumption that those transitions predicted with small intensities could not be observed.

While for model *E* and *F* standard parameters for all sulfur atoms could be used this was not the case for model *D*. Here the outer sulfur centers contribute formally $1\frac{1}{2}$ electrons to the π-system. A variation of the sulfur parameters shows a reasonable agreement between experiment and calculation concerning the band positions and intensities. The polarization direction of the intense band at 38 kK is not reproduced correctly.

A correction of this result can be achieved by using a C_S symmetry. This leads to model *F*.

In Table 4 the results on *Ia* for the three different models are compared by experiment. These results do not imply that the geometry of *Ia* necessarily has

Table 4. Calculated transitions for *Ia* according to models *D*, *E* and *F*

Model	Transition $\tilde{\nu}$[kK]	Oscillator strength *f*	Direction of polarization δ
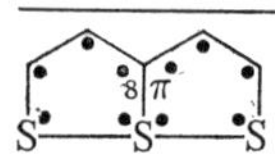	33.2	0.75	90°
D	39.5	0.02	0°
	44.4	0.80	0°
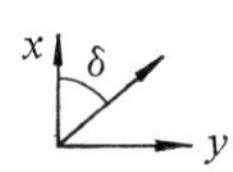	50.2	0.13	0°
	32.9	0.50	90°
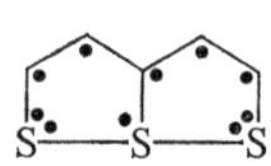	35.1	0.26	0
	44.5	0.50	0
E	46.4	1.30	90°
	20.4	0.39	85°
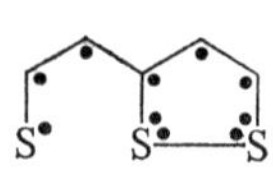	34.5	0.02	51°
	37.6	0.56	—1°
F	40.7	0.45	68°

C_S symmetry since the direction of the transition moment depends not only on the potential curve of the ground state but also on the corresponding excited state [70,91].

A CNDO/CI calculation [88] favours also a model with C_S symmetry for *I*. The accord of this calculation with the experiment depends very much on how many of the predicted $\pi^* \leftarrow n$ and $\sigma^* \leftarrow \pi$ transitions can be observed. At least there are enough — probably too many — predicted transitions to choose from.

The quantum mechanical interpretations of the electronic spectrum of *I* are not at all satisfactory. This is partly due to the fact that there are no analogous compounds around to adjust the heteroatom parameters and there are more parameters to vary (model *D*) than observable ones. An improved all valence treatment for sulfur compounds is desirable.

2. Electronic Spectra of N- and O-Analogues of Trithiapentalenes

As an example for the absorption spectra of O- and N-analogues of trithiapentalenes we have compared the electronic spectra of *IVb*, *Vb*, *VIb* and *VIIb* with the one of *Ib* in Fig. 10 and Table 5. The stretched film spectra of *IVb* and *Vb* are shown in Fig. 11 [90].

Table 5. Observed transitions for some N- and O-analogues of trithiapentalene. For the definition of δ see Table 3

Compound	Band	$\tilde{\nu}$[kK][1])	Direction of polarization δ[1])	log ε[2])
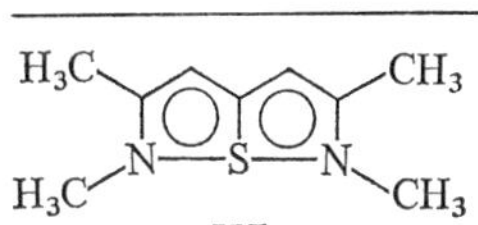 *IVb*	①	26.7	90°	4.16
	②	37.0	0°	3.61
	③	44.6	—	3.99
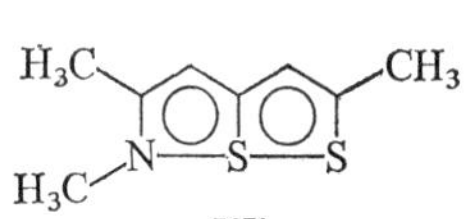 *VIb*	①	24.1	—	3.95
	②	38.0	—	3.89
	③	42.2	—	4.24
	④	46.7	—	4.05
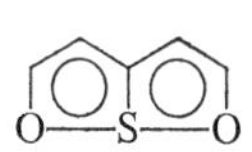 *Va*	①	29.5	90°	4.06
	②	38.9	0°	3.35
	③	45.0	—	3.38
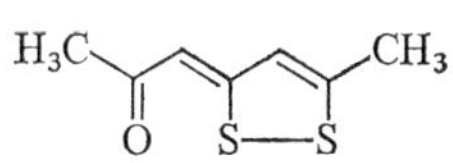 *VIIb*	①	23.9	—	4.11
	②	38.5	—	3.40
	③	43.7	—	4.28

[1]) In polyethylene.
[2]) In solution.

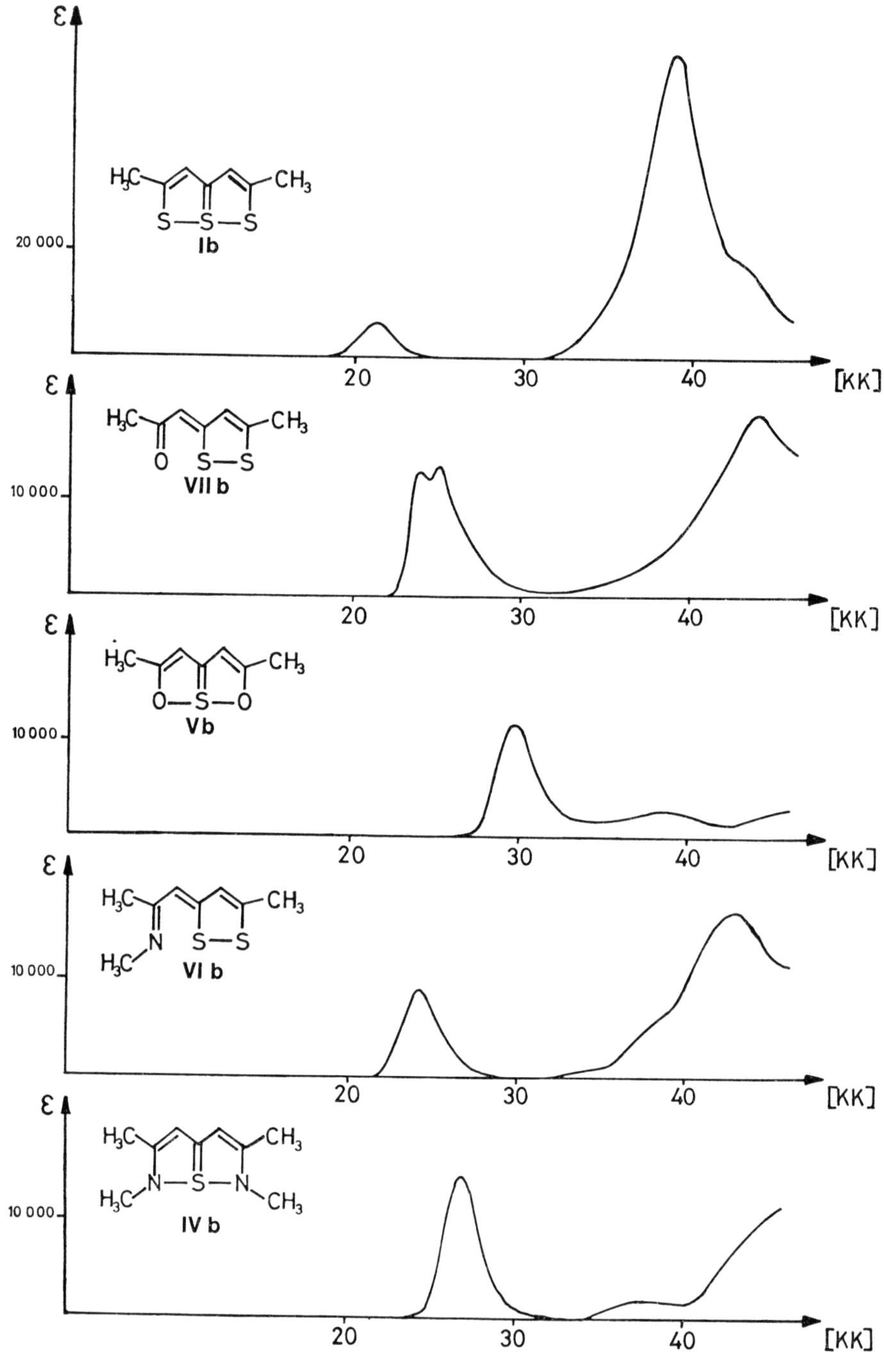

Fig. 10. Comparison between the electronic spectra of *IVb, Vb, VIb* and *VIIb* with *Ib*

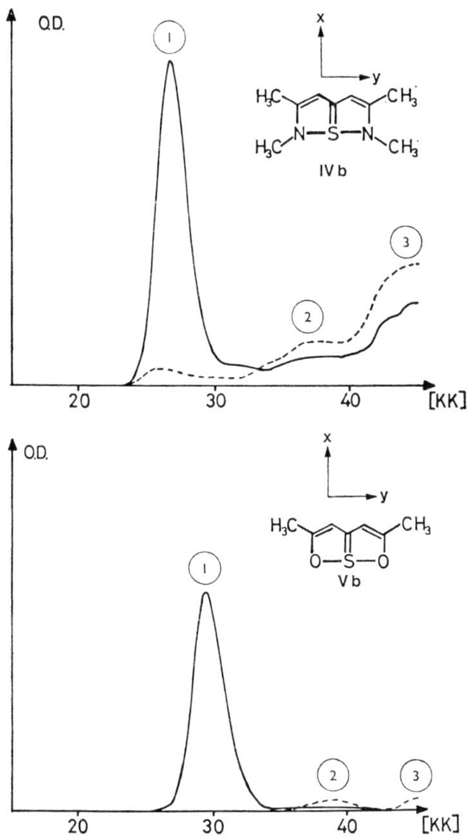

Fig. 11. Electronic spectrum of *IVb* and *Vb* in stretched polyethylene film measured with light parallel (——) and perpendicular (----) to the *y*-axis of the molecule (compare legend to Fig. 9)

As anticipated in the foregoing chapter both models *E* and *F* predict that the first band is due to a $\overset{*}{\pi}_1 \leftarrow \pi_1$ transition. The observed blue shift (see Fig. 10) of the first band is also consistent with both models. The intense first band is followed by a second less intense peak towards the short wave length region. From the stretched film experiment we conclude that the transition moment is perpendicular to the long axis of the molecule. For a more detailed picture more research is necessary.

3. Photoelectron Spectra (584 Å) of Trithiapentalenes

The HeI photoelectron (PE) spectra of trithiapentalene and substitution products have already been reported [92]. The comparison between PE experiment and model calculation is subject to several limitations:

1. The most serious limitation is the assumption of Koopmans' theorem [93]: $-\varepsilon_J = I_{V,J}$. This implies that the MO's of the generated cation are the same as the

MO's of the neutral molecule and that the correlation energy in the ion is the same as in the molecule.

2. The current methods of calculations are unable to predict ionization potentials of molecules as complex as considered in this article. Moreover, most of the calculations are based on approximated geometries.

From this it is evident that PE spectroscopy is not suited to solve the problem of the structure of these compounds. Nevertheless, it gives a better understanding of the highest occupied MO's of these compounds.

The PE spectra of *Ib*, *IVb*, *Vb*, *VIb* and *VIIIb* are shown in Fig. 12. The first ionization potentials are collected in Table 6 and compared with orbital energies. In Fig. 13 the first ionization potentials are compared with each other.

The assignment given in Fig. 13 and Table 6 is based on the assumption [92]: 1) that the ionization potential is shifted monotonously by replacing the S atom subsequently by one or two other heteroatoms and 2) that the comparison with the MO calculations (Koopmans' theorem [93]) is valid.

It is interesting to note that the π orbital difference between the first three π-orbitals of *Ia* is in good accord with PPP calculations, using model *F* [90]. Also the CNDO/CI calculations [88] account reasonably well for the electronic spectra and support the reported PE assignment for *Ia*.

4. ESCA-Spectra of Trithiapentalenes

Three ESCA studies in the solid state have been published [92a,94,95]. However, the unresolved S_{2p} peak obtained in these studies had such an anomalous shape that an unequivocal assignment was not possible. As a result of this, different deconvolutions were reported. A fourth experiment carried out in the gas phase has been published [96]. The spectrum consists of one sharp and one broad doublet (see Fig. 14a).

The sharp doublet has been described as due to an ionization from the central sulfur atom, the broad doublet as due to an ionization from the outer sulfur centers. The broadening (vibrational broadening) is explained by assuming a U-shaped potential for the ground state and that ionic state (S_1^*) which corresponds to ejection of an electron from the $2p$ shell of the central sulfur atom. The ejection of an electron out of the $2p$ shell on the outer sulfur atoms will cause a change in the geometry in the corresponding state (S_2^*).

This will be due to the effect that the ejection of an inner electron from the outer sulfur atoms will cause this atom to be much more electronegative, giving about the same effect as if it were substituted by an oxygen. In Fig. 14b the potential curves for the ground state and the two ionic states S_1^* and S_2^* are shown.

These arguments support but do not prove a U-shaped potential for the trithiapentalene in the gas phase.

Besides the ESCA spectra of *I* also those of oxygen analogues have been reported [95,97].

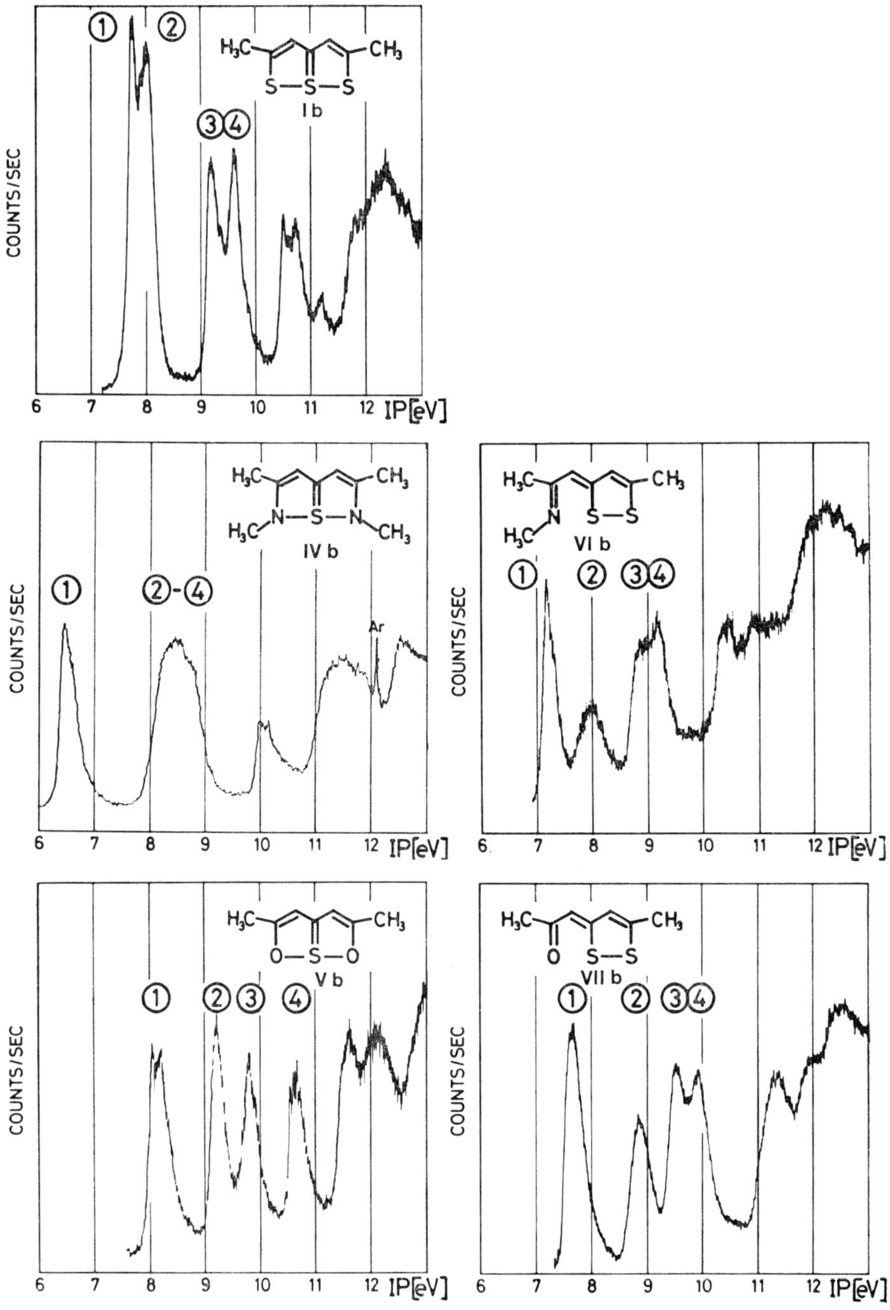

Fig. 12. Photoelectron spectra of *Ib*, *IVb*, *Vb*, *VIb* and *VII b*

Table 6. Comparison between measured vertical ionisation potentials $I_{V,J}$ and calculated orbital energies of trithiapentalene and analogous compounds. The calculations have been carried out on the unsubstituted compounds except for the ab initio calculation on *IVb*. All values in eV

Compound	Band	$I_{V,J}$	Assignment	Ab initio [75]	CNDO/2	EH
	①	8.11	19 $a_1(\sigma,n)$	— 8.51(19 a_1)	— 8.91(19 a_1)	—10.71(19 a_1)
	②	8.27	3 $a_2(\pi)$	— 8.54(3 a_2)	—10.34(3 a_2)	—11.84(3 a_2)
	③	9.58	5 $b_1(\pi)$	—10.84(5 b_1)	—12.30(5 b_1)	—12.94(5 b_1)
Ia	④	10.01	4 $b_1(\pi)$	—11.37(4 b_1)	—14.07(4 b_1)	—13.26(4 b_1)
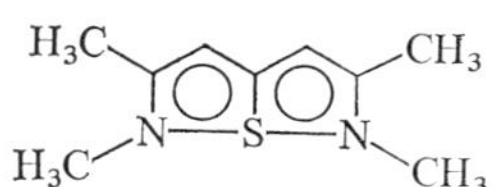	①	6.44	2 $a_2(\pi)$	— 6.64(2 a_2)	— 9.56(2 a_2)	—10.68(15 a_1)
	②	8.10	15 $a_1(\sigma,n)$	— 8.90(15 a_1)	—10.47(15 a_1)	—11.96(2 a_2)
	③	8.44	4 $b_1(\pi)$	— 9.66(4 b_1)	—13.34(4 b_1)	—12.75(4 b_1)
IVb	④	8.95	3 $b_1(\pi)$	— 9.81(3 b_1)	—14.27(3 b_1)	—13.34(3 b_1)
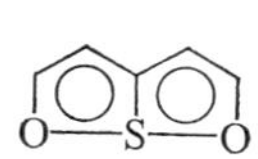	①	8.58	2 $a_2(\pi)$	— 9.16(2 a_2)	—11.00(2 a_2)	—12.47(2 a_2)
	②	9.76	15 $a_1(\sigma,n)$	—11.78(15 a_1)	—11.42(15 a_1)	—12.49(15 a_1)
	③	10.28	4 $b_1(\pi)$	—12.05(4 b_1)	—13.74(4 b_1)	—12.74(4 b_1)
Va	④	10.98	3 $b_1(\pi)$	—13.54(11 b_2)	—14.28(11 b_2)	—13.28(11 b_2)
				—14.05(3 b_1)	—15.36(3 b_1)	—13.98(3 b_1)
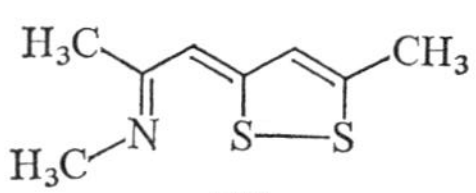	①	7.17	7 $a''(\pi)$	— 8.02(7 a'')	— 9.36(30 a')	—10.77(30 a')
	②	7.97	30 $a'(\sigma,n)$	— 8.80(30 a')	— 9.83(7 a'')	—11.91(7 a'')
	③	8.83	6 $a''(\pi)$	—10.83(6 a'')	—12.78(6 a'')	—12.86(6 a'')
VIb	④	9.17	5 $a''(\pi)$	—11.45(5 a'')	—13.90(5 a'')	—13.27(5 a'')
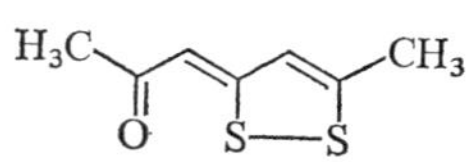	①	7.68	7 $a''(\pi)$	— 9.01(7 a'')	—10.66(7 a'')	—11.95(7 a'')
	②	8.88	30 $a'(n,\sigma)$	—11.08(30 a')	—12.56(30 a')	—12.96(30 a')
	③	9.60	6 $a''(\pi)$	—11.80(6 a'')	—13.71(6 a'')	—13.02(6 a'')
VIIb	④	9.94	5 $a''(\pi)$	—12.96(5 a'')	—14.06(5 a'')	—13.45(5 a'')

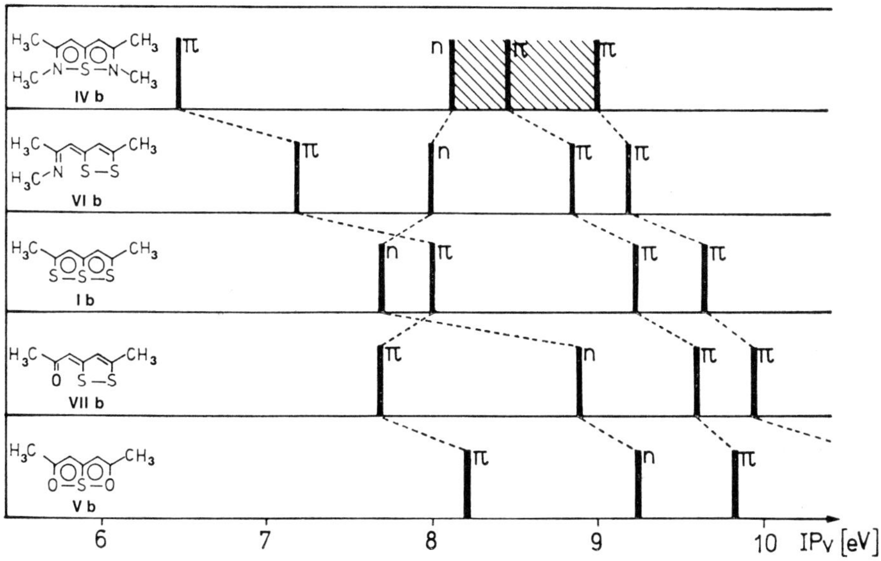

Fig. 13. Correlation between the first ionization potentials of the PE spectra of *Ib*, *IVb*, *Vb*, *VIb* and *VII b*

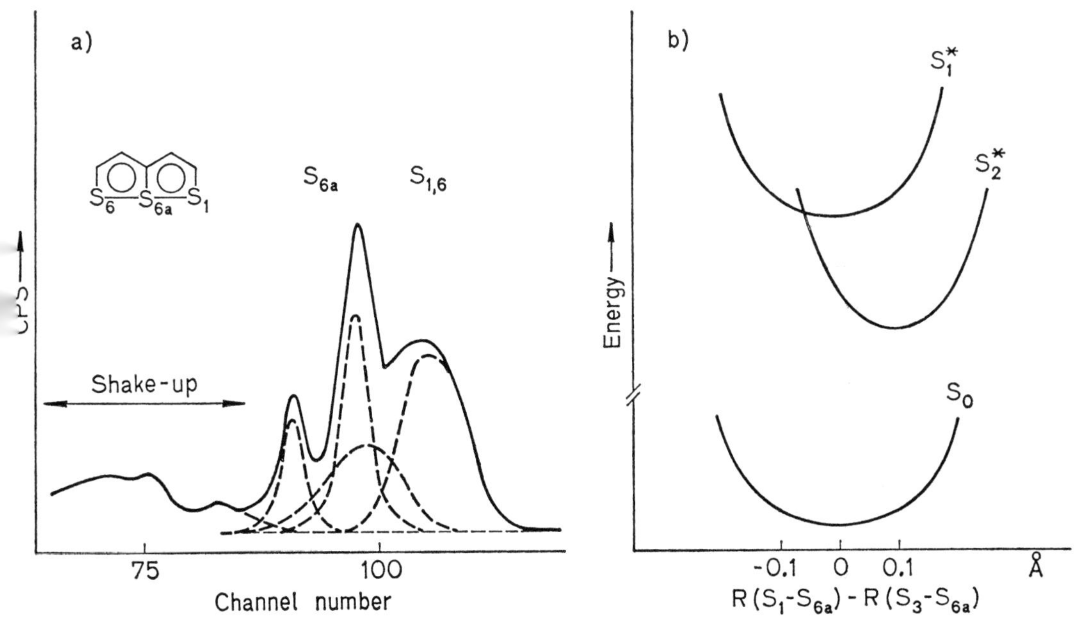

Fig. 14. (a) The S_{2p} photoelectron spectrum of *I a* in the gasphase. The deconvolution is indicated. (b) Schematic potential curves for the ground state (S_0) and the ionization of an S_{2p} electron of the central (S_1^*) and peripheral (S_2^*) sulfur center(s)

5. ESR-Spectra of Trithiapentalene Radical Anions

The ESR spectra of Ia^-, Ib^-, Id^- and some bridged compounds have been studied in solution [98,99]. The results obtained are consistent only with a structure with C_{2v} symmetry in a temperature range between +25 to −60 °C. The π-spinpopulation at center 2 of I^- is found relatively large while the spinpopulation at center 3 is relatively small. This is qualitatively in agreement with $a_2(\pi^*)$ as the LUMO assuming the same topology in the anion as in the neutral species. In Table 7 we have listed the coupling constants for Ia^-, Ib^-, Im^- and Id^-.

Table 7. Proton coupling constants ($a_{H\mu}$ in Gauss) for the radical anions Ia^- and substitution products [98]. The coupling constants are given for 25 °C

Radical anion	a_{H2}	a_{H3}	a_{CH_3}	a_{CH_2}	Phenyl a_H^o	a_H^m	a_H^p
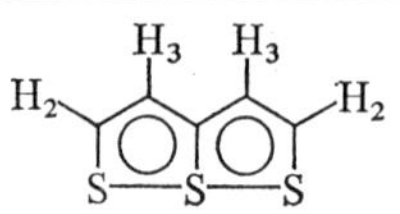 *Ia*	7.4	2.6	—	—	—	—	—
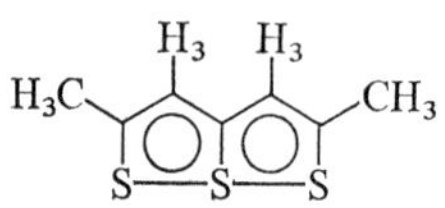 *Ib*	—	2.17	6.35	—	—	—	—
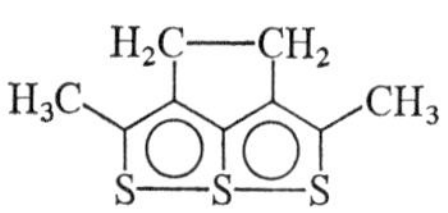 *Im*	—	—	6.55	2.52	—	—	—
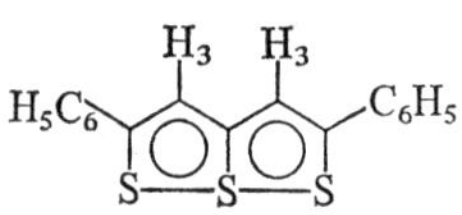 *Id*	—	1.58	—	—	1.37	0.45	1.58

ESR studies on the radical cations and polarographic studies could provide supplementary information concerning the process of reduction (*e.g.* reversibility).

V. Chemical Properties of Trithiapentalenes

From the many reaction types observed with π-systems (substitution, addition, cycloaddition...) only a few reactions have been studied [4,8,25]. The results are qualitative in character and detailed studies concerning the mechanism are missing. Reports on the reactivity of related heterocycles are sporadic [18,22,23,103].

1. Electrophilic Substitution

It is found that bromination [100] and Vilsmeier formylation [101] of trithiapentalenes proceed normally to give the substitution in 3 position.

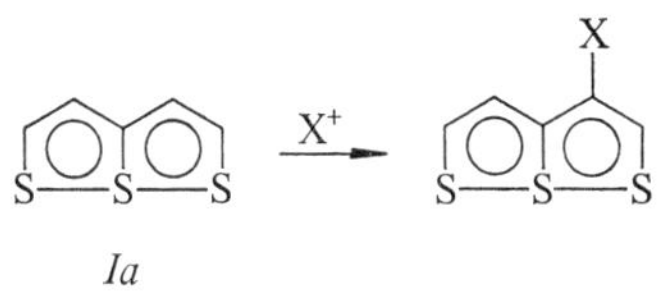

(15)

However, nitrosation and nitration [100] of *Id* gives the oxadithiaazapentalene shown below. A similar change is observed [101,102] by reaction with arenedia-

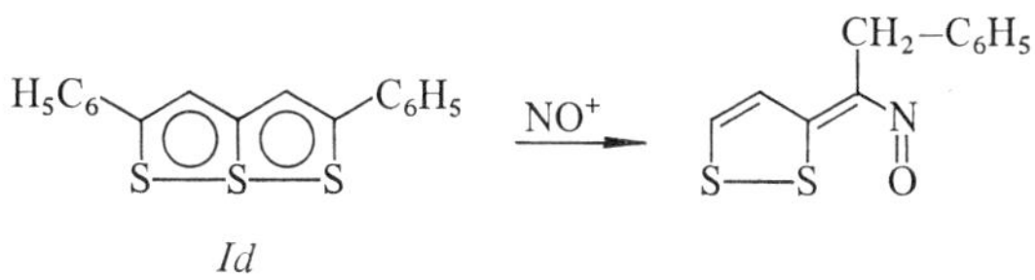

(16)

zonium fluorborates. This has also been reported for the derivatives of *V* and *VII* [103].

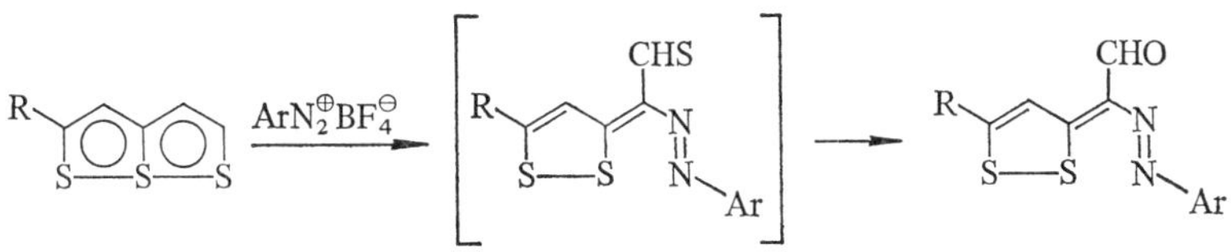

(17)

Alkylation occurs at sulfur [22,104]. The reaction with $Hg(OAc)_2$ and with strong acids [11] probably starts by attack at the sulfur center.

2. Nucleophilic Substitution

The reaction with hydrogensulfide or sulfide [105] is assumed to take place at the 2 position.

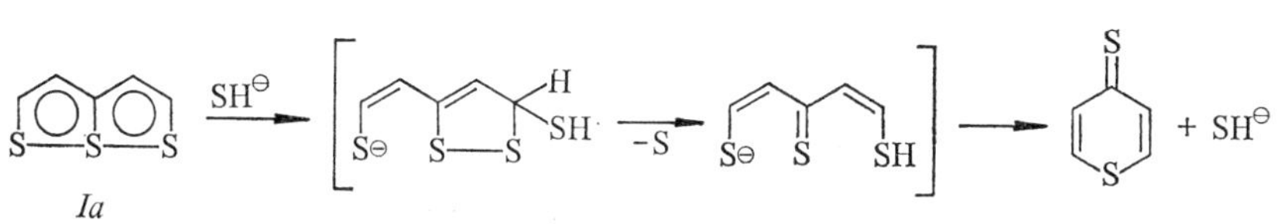

(18)

Methylthiosubstituents in position 2 or 5 are replaced by ethoxy groups [106] or aliphatic amines [107].

3. Model Calculations

These experimental facts have been rationalized using semiempirical calculations.

Based on a HMO model with a charge iteration procedure it was found [107] that the 2 position of *I* shows a surplus of positive charge while for the 3 position a surplus of negative charge is found.

Using a CNDO/2 model the energies for adding H^+ and H^- to *Ia* have been calculated [71]. No geometry optimization of these reaction products was made. Therefore these results have also a more or less qualitative character. It is found that the 3 position can be seen to be the preferred position for electrophilic attack. It is predicted that it will take place as facile as with benzene.

Another interesting point is the similarity of the localization energies for an attack at sulfur and for the attack at carbon.

From the calculations it is concluded that nucleophiles add favourably at position 2. This attack should occur more facile than in benzene. The reaction at the peripheral sulfur atom is predicted to be competitive with the one at position 2.

Both models mentioned explain qualitatively the experimental results.

4. Photochemistry

By irradiation into the long wavelength band of compounds of the type *I, V, VI* and *VII* photochromism is observed [58,70,108–110]. In most cases the reaction shows a high degree of reversibility. The assumed *cis-trans* isomerization [108, 109] has been confirmed by IR and NMR spectroscopy [58,109,110].

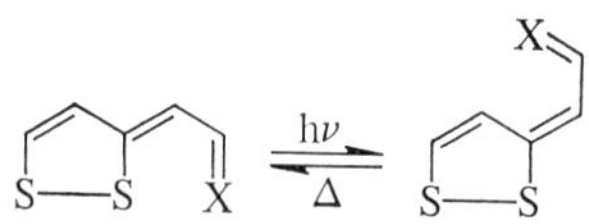

X=O, S, N–R (19)

Kinetic data for the thermal back reaction have been reported in different solvents [108–110]. A rationalization of this reaction has been given using semiempirical calculations [109]. A polar transition state *XVII* for the thermal back reaction has been suggested.

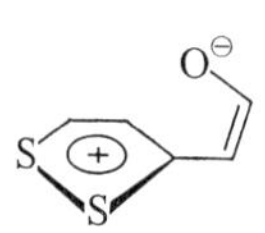

XVII

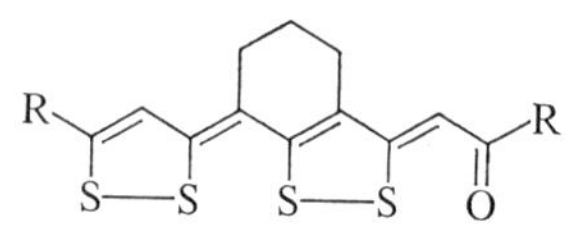

XVIII

Recently photochromism has also been detected for large systems such as *XVIII* [111]. Further questions have to be solved such as, from which excited state these reactions occur, the quantum yield of these isomerizations and the reaction products in case of irreversibility.

VI. Other Compounds Related to Trithiapentalenes

In Chapter III.3. the model compounds *IX—XI, XIII* and *XVI* were discussed as examples of the extension of the no-bond resonance concept. Derivatives of these model systems have been synthesized [4,8,81,112–119] (see Scheme 2) and

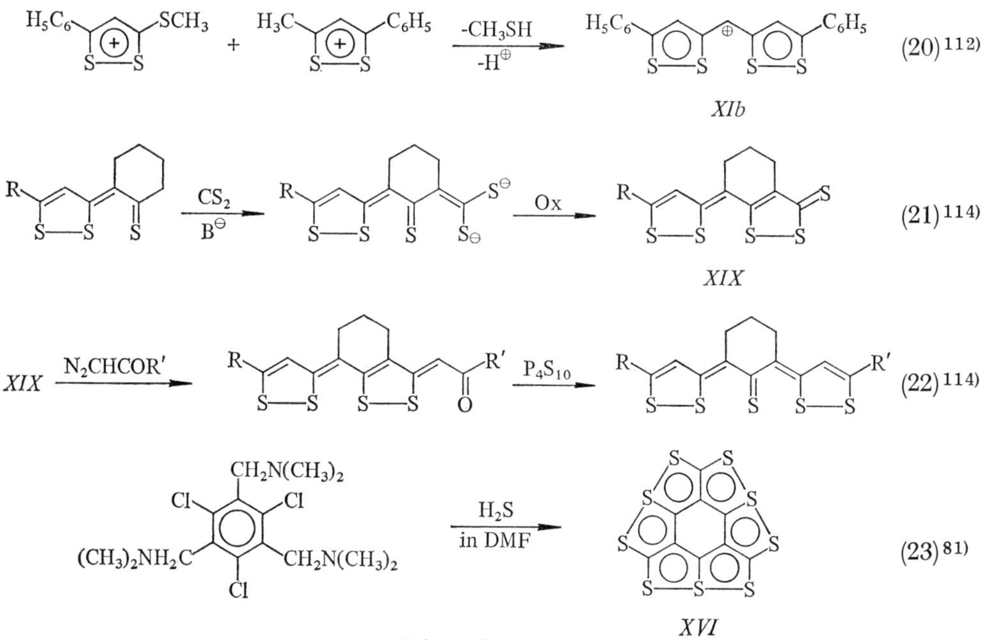

Scheme 2

X-ray results have been reported [78–80,120,121]. The S··S bond distances for some derivatives are shown below.

A comparison between the structures of *IXc* and *IXd* reveals only a minor sensitivity towards substituents in the carbon skeleton in contrast to the large sensitivity in the trithiapentalene case (see Chapter II.1.).

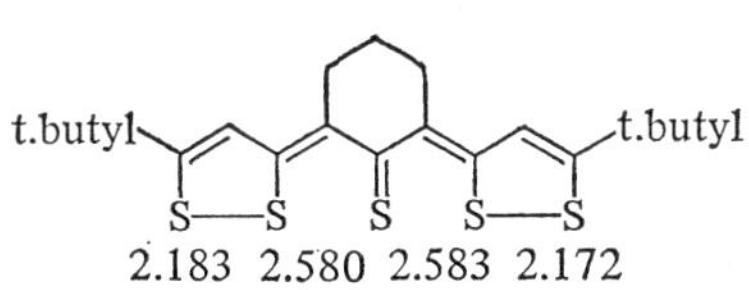

IXc [78]

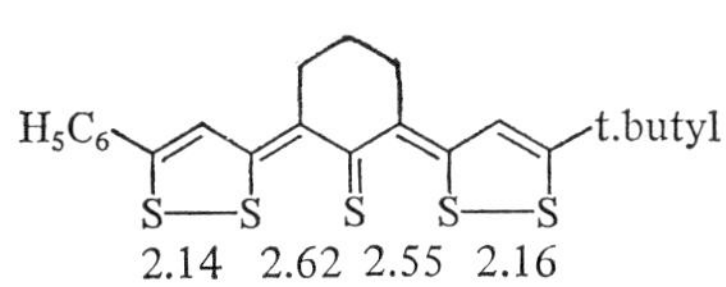

IXd [120]

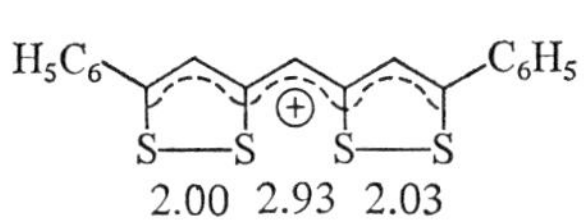

XIb [79]

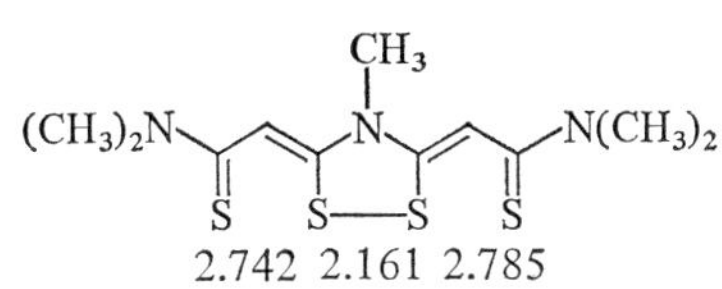

XIIIc [80]

Xb [121]

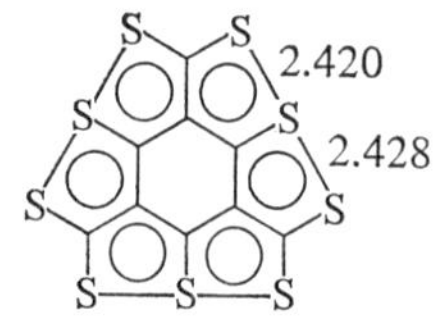

XVI [82]

The electronic spectra of *IXc* [117], *Xb* [122,123] and *XIb* [124,125] as well as related systems [113,117] have been reported. The good accord between calculation on *XIb* and measurements concerning position [124], intensity and polarization [126] direction of the transition moment shows that this system can be treated using the usual sulfur parameters. A similarly good accord has been found for derivatives of *X* [122,123].

When the electron-rich three center bond of trithiapentalenes was discussed the carbon skeleton of these species was disregarded. Therefore it seems interesting to look for systems where this skeleton is not present.

One thoroughly studied example is the triselenocyanate ion which has been studied in detail [127]. The geometry reported for its cesium salt is shown below [128]. The reported Se—Se bond length of 2.65 Å is 0.32 Å longer than the Se—Se "single-bond length" in Se $(SeCN)_2$ [129].

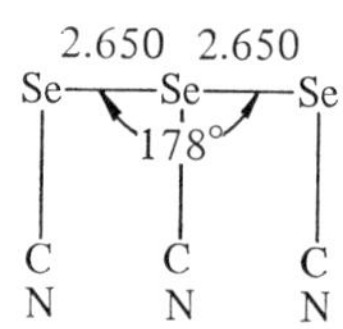

Linear chains with three selenium centers occur also in the potassium [130] and rubidium [131] salts as well as in the trisselenoureadichlorides and dibromides [132].

A large number of tellurium(II)-complexes exist also with an electron-rich three-center bond [127]. In most complexes the tellurium(II) shows square planar four coordination.

Another example are the triiodide anions [133]. Here as well as in the selenocyanates the distance depends on the counterion and shows a similar fluctuation as do the trithiapentalenes concerning the interatomic distances [134].

Apart from the examples mentioned so far, where at least the central atoms belonged to the second row, one should look for examples where the central atom is a first row atom [135].

So far there are no reports on the isolation of a system with C_{2V} symmetry where the central atom is a first row atom.

Examples pertaining to this are nitrobenzofuroxanes (*XX*) [136], 7-acetyl-3-methylanthranil (*XXI*) [137] and o,o′ disubstituted benzylcations (*XXII*) [138,139].

All examples studied so far show an equilibrium but no indication of a structure with C_{2V} symmetry. An electron-rich three-center bond is discussed for the transition state of *XX*, *XXI* [137] and *XXII* [60]. In this context an isoster of *XXI*, the cation *XXIII* [140] deserves special interest concerning its structure.

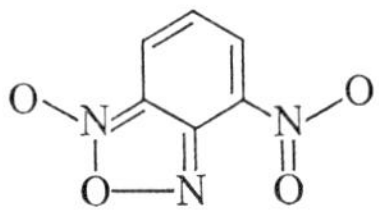

XX

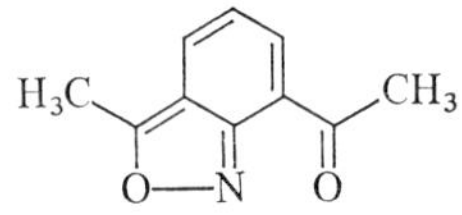

XXI

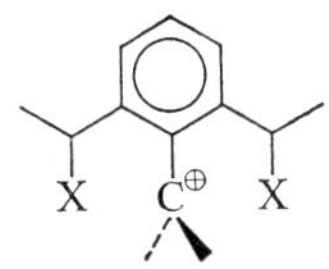

XXII

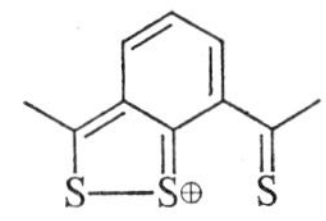

XXIII

The observation that at least the central atom has to be a second or third row atom to stabilize a C_{2V} symmetry does not necessarily mean that participation of empty *d* or *s* orbitals is essential. It could just as well mean that the overlap provided by the $2p$ orbitals of the central atom is not sufficient for a stabilization of the three-center bond.

There are many reports in the literature on structures with bond distances between S···O and S···S respectively between a single bond and the sum of the Van-der-Waals radii. A few examples are listed below.

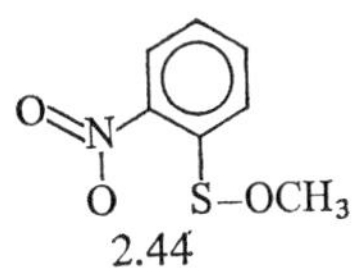

XXIV [141]

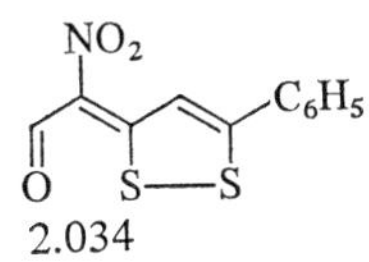

XXV [142]

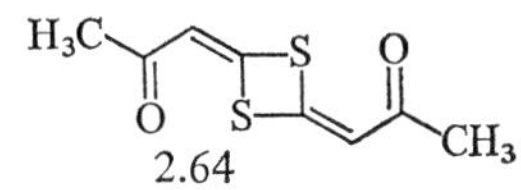

XXVI [143]

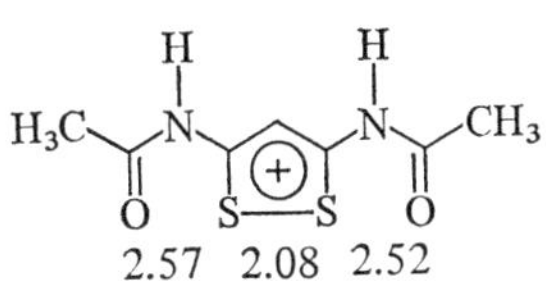

XXVII [144]

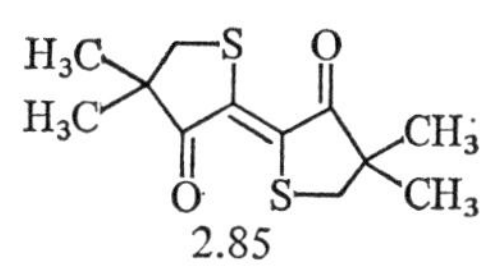

XVIII [145]

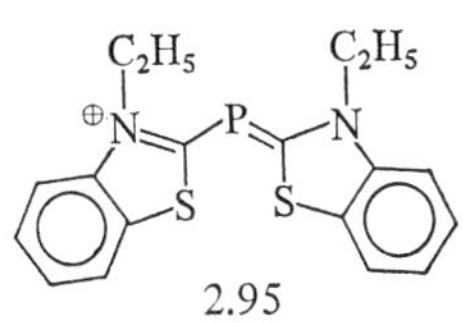

XXIX [146]

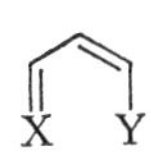

XXX

The U-shaped arrangement of the part (*XXX*) which is isovalence electronic to the pentadienyl anion, is common to all structures. This suggests that the π-interaction between X and Y, although not very large, rules the geometry [123, 147]. This suggestion could be easily checked by looking at the corresponding saturated species.

Probably, the preponderance of the U shape of *XXX* is a delicate balance between the repulsive forces (repulsion of the nuclei and the lone pairs) and attractive forces (p_π—p_π interaction and p_σ—$3d$ or $4s$ interaction).

VII. Outlook

The electron diffraction and ESCA experiment have clarified the picture concerning the structure of trithiapentalene in giving strong evidence for a broad U-shaped potential (see Fig. 14). However, there are still many questions to be answered until a full understanding of the bonding and chemical properties is achieved. In many chapters we indicated some of the questions.

For us the most interesting experiments are those concerned with structures related to *XXII* and efforts to clarify the picture of the partial bonds present in *XXIV* to *XXIX*.

Acknowledgments. Financial support has been obtained by Ciba Geigy S. A. and Hoffmann La Roche, both Basel, the Schweizerische Nationalfonds zur Förderung der wissenschaftlichen Forschung (Nr. 2.159.74), the Deutsche Forschungsgemeinschaft, the Fonds der Chemischen Industrie and the Otto Röhm Stiftung. R. Gygax is indepted to the Camille and Henry Dreyfus- Foundation for a stipend.

We would like to express our most sincere thanks to the many collaborators whose names appear in the References, for their contributions.

VIII. References

1) Arndt, F., Nachtwey, P., Pusch, J.: Chem. Ber. *58*, 1633 (1925).
2) Bezzi, S., Mammi, M., Garbuglio, C.: Nature *182*, 247 (1958).
3) Pfister-Guillouzo, G.: Bull. Soc. Chim. France *1958*, 1316.
4) Klingsberg, E.: Quart. Rev. *23*, 537 (1969) and references therein.
5) Wheland, G. W.: Resonance in organic chemistry, p. 151. New York: Wiley 1955.
6) Pfister-Guillouzo, G., Lozac'h, N.: Bull. Soc. Chim. France *1963*, 153.
7) Behringer, H., Ruff, M., Wiedemann, R.: Chem. Ber. *97*, 1732 (1964).
8) Lozac'h, N.: Advan. Heterocyclic Chem. *13*, 161 (1971).
9) Klingsberg, E.: J. Am. Chem. Soc. *85*, 3244 (1963).
10) Stavaux, M., Lozac'h, N.: Bull. Soc. Chim. France *1967*, 2082.
11) Arndt, F.: Rev. Fac. Sci. Univ. Istanbul *A 13*, 57 (1948).
12) Traverso, G.: Ann. Chim. (Rome) *45*, 687 (1955).
13) Behringer, H., Grimm, A.: Liebigs Ann. Chem. *682*, 188 (1965).
14) Klingsberg, E.: J. Am. Chem. Soc. *83*, 2934 (1961); J. Org. Chem. *31*, 3489 (1966).
15) Coulibaly, O., Mollier, Y.: Bull. Soc. Chim. France *1969*, 3208.
16) Leaver, D., McKinnon, D. M.: Chem. Ind. *1964*, 461.
17) Beer, R. J. S., Carr, R. P., Cartwright, C., Harris, D., Slater, R.A.: J. Chem. Soc. (C). *1968*, 2490.
18) Reid, D. H., Webster, R. G.: Chem. Commun. *1972*, 1283.
19) Dingwall, J. G., Reid, D. H., Symon, J. D.: J. Chem. Soc. (C) *1970*, 2412; Chem. Commun. *1969*, 466.
20) Reid, D. H., Webster, R. G.: J. C. S. Perkin I *1972*, 1447.
21) Dingwall, J. G., McKenzie, S., Reid, D. H.: J. Chem. Soc. (C) *1968*, 2543.
22) Ingram, A. S., Reid, D. H., Symon, J. D.: J. C. S. Perkin I *1974*, 242.
23) Dingwall, J. G., Ingram, A. S., Reid, D. H., Symon, J. D.: J. C. S. Perkin I *1973*, 2351.
24) Klingsberg, E.: J. Org. Chem. *33*, 2915 (1968).
25) Reid, D. H.: In: Organic compounds of sulphur, selenium and tellurium, Vol. 1 (1970). — Beer, R. J. S.: In: Organic compounds of sulphur, selenium and tellurium, Vol. 2 (1973).
26) Hansen, L. K., Hordvik, A.: Acta Chem. Scand. *27*, 411 (1973) and references given therein.
27) Leung, F., Nyburg, S. C.: Chem. Commun. *1969*, 137.
28) Hordvik, A., Saethre, L. J.: Acta Chem. Scand. *26*, 3114 (1972).
29) Hordvik, A.: Acta Chem. Scand. *25*, 1583 (1971).
30) Johnson, P. L., Paul, I. C.: Chem. Commun. *1969*, 1014.
31) Hordvik, A., Sletten, E., Sletten, J.: Acta Chem. Scand. *23*, 1852 (1969).
32) Hordvik, A., Saethre, L. J.: Acta Chem. Scand. *24*, 2261 (1970); *26*, 1729 (1972).
33) Birknes, B., Hordvik, A., Saethre, L. J.: Acta Chem. Scand. *26*, 2140 (1972).
34) Hordvik, A., Julshamn, K.: Acta Chem. Scand. *25*, 1835 (1971).
35) Hordvik, A., Sjølset, O., Saethre, L. J.: Acta Chem. Scand. *26*, 1297 (1972).
36) Pauling, L.: The nature of the chemical bond, 3rd edit. New York: Cornell University Press 1960.
37) Hordvik, A.: Acta Chem. Scand. *20*, 1885 (1966); Quart. Rev. Sulphur Chem. *5*, 21 (1970)
38) Hordvik, A., Julshamn, K.: Acta Chem. Scand. *25*, 2507 (1971).
39) Hordvik, A., Porten, J. A.: Acta Chem. Scand. *27*, 485 (1973).
40) Hordvik, A., Julshamn, K.: Acta Chem. Scand. *25*, 1895 (1971).
41) Hordvik, A., Rimala, T. S., Saethre, L. J.: Acta Chem. Scand. *26*, 2139 (1972); *27*, 360 (1973).
42) Hordvik, A., Julshamn, K.: Acta Chem. Scand. *26*, 343 (1972).
43) Gilardi, R. P., Karle, I. K.: Acta Cryst. *B 27*, 1073 (1971).
44) Leung, F., Nyburg, S. C.: Chem. Commun. *1970*, 707; Can. J. Chem. *49*, 167 (1971).
45) Mammi, M., Bardi, R., Traverso, G., Bezzi, S.: Nature *192*, 1282 (1961).
46) Hordvik, A., Sletten, E., Sletten, J.: Acta Chem. Scand. *23*, 1377 (1969).
47) Pinel, R., Mollier, Y., Llaguno, E. C., Paul, I. C.: Chem. Commun. *1971*, 1352.
48) Hordvik, A., Saethre, L. J.: Acta Chem. Scand. *26*, 849 (1972).

49) Beer, R. J. S., Hatton, J. R., Llaguno, E. C., Paul, I. C.: Chem. Commun. *1971*, 594. — Llaguno, E. C., Paul, I. C.: J. C. S. Perkin II, *1972*, 2001; Tetrahedron Letters *1973*, 1565.
50) Leading references: Muetterties, E. L., Schum, R. A.: Quart. Rev. *20*, 245 (1966). — Perozzi, E., Martin, J. C., Paul, I. C.: J. Am. Chem. Soc. *96*, 6735 (1974).
51) Shen, Q., Hedberg, K.: J. Am. Chem. Soc. *96*, 289 (1974).
52) Hertz, H. G., Traverso, G., Walter, W.: Liebigs Ann. Chem. *625*, 43 (1959).
53) Pfister-Guillouzo, G., Lozac'h, N.: Bull. Soc. Chim. France *1964*, 3254.
54) Bohlmann, R., Bresinsky, E.: Chem. Ber. *100*, 107 (1967).
55) Lapper, R. D., Poole, A. J.: Tetrahedron Letters *1974*, 2783.
56) Pinel, R., Mollier, Y., Lozac'h, N.: Bull. Soc. Chim. France *1967*, 856.
57) Festal, D., Mollier, Y.: Tetrahedron Letters *1970*, 1259.
58) Gleiter, R., Knauer, K. H., Schmidt, E., Mollier, Y., Pinel, R.: Tetrahedron Letters *1973*, 1257.
59) Hach, H. J., Rundle, R. E.: J. Am. Chem. Soc. *73*, 4321 (1951). — Rundle, R. E.: J. Am. Chem. Soc. *85*, 112 (1963).
60) Gleiter, R., Hoffmann, R.: Tetrahedron *24*, 5899 (1968).
61) a) Hund, F.: Z. Physik *43*, 805 (1927);
b) Harmony, M. D.: Chem. Soc. Rev. *1*, 211 (1972);
c) Pitzer, K. S.: J. Chem. Phys. *7*, 314 (1939);
d) Dennison, D. M., Uhlenbeck, G. E.: Phys. Rev. *41*, 313 (1932).
62) Wilson, E. B., Decius, J. C., Cross, P. C.: Molecular vibrations. New York: McGraw Hill 1955.
63) Herzberg, G.: Infrared and raman spectra, p. 131 (a) and p. 220 (b). Princeton: D. Van Nostrand 1945.
64) Giacometti, G., Rigatti, G.: J. Chem. Phys. *30*, 1633 (1959).
65) Shustorovich, E. M.: J. Gen. Chem. USSR (Engl. Transl.) *29*, 2424 (1959).
66) Maeda, K.: Bull. Chem. Soc. Japan *34*, 785, 1166 (1961).
67) Hoffmann, R.: J. Chem. Phys. *39*, 1397 (1963). — Hoffmann, R., Lipscomb, W. N.: J. Chem. Phys. *36*, 2179, 3489 (1962); *37*, 2872 (1962).
68) Pople, J. A., Beveridge, D. L.: Approximate molecular orbital theory. New York: McGraw-Hill 1970. — Santry, D. P., Segal, G. A.: J. Chem. Phys. *47*, 158 (1967).
69) Bingham, R., Dewar, M. J. S., Lo, D. H.: J. Am. Chem. Soc. *97*, 1285 (1975).
70) Gleiter, R., Werthemann, D., Behringer, H.: J. Am. Chem. Soc. *94*, 651 (1972).
71) Clark, D. T., Kilcast, D.: Tetrahedron *27*, 4367 (1971).
72) Hansen, L. K., Hordvik, A., Saethre, L. J.: Chem. Commun. *1972*, 222.
73) Bischof, P.: private communication.
74) Clark, D. T., Intern. J. Sulfur Chem. *C7*, 11 (1972).
75) Palmer, M. H., Findlay, R. H.: Tetrahedron Letters *1972*, 4165; J. C. S. Perkin II *1974*, 1885. — Palmer, M. H.: private communication.
76) Calzaferri, G., Gleiter, R.: J. C. S. Perkin II *1975*, 559.
77) Brown, E. I. G., Leaver, D., Rawlings, T. J.: Chem. Commun. *1969*, 83.
78) Kristensen, R., Sletten, J.: Acta Chem. Scand. *27*, 2517 (1973).
79) Hordvik, A.: Acta Chem. Scand. *19*, 1253 (1965).
80) Flippen, J. L.: J. Am. Chem. Soc. *95*, 6073 (1973). — Sletten, J.: Acta Chem. Scand. *A28*, 989 (1974); *A29*, 317 (1975).
81) Brown, J. P., Gay, T. B.: J. C. S. Perkin I *1974*, 866.
82) Hansen, L. K., Hordvik, A.: Chem. Commun. *1974*, 800.
83) Calzaferri, G., Gleiter, R.: unpublished results.
84) Leading references: Clark, A. H.: In: Internal rotation in molecules (ed. W. J. Orville-Thomas). London-New York-Sydney-Toronto: J. Wiley and Sons 1974.
85) Brown, E. I. G.: Leaver, D., McKinnon, D. M.: J. Chem. Soc. (C) *1970*, 1202.
86) Johnstone, R. A. W., Ward, S. D.: Theoret. Chim. Acta *14*, 420 (1969).
87) Gleiter, R., Schmidt, D., Behringer, H.: Chem. Commun. *1971*, 525.
88) Kroner, J., Proch, D.: Tetrahedron Letters *1972*, 2537.
89) Fabian, J.: Z. Chem. *13*, 26 (1973).

90) Gleiter, R., Gygax, R.: unpublished. — Gygax, R.: planned dissertation at the University of Basel, 1976.
91) Heilbronner, E., Gerdil, R.: Helv. Chim. Acta *39*, 1996 (1956). — Jaffé, H. H.: Theory and applications of ultraviolet spectroscopy, 392f. New York: Wiley 1962.
92) Gleiter, R., Hornung, V., Lindberg, B., Högberg, S., Lozac'h, N.: Chem. Phys. Letters *11*, 401 (1971). — b) Gleiter, R., Gygax, R., Reid, D. H.: Helv. Chim. Acta *58*, 1591 (1975).
93) Koopmans, T.: Physica *1*, 104 (1934).
94) Clark, D. T., Kilcast, D., Reid, D. H.: Chem. Commun. *1970*, 638.
95) Lindberg, B., Högberg, S., Malmsten, G., Bergmark, J., Nilsson, Ö., Karlsson, S.-E., Fahlman, A., Gelius, U., Pinel, R., Stavaux, M., Mollier, Y., Lozac'h, N.: Chem. Scripta *1*, 183 (1971).
96) Gelius, U.: J. Electron Spectr. Rel. Phen. *5*, 985 (1974).
97) Lindberg, B. J., Pinel, R., Mollier, Y.: Tetrahedron *30*, 2539 (1974).
98) Gerson, F., Gleiter, R., Heinzer, J., Behringer, H.: Angew. Chem. *82*, 294 (1970); Angew. Chem. Intern. Ed. Engl. *9*, 306 (1970). — Gerson, F., Gleiter, R., Heinzer, J.: unpublished results.
99) Gerson, F., Heinzer, J., Stavaux, M.: Helv. Chim. Acta *56*, 1845 (1973).
100) Beer, R. J. S., Cartwright, D., Gait, R. J., Harris, D.: J. Chem. Soc. (C) *1971*, 963, and references therein.
101) Dingwall, J. G., Reid, D. H., Wade, K. O.: J. Chem. Soc. (C) *1969*, 913. — Duguay, G., Reid, D. H., Wade, K. O., Webster, R. G.: J. Chem. Soc. (C) *1971*, 2829.
102) Bignebat, J., Quiniou, H.: Compt. Rend. *267 C*, 180 (1968); *269 C*, 1129 (1969).
103) Christie, R. M., Ingram, A. S., Reid, D. H., Webster, R. G.: Chem. Commun. *1973*, 92.
104) Behringer, H., Falkenberg, J.: Chem. Ber. *102*, 1585 (1969).
105) Dingwall, J. G., Reid, D. H.: Chem. Commun. *1968*, 863.
106) Beer, R. J. S., Carr, R. P., Cartwright, D., Harris, D., Slater, R. A.: J. Chem. Soc. (C) *1968*, 2490.
107) Beer, R. J. S., Cartwright, D., Gait, R. J., Johnstone, R. A. W., Ward, S. D.: Chem. Commun. *1968*, 688.
108) Pedersen, C. Th., Lohse, C.: Chem. Commun. *1973*, 123; J. C. S. Perkin I *1973*, 2837.
109) Calzaferri, G., Gleiter, R., Knauer, K. H., Rommel, E., Schmidt, E., Behringer, H.: Helv. Chim. Acta *56*, 597 (1973).
110) Calzaferri, G., Gleiter, R., Gygax, R., Knauer, K. H., Schmidt, E., Behringer, H.: Helv. Chim. Acta *56*, 2584 (1973).
111) Pedersen, C. Th., Lohse, C., Stavaux, M.: J. C. S. Perkin I *1974*, 2722.
112) Klingsberg, E.: J. Heterocycl. Chem. *3*, 243 (1966).
113) Klingsberg, E.: Chem. Ind. (London) *1968*, 1813.
114) Stavaux, M., Lozac'h, N.: Bull. Soc. Chim. France *1967*, 3557; *1968*, 4273; *1969*, 4184.
115) Stavaux, M.: Bull. Soc. Chim. France *1971*, 4418.
116) Stavaux, M., Lozac'h, N.: Bull. Soc. Chim. France *1971*, 4419, 4423.
117) Stavaux, M.: Bull. Soc. Chim. France *1971*, 4426, 4429.
118) Goerdeler, J., Ulmen, J.: Chem. Ber. *105*, 1568 (1972).
119) Oliver, J. E., Stokes, J. B.: Intern. J. Sulfur Chem. *A2*, 105 (1972).
120) Sletten, J.: Acta Chem. Scand. *24*, 1464 (1970); *25*, 3577 (1971).
121) Grundvig, E., Hordvik, A., Acta Chem. Scand. *25*, 1567 (1971).
122) Lüttringhans, A., Mohr, M., Engelhard, N.: Liebigs Ann. Chem. *661*, 84 (1963).
123) Fabian, K., Hartmann, H., Fabian, J., Mayer, R.: Tetrahedron *27*, 4705 (1971).
124) Fabian, J., Hartmann, H.: Tetrahedron *29*, 2597 (1973).
125) Fabian, J., Hartmann, H., Fabian, K.: Tetrahedron *29*, 2609 (1973).
126) Gleiter, R., Klingsberg, E., Schmidt, D.: unpublished. The first band in *XIb* and 1,3-diphenyldithioliumperchlorate is found to be polarized parallel to the long axis of the molecule.
127) Foss, O.: Pure Appl. Chem. *24*, 31 (1970). In: Selected topics in structure chemistry, p. 145 (P. Andersen, O. Bastiansen, and S. Furberg, ed.). Universitetsforlaget, Oslo 1967, and references therein.
128) Hauge, S.: Acta Chem. Scand. *A29*, 163 (1975).
129) Aksnes, O., Foss, O.: Acta Chem. Scand. *8*, 1787 (1954).

130) Hauge, S., Sletten, J.: Acta Chem. Scand. *25*, 3094 (1971).
131) Hauge, S.: Acta Chem. Scand. *25*, 3103 (1971).
132) Hauge, S., Opedal, D., Årskog, J.: Acta Chem. Scand. *A 29*, 225 (1975).
133) Mooney, R. C. L.: Phys. Rev. *53*, 851 (1938). — Mooney, R. C. L., Slater, J. C.: Acta Cryst. *12*, 187 (1959). — Mooney, R. C. L.: Z. Krist. *90*, 143 (1935). — Tasman, H. A., Boswijk, K. H.: Acta Cryst. *8*, 59, 857 (1955).
134) Bürgi, H. B.: Angew. Chem. *87*, 461 (1975); Angew. Chem. Intern. Ed. Engl. *14*, 460 (1975).
135) Klingsberg, E.: Lectures in heterocyclic chemistry, Vol. 1, S. 19 (1972); A supplementary issue of the Journal of Heterocyclic Chemistry (R. N. Castle and E. F. Elslager, ed.).
136) Boulton, A. J.: Lectures in heterocyclic chemistry, Vol. 2, S. 45 (1975). — Boulton, A. J., Ghosh, P. B.: Advan. Heterocycl. Chem. *10*, 1 (1969), and literature cited therein.
137) Parry, K. P., Rees, C. W.: Chem. Commun. *1971*, 833.
138) Breslow, R., Kaplan, L., LaFollette, D.: J. Am. Chem. Soc. *90*, 4056 (1968), and references therein.
139) Martin, J. C., Basaly, R. J.: J. Am. Chem. Soc. *95*, 2572 (1973).
140) Brown, E. I. G., Leaver, D., Rawlings, T. J.: Chem. Commun. *1969*, 83.
141) Hamilton, W. C., LaPlaca, S. J.: J. Am. Chem. Soc. *86*, 2289 (1964).
142) Johnson, P. L., Reid, K. I. G., Paul, I. C.: J. Chem. Soc. (B) *1971*, 946; see also Johnson, P. L., Paul, I. C.: J. Am. Chem. Soc. *91*, 781 (1969).
143) Lynch, T. R., Melor, I. P., Nyburg, S. C., Yates, P.: Tetrahedron Letters *1967*, 373. — Kapecki, J. A., Baldwin, J. E.: J. Am. Chem. Soc. *91*, 1120 (1969).
144) Hordvik, A., Kjøge, H. M.: Acta Chem. Scand. *20*, 1923 (1966).
145) Hermann, H. J. A., Amon, H. L., Gibson, R. E.: Tetrahedron Letters *1969*, 2559; thioindigo see Gribova, E. A., Zhdanov, G. S., Gol'der, G. A.: Soviet Phys.-Cryst. (English Transl.) *1*, 39 (1956).
146) Allmann, R.: Chem. Ber. *99*, 1332 (1966).
147) Hoffmann, R., Olofson, R. A.: J. Am. Chem. Soc. *88*, 943 (1966).

Received October 23, 1975

The Molecular Zeeman Effect

Prof. Dr. D. H. Sutter

Institute for Physical Chemistry, University of Kiel, D-2300 Kiel, Germany

Prof. Dr. W. H. Flygare

Noyes Chemical Laboratory University of Illinois Urbana, Illinois 61801, USA

Contents

Introduction

Chemical behavior depends on the geometry and charge distribution of interacting molecules. Thus, from the beginning chemists were interested in accurate information on molecular structures (including electronic structures). Since its introduction about thirty years ago, rotational microwave spectroscopy has proved to be one of the most powerful tools available for the investigation of small molecules. Extensive work has been reported in determining molecular structures, investigating low frequency vibrations including internal rotation, measuring nuclear quadrupole coupling constants, and in determining the square of the permanent electric dipole moment. About six years ago the use of the molecular Zeeman effect was extensively developed to study diamagnetic molecules which provides significant new information concerning the electronic structure of molecules. This new development in microwave spectroscopy has led to accurate values for the diagonal elements in the molecular *g*-value tensor, anisotropies in the diagonal values of the magnetic susceptibility tensor and the diagonal elements in the molecular electric quadrupole moment tensor for a wide range of diamagnetic molecules.

These new results have stimulated new interest in understanding the nature of the electronic structure in molecules in reference to understanding the nature of molecular electric quadrupole moments and the field stimulated electric currents which lead to the magnetic susceptibility anisotropies. The molecular electronic quadrupole moments are an electronic ground state property which can be quite effectively calculated by semiempirical and *ab initio* quantum mechanical methods. The magnetic susceptibility is a sum of a diamagnetic term, which is a ground state property, and a paramagnetic term, which depends on all of the excited electronic states. The diamagnetic term can be calculated by quantum theory in a manner similar to the methods used to calculate quadrupole moments. However, the paramagnetic term which includes a sum over all electronic states has not been as easy to understand. *Ab initio* methods of calculating the paramagnetic susceptibility have not been generally successful. However, semiempirical methods can be employed to estimate the diagonal elements in the magnetic susceptibility of new molecules in the absence of the experimental results.

In the first Chapter of this contribution we will familiarize the reader with the basic principles of rotational Zeeman spectroscopy. In the second Chapter some recent results on molecular quadrupole moments and magnetic susceptibility anisotropies will be presented in order to demonstrate the possibilities of the method. In the third Chapter the instrumentation and the analysis of the spectra will be described in detail. In this Chapter the effects which limit the accuracy of the measurements will also be discussed. The detailed theory starting from the Lagrangian of the molecular system is given in the final Chapter. In the Appendix the latest results covering the period from 1972 to spring 1975 are compiled in the form of a table. For earlier results the reader is referred to the review paper by Flygare and Benson [1] and to the compilation of molecular magnetic data by Hüttner and Tischer which appeared in Landolt-Börnstein [2].

I. Basic Principles

We limit our discussion to diamagnetic molecules, molecules with no net electronic angular momentum in the non-rotation electronic ground state. The rigid rotor model has been very successful in describing the major share of the rotational properties which are observed by spectroscopy. Within this model the molecule is approximated by a rigid frame of atomic point masses fixed at the equilibrium positions of the nuclei (minima of the potential hypersurface). Small positive and negative point charges in the order of 1/10 of the electron charge, e, may account for the electric dipole moment of the molecule. If the "molecule" then rotates, the rotating point charge distribution necessarily will produce a magnetic field, which to a first approximation corresponds to that of a magnetic dipole moment. The larger the speed of rotation, the larger is this magnetic dipole moment. Obviously, it must be proportional to the angular momentum and it will depend on the axis of rotation within the molecule. For a molecule rotating about its a-axis the amount of mutual cancellation of rotating positive and negative currents will be usually different from the amount of cancellation in the case of a rotation about its b- or its c-axis as shown, for instance, in Fig. I.1. Thus, in general we will expect a

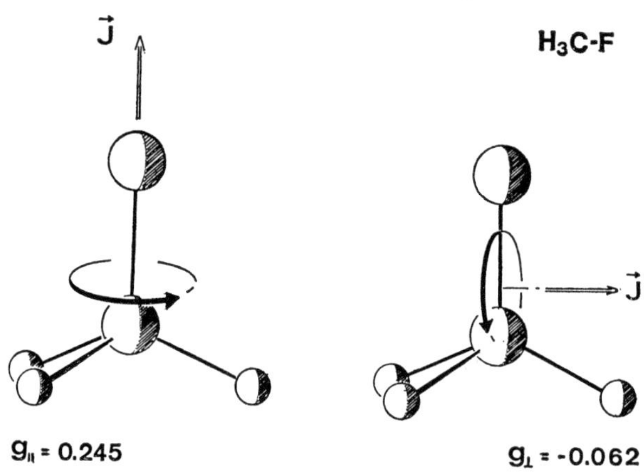

Fig. I.1. Schematic showing the rotational motion about axes parallel and perpendicular to the C—F bond in CH_3F. The currents for the parallel and perpendicular rotations will lead to different magnetic moments and therefore a tensorial relationship between the angular momentum and the induced moment is appropriate

tensorial relationship between the rotational magnetic moment, $\boldsymbol{\mu}_{\text{rot}}$, and the angular momentum vector of the overall rotation, $\hbar \boldsymbol{J}$. If $\boldsymbol{\mu}_{\text{rot}}$ is measured in units of the nuclear magneton, $\mu_N = |e|\hbar/(2M_p c)$, this relation may be written in matrix notation as

$$\begin{pmatrix} \mu_{a,\text{ rot}} \\ \mu_{b,\text{ rot}} \\ \mu_{c,\text{ rot}} \end{pmatrix} = \mu_N \begin{pmatrix} g_{aa} & g_{ab} & g_{ac} \\ g_{ba} & g_{bb} & g_{bc} \\ g_{ca} & g_{cb} & g_{cc} \end{pmatrix} \cdot \begin{pmatrix} J_a \\ J_b \\ J_c \end{pmatrix} \tag{I.1}$$

or more compactly with $\underset{\sim}{g}$ the molecular $\underset{\sim}{g}$-tensor:

$$\mu_{\text{rot}} = \mu_{\text{N}} \underset{\sim}{g} \cdot J \tag{1.1'}$$

In Eq. (I.1) J_a, J_b, and J_c stand for the a, b, and c-components of the overall angular momentum measured in units of $\hbar$. They are referred to the principal axis system of the molecular moment of inertia tensor. Because of the mutual compensation of positive and negative contributions, the absolute values of the g-tensor elements are usually smaller than 1 (typically on the order of 0.01 to 0.1 as may be checked in Table AI of the Appendix). In many cases the off-diagonal elements of the g-tensor in Eq. (I.1) will be zero because of molecular symmetry. Formaldehyde or 1,2-difluorobenzene may serve as examples.

In Chapter IV it will be shown that the theoretical expressions for the diagonal elements of the g-tensor are given by [3]

$$g_{aa} = \underbrace{\frac{M_{\text{p}}}{I_{aa}^{(\text{n})}}\left(1 + \frac{2}{I_{aa}^{(\text{n})}}\left(\frac{L_aL_a}{\Delta}\right)\right)\sum_{\nu}^{\text{nuclei}} Z_\nu(b_\nu^2 + c_\nu^2)}_{\text{"nuclear contribution"}} + \underbrace{\frac{M_{\text{p}}}{I_{aa}^{(\text{n})}}\frac{2}{m}\left(\frac{L_aL_a}{\Delta}\right)}_{\text{"electronic contribution"}} \tag{I.2}$$

where M_{p} is the proton mass, m is the electron mass, and L_a is the a-component of the electronic angular momentum operator defined as

$$L_a = \frac{\hbar}{\text{i}} \sum_{\varepsilon} \left(b_\varepsilon \frac{\partial}{\partial c_\varepsilon} - c_\varepsilon \frac{\partial}{\partial b_\varepsilon}\right).$$

a_ν, b_ν, c_ν and a_ε, b_ε, c_ε are the coordinates of the ν-th nucleus and ε-th electron respectively referred to the principal inertia axis system. The sum over ε is over all of the electrons.

The perturbation sum is written as a sum over all excited states, n:

$$\left(\frac{L_aL_a}{\Delta}\right) = \sum_n \frac{|\langle n|L_a|0\rangle|^2}{E_0 - E_n}.$$

Since $E_0 < E_n$, $\left(\frac{L_aL_a}{\Delta}\right)$ is negative. $\text{I}_{aa}^{(\text{n})}$ etc. are the nuclear contributions to the moment of inertia.

A second source for the molecular magnetic moment lies in the intramolecular electronic currents which are induced by the exterior magnetic field as demonstrated in Fig. I.2. Again one expects a tensorial relationship, in this case between the field induced magnetic moment, μ_{ind}, and the exterior filed, H. This may be expressed as:

$$\begin{pmatrix} \mu_{a,\text{ind}} \\ \mu_{b,\text{ind}} \\ \mu_{c,\text{ind}} \end{pmatrix} = \begin{pmatrix} \chi_{aa} & \chi_{ab} & \chi_{ac} \\ \chi_{ba} & \chi_{bb} & \chi_{bc} \\ \chi_{ca} & \chi_{cb} & \chi_{cc} \end{pmatrix} \cdot \begin{pmatrix} H_a \\ H_b \\ H_c \end{pmatrix} \tag{I.3}$$

or in tensorial form:

$$\mu_{\text{ind}} = \underset{\sim}{\chi} \cdot H \tag{I.3'}$$

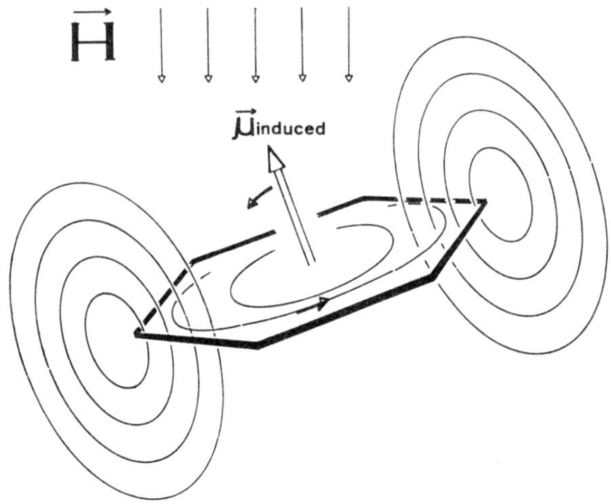

Fig. I.2. The field induced magnetic moment is depicted schematically in this drawing. This effect is most pronounced in aromatic molecules such as fluorobenzene, where comparatively strong electron ring currents may be induced, leading to a field induced, molecular magnetic dipole moment which opposes the exterior field. Trying to align the induced moment, the exterior field will exert a torque $\boldsymbol{\mu}_{\mathrm{ind}} \times \boldsymbol{H}$ on the molecule and will thus perturb the overall rotation. This perturbation is seen as a splitting in the rotational spectra. Since there will be a torque only in the case that $\boldsymbol{\mu}_{\mathrm{ind}}$ and $\boldsymbol{H}$ are not aligned, *i.e.*, if $\underset{\sim}{\chi}$ is anisotropic, only the anisotropies of the molecular susceptibility tensor can be obtained from the splittings of the rotational lines

with χ the molecular magnetic susceptibility tensor. As will be shown in Chapter IV, the theoretical expressions for the diagonal elements of the susceptibility tensor may be approximated as [4)]

$$\chi_{aa} = \underbrace{-\frac{e^2}{4mc^2}\left\langle 0 \left| \sum_{\varepsilon} (b_\varepsilon^2 + c_\varepsilon^2) \right| 0 \right\rangle}_{\text{diamagnetic susceptibility}} \underbrace{- \frac{e^2}{2m^2c^2}\left(\frac{L_a L_a}{\Delta}\right)}_{\text{paramagnetic susceptibility}} \tag{I.4}$$

where $|e|$ is the absolute value of the electron charge and c is the speed of light.

Eqs. (I.2) and (I.4) are the leading terms following from a second order perturbation treatment within the electronic states assuming a rigid rotor (neglect of vibrations). In both equations the two contributions ("nuclear" and "electronic" in the case of the g-values, and "diamagnetic" and "paramagnetic" in the case of the χ-values), have the same order of magnitude but opposite signs. Thus, a rather high numerical accuracy is required if these quantities are to be calculated by quantum chemical methods.

From Eqs. (I.1′) and (I.3′) the potential energy of a molecule at rest within the exterior magnetic field may be written as:

$$V(H) = -\int_{H=0}^{H} d\boldsymbol{H} \cdot (\boldsymbol{\mu}_{\mathrm{rot}} + \boldsymbol{\mu}_{\mathrm{ind}}(\boldsymbol{H})) = -\mu_{\mathrm{N}}\boldsymbol{H} \cdot \underset{\sim}{\boldsymbol{g}} \cdot \boldsymbol{J} - \tfrac{1}{2}\boldsymbol{H} \cdot \underset{\sim}{\chi} \cdot \boldsymbol{H} \tag{I.5}$$

In addition to Eq. (I.5) a further contribution to the energy arises if the molecule is in translational motion with a center of mass velocity $\boldsymbol{V}_0$. In this case the Lorentz forces corresponding to $\boldsymbol{V}_0$ force the positive nuclei in one direction and the negative electrons in the opposite direction.

$$\boldsymbol{F}_{\text{Lorentz}} = \frac{q}{c}(\boldsymbol{V}_0 \times \boldsymbol{H}) \qquad q = \begin{array}{ll} -|e| & \text{for the electrons} \\ +Z_\nu|e| & \text{for the nuclei} \end{array}$$

This force has the same effect as an electric field $\boldsymbol{E}_{\text{TS}} = \frac{1}{c}(\boldsymbol{V}_0 \times \boldsymbol{H})$ which leads to a translational Stark effect with potential energy given by

$$V_{\text{TS}} = -\boldsymbol{\mu}_{\text{el}} \cdot \boldsymbol{E}_{\text{TS}} = -\boldsymbol{\mu}_{\text{el}} \cdot \frac{1}{c}(\boldsymbol{V}_0 \times \boldsymbol{H}) \tag{I.6}$$

($\boldsymbol{\mu}_{\text{el}}$ = molecular electric dipole moment). This effect may become important in symmetric top molecules. This point will be discussed in detail in Chapter III.

After addition of the kinetic energy of the overall molecular rotation, the phenomenological "derivation" of the effective rotational Hamiltonian is complete:

$$\mathcal{H} = \underbrace{\mathrm{h}(AJ_a^2 + BJ_b^2 + CJ_c^2)}_{\mathcal{H}_{\text{rot}}} - \underbrace{\mu_{\mathrm{N}}\boldsymbol{H}\cdot\boldsymbol{g}\cdot\boldsymbol{J}}_{+\ \mathcal{H}_{\mathrm{g}}} - \underbrace{\frac{1}{2}\boldsymbol{H}\cdot\underset{\sim}{\chi}\cdot\boldsymbol{H}}_{+\ \mathcal{H}_\chi} - \underbrace{\boldsymbol{\mu}_{\text{el}}\cdot(\boldsymbol{V}_0 \times \boldsymbol{H})/\mathrm{c}.}_{+\ \mathcal{H}_{\text{TS}}} \tag{I.7}$$

For completeness we also give the theoretical expressions for the rotational constants A, B, and C which are essentially given by the inverse moments of inertia and which enter into the rotational energy expression, $\mathcal{H}_{\text{rot}}$.

$$A = \frac{h}{8\pi^2 I_{aa}} = \frac{h}{8\pi^2 I_{aa}^{(\mathrm{n})}}\left(1 + \frac{2}{I_{aa}^{(\mathrm{n})}}\left(\frac{L_a L_a}{\Delta}\right)\right) + \ldots \tag{I.8}$$

and cyclic permutations for B and C. The small corrections $\frac{2}{I_{aa}^{(\mathrm{n})}}\left(\frac{L_a L_a}{\Delta}\right)$ correspond to the electronic contributions to the molecular moments of inertia [5]. In quantum mechanics, the components of the angular momentum and the direction cosines between the space fixed direction of the exterior field and the basis vectors of the rotating molecular principal inertia axis system both become operators and the corresponding eigenvalue problem has to be solved. In practice this is usually done by setting up the Hamiltonian matrix corresponding to Eq. (I.7) within the well known eigenfunction basis of the limiting symmetric top and subsequent diagonalization. A detailed discussion of this procedure will be postponed to Chapter III. At present we will merely state the following facts:

1. The perturbation of the rotational energy by the exterior field, $\mathcal{H}_{\mathrm{g}} + \mathcal{H}_\chi + \mathcal{H}_{\text{TS}}$ in Eq. (I.7), is small. If expressed in frequency units the spacing of the rotational energy levels is on the order of GHz while the splittings due to the magnetic field are only on the order of MHz (at $H \simeq 20{,}000$ G).

2. The magnetic field strength and the direction cosines between the field and the molecular axis system enter linearly in the $\mathscr{H}_g$ term ("linear Zeeman effect contribution"). Since the direction cosines are proportional to M, the quantum number for the component of the angular momentum in the direction of the exterior field ($-J \leq M \leq +J$, M integer), $\mathscr{H}_g$ will cause a field dependent splitting of the rotational level into a pattern of $2J+1$ sublevels which are symmetrically arranged around the zero-field position.
3. The magnetic field and the direction cosines enter quadratically in $\mathscr{H}_\chi$ ("quadratic Zeeman effect contribution"). As a consequence this term will be negligible at low fields, but may play the dominant role at very high fields. The quadratic dependence on the direction cosines leads to a quadratic dependence on the quantum number M and causes a shift of the $\pm M$-doublets with respect to each other. Thus, at higher fields the initial symmetry of the Zeeman multiplets is destroyed.

Figure I.3, which shows a recording of the $2_{12} \rightarrow 2_{21}$ and $4_{31} \rightarrow 4_{40}$ rotational transitions of ethyleneoxide, demonstrates the effect of the linear and quadratic Zeeman contributions in a medium-sized molecule. In light and medium-sized molecules, the linear Zeeman effect usually dominates, leading to rather sym-

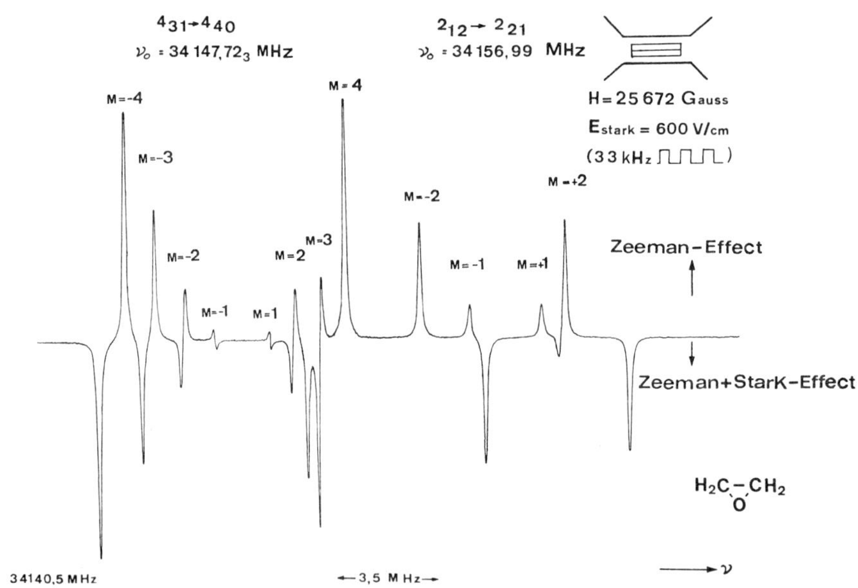

Fig. I.3. This figure shows a small section of the rotational spectrum of ethyleneoxide in the presence of a magnetic field of 25.672 kG. A Stark effect modulated microwave spectrometer operated with $\Delta M = 0$ selection rule was used for this recording, which actually consists of two superimposed absorption spectra. One of these spectra is observed in the absence of the modulating Stark field (above the horizontal line) and the other is observed during the periods when the modulating field is switched on (below the horizontal line). In most investigations only the upper part (pure Zeeman effect) is used for the analysis, since calibration uncertainties and the inhomogeneity of the modulating Stark-field lead to a reduced accuracy of Zeeman data derived from the splittings observed in the simultaneous presence of both fields metric-looking Zeeman patterns. In heavier molecules, however, especially in aromatic rings, the quadratic Zeeman effect becomes more important due to the bigger anisotropies of the magnetic susceptibility tensor. This is illustrated in Fig. I.4 where calculated line spectra are shown for propene, methylenecyclopropene, cyclopentadiene, and fluorobenzene.

It is well known that nuclei also couple with the exterior magnetic field. (Of course, the electronic spins may be neglected because there is no net electronic spin moment in the closed shell molecules considered here). In many cases of interest, the nuclear spin orientations are so loosely coupled to the overall rotation that to a good approximation the nuclei may be assumed to be fixed in orientation with respect to the exterior field. In other words — within the experimental resolution of microwave spectroscopy — the nuclear spin system may be treated as an independent system which does not affect the rotational states or the rotational Zeeman-effect. An exception of this generale rule occurs in molecules with strong nuclear-molecule coupling. For instance, nuclei with electric quadrupole moments have preferred orientations with respect to the intramolecular electric field and the coupling between the nuclear spin orientation and the overall rotation must be included in the Hamiltonian. This interesting situation will be treated in detail in Chapter III.

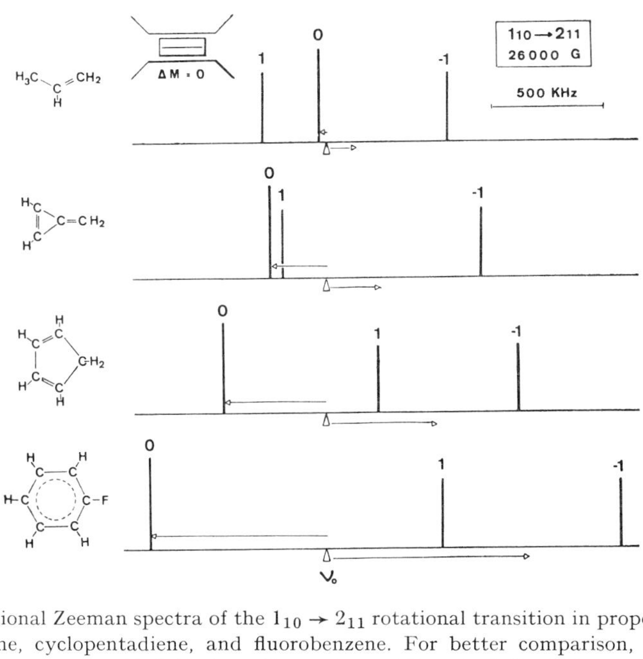

Fig. I.4. Rotational Zeeman spectra of the $1_{10} \rightarrow 2_{11}$ rotational transition in propene, methylenecyclopropene, cyclopentadiene, and fluorobenzene. For better comparison, spectra calculated for the same magnetic field strength are shown. The calculation is based on the experimentally determined g-values and susceptibility anisotropies. While the order of magnitude of the $M \pm 1$ splitting ($\underset{\sim}{g}$-tensor contribution) remains essentially the same, the shifts of the $M = 0$ satellite and of the $M = \pm 1$ doublet due to the $\underset{\sim}{\chi}$-tensor contribution increase almost by a factor of ten when going from the small open chain molecule propene to the aromatic ring fluorobenzene. These susceptibility shifts are indicated by the horizontal arrows to the right for $M = \pm 1$ shifts and to the left for $M = 0$ shifts.

II. Results

A. General Remarks

Only the diagonal elements of the molecular g-tensor, g_{aa}, g_{bb}, and g_{cc}, and the anisotropies of the diagonal elements of the magnetic susceptibility tensor, $2\chi_{aa}-\chi_{bb}-\chi_{cc}$, $2\chi_{bb}-\chi_{cc}-\chi_{aa}$, and $2\chi_{cc}-\chi_{aa}-\chi_{bb}$, are usually obtained from an analysis of the rotational Zeeman effect (see Chapter III). From the theoretical expressions for the $\underset{\sim}{g}$- and $\underset{\sim}{\chi}$-tensor elements in Eqs. (I.2) and (I.4), we know that information on the second moments of the molecular charge distribution as well as on the individual components of the diamagnetic and paramagnetic susceptibilities is not obtained directly from this experimental data. Additional information such as the rotational constants, the structure of the nuclear frame, and the bulk magnetic susceptibility is also needed. In the following we will list several useful relations which may be obtained by simple manipulations of Eqs. (I.2), (I.4), and (I.8).

Molecular quadrupole moments [6]:

$$
\begin{aligned}
Q_{aa} &= \frac{|e|}{2}\left(\sum_{\nu}^{\text{nuclei}} Z_\nu(2a_\nu^2 - b_\nu^2 - c_\nu^2) - \langle 0|\sum_{\varepsilon}^{\text{electrons}} 2a_\varepsilon^2 - b_\varepsilon^2 - c_\varepsilon^2\,|0\rangle\right) \\
&= -\frac{h|e|}{16\pi^2 M_\mathrm{p}}\left\{\frac{2g_{aa}}{A} - \frac{g_{bb}}{B} - \frac{g_{cc}}{C}\right\} - \frac{2mc^2}{|e|}\left\{2\chi_{aa} - \chi_{bb} - \chi_{cc}\right\}.
\end{aligned} \tag{II.1}
$$

Paramagnetic susceptibilities:

$$
\chi_{aa}^{\mathrm{P}} = -\frac{e^2}{2m^2c^2}\left(\frac{L_a L_a}{\Delta}\right) = -\frac{e^2}{4mc^2}\left(\frac{h}{8\pi^2 M_\mathrm{p}}\frac{g_{aa}}{A} - \sum_{\nu}^{\text{nuclei}} Z_\nu(b_\nu^2 + c_\nu^2)\right). \tag{II.2}
$$

Anisotropies of the second moments of the electronic charge distribution:

$$
\begin{aligned}
\left\langle 0\left|\sum_{\varepsilon}^{\text{electrons}} (a_\varepsilon^2 - b_\varepsilon^2)\right|0\right\rangle &= \sum_{\nu}^{\text{nuclei}} Z_\nu(a_\nu^2 - b_\nu^2) + \frac{h}{8\pi^2 M_\mathrm{p}}\left(\frac{g_{aa}}{A} - \frac{g_{bb}}{B}\right) \\
&\quad + \frac{4mc^2}{3e^2}\left\{(2\chi_{aa} - \chi_{bb} - \chi_{cc}) - (2\chi_{bb} - \chi_{cc} - \chi_{aa})\right\}
\end{aligned} \tag{II.3}
$$

Second moments of the electronic charge distribution:

$$
\begin{aligned}
\left\langle 0\left|\sum_{\varepsilon}^{\text{electrons}} a_\varepsilon^2\right|0\right\rangle &= -\frac{2mc^2}{|e|^2}(\chi_{bb} + \chi_{cc} - \chi_{aa}) - \frac{h}{16\pi^2 M_\mathrm{p}}\left(\frac{g_{bb}}{B} + \frac{g_{cc}}{C} - \frac{g_{aa}}{A}\right) \\
&\quad + \sum_{\nu}^{\text{nuclei}} Z_\nu a_\nu^2
\end{aligned} \tag{II.4}
$$

with cyclic permutations of a, b, and c.

In the above equations, the inverse of the moments of inertia are replaced by the rotational constants and the electronic corrections $\frac{2}{I_{aa}^{(\mathrm{n})}}\left(\frac{L_a L_a}{\Delta}\right)$ in Eqs.

(I.2) and (I.8) are neglected. This neglection will be discussed later in this section in connection with the discussion on molecular quadrupole moment determinations. In Fig. (II.1) we give a diagram which illustrates where the additional information enters into the evaluation of the Zeeman data. The determination of the sign of the electric dipole moments which is also mentioned in this diagram will be discussed shortly.

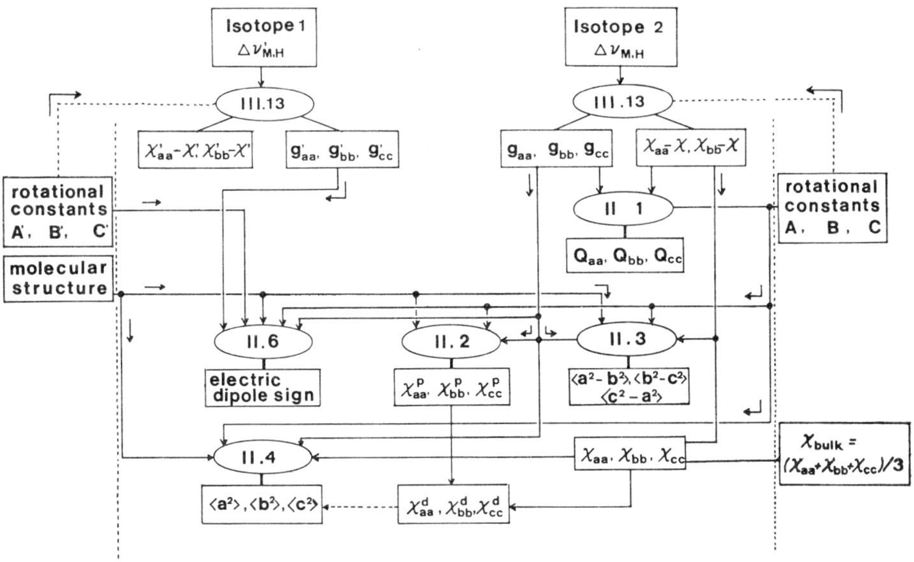

Fig. II.1. Flow diagram showing near the top the experimental measurements of the field, H, and frequency shifts, $\Delta\nu$. The g-values and magnetic susceptibility anisotropies are extracted directly from the experimental data. The remaining quantities such as χ^{p}_{aa}, χ_{aa}, and so on, require additional information to the right and left of the dotted lines. The numbers at the bottom of the diagram require the greatest input of information from different sources

Before we enter into a more detailed discussion on the determination of the molecular electric quadrupole moments and on additivity rules for atom susceptibilities, we will draw some general conclusions from the theoretical expressions for the g- and χ-values given in Eqs. (I.2) and (I.4), respectively. We first restate that the perturbation sums are necessarily zero if the total electronic wavefunction (for simplicity we may think of a Slater determinant) has cylindrical symmetry with respect to the rotational axis in consideration. To see this, we recall that in cylindrical coordinates with a as the symmetry axis:

$$L_{a_\varepsilon} = \frac{\hbar}{\mathrm{i}}\left(b_\varepsilon \frac{\partial}{\partial c_\varepsilon} - c_\varepsilon \frac{\partial}{\partial b_\varepsilon}\right) = \frac{\hbar}{\mathrm{i}} \frac{\partial}{\partial \alpha_\varepsilon}$$

where α_ε stands for the polar angle describing the instantaneous position of the ε-th electron. Using this relation we have:

$$\sum_\varepsilon L_{a_\varepsilon}\,\psi = \frac{\hbar}{\mathrm{i}} \sum_\varepsilon \frac{\partial \psi}{\partial \alpha_\varepsilon} = \frac{\hbar}{\mathrm{i}} \lim_{\Delta\alpha \to 0} \frac{\psi(\alpha_1 + \Delta\alpha, \alpha_2 + \Delta\alpha, \ldots) - \psi(\alpha_1, \alpha_2, \ldots)}{\Delta\alpha} = \frac{\hbar}{\mathrm{i}} \frac{\partial \psi}{\partial \alpha}$$

From this we expect to find negative susceptibility anisotropies $\chi_{||} - \chi_\perp$ in diatomic molecules where $\chi_{||} = \chi_{aa}$ (a is molecular axis) because of the cylindrical symmetry of the electronic wavefunction about the molecular axis giving $L_a|0\rangle = 0$ and $\left(\frac{L_a L_a}{\Delta}\right) = 0$. Therefore, in χ_{aa} the diamagnetic contribution (from $\langle 0|\sum_\varepsilon (b_\varepsilon^2 + c_\varepsilon^2)|0\rangle$) is not balanced by a paramagnetic term while in $\chi_{bb} = \chi_{cc}$, the diamagnetic and paramagnetic contributions will largely cancel. This expected behavior, evident from the results in Table A1 of Appendix I, is indeed quite generally found. On the contrary, we may expect to find fairly strong positive anisotropies $\bar{\chi}_{||} - \chi_\perp$ in planar aromatic ring molecules where $\chi_\perp = \chi_{cc}$ with the c-axis perpendicular to the molecular plane and $\bar{\chi}_{||}$ is the average in plane susceptibility. In this case, the anisotropy will be mainly due to delocalized π-type electrons which, at least to some extent, have the cylindrical symmetry about the c-axis required to give a comparatively small value for $\left(\frac{L_c L_c}{\Delta}\right)$ and thus a comparatively big negative value for $\chi_{cc} = \chi_\perp$. In this context, we note further that substituents which disturb the quasicylindrical symmetry of the electronic system should effectively quench part of the negative χ_{cc}-value and should thus lead to smaller anisotropies, $\bar{\chi}_{||} - \chi_\perp$. Again these expectations are substantiated by the experimental findings as will be discussed later in this Chapter.

Up until now we have discussed the perturbation sums only in reference to the matrix elements or with respect to delocalization of the electronic wavefunctions. Of course, the energy differences in the denominator also play an important role. The series of diatomics CO, CS, CSe may serve as an example. In this series, the observed increase of the paramagnetic susceptibility, $\chi_\perp^{\mathrm{para}}$, is paralleled by a decrease in the lowest energy transition ($A'\pi - X'\Sigma^+$) and also by a decrease in dissociation energy [7]. Thus, in this series the change in the energy differences may be assumed to play the dominant role in the systematic change in $\chi_\perp^{\mathrm{para}}$.

B. Molecular Electric Quadrupole Moments and Molecular Electric Dipole Moment Signs

We will now enter into a more detailed discussion on the determination of the molecular electric quadrupole moments from Eq. (II.1). Quadrupole moments are important in the calculation of intermolecular interaction potentials [8] and as test quantities for quantum chemical calculations. Several quadrupole moments calculated according to Eq. (II.1) from experimentally determined ground state rotational constants, g-values, and magnetic susceptibility anisotropies are listed in Table II.1. Also listed for comparison are quadrupole moments calculated from empirically determined atom dipoles [9] and from INDO-wavefunctions [10].

Table II.1. Molecular quadrupole moments calculated with the molecular Zeeman parameters and Eq. (II.1) [1]. The experimental uncertainties follow from standard error propagation and do not reflect systematic errors introduced, for instance, through the neglect of vibrations. Also listed for comparison are values calculated from atom dipoles [9] and values calculated from INDO-wavefunctions [10]. Only the quadrupole moments of the most abundant isotopic species are listed in each case. The values are given in units of 10^{-26} esu cm^2 and are referred to the principal axis system of the moment of inertia tensor. The structure references are given in Refs. [9] and [10]

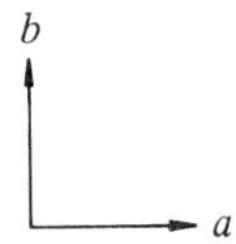	$Q_{aa,\mathrm{exp}}$ $Q_{bb,\mathrm{exp}}$ $Q_{cc,\mathrm{exp}}$	Calc. from Atom Dipoles	Calc. from INDO-Wavefunctions
F–C≡N	−4.2 +2.1 +2.1	−3.8 +1.9 +1.9	−5.6 +2.8 +2.8
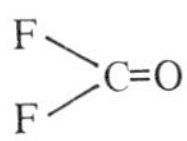	−3.7 ± 0.7 −0.2 ± 0.5 +3.9 ± 1.1	−3.1 −0.8 +3.5	−3.7 −1.0 +4.7
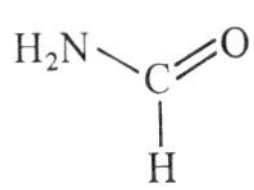	−0.3 ± 0.5 +3.4 ± 0.4 −3.1 ± 0.8	−0.5 +3.2 −2.7	−0.91 +2.64 −1.73
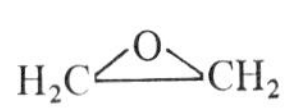	+2.6 ± 0.10 −3.7 ± 0.14 +1.1 ± 0.22	+2.3 −5.5 +2.2	+2.17 −3.77 +1.6
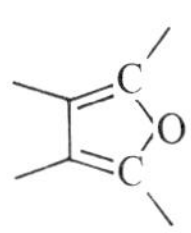	+0.2 ± 0.4 +5.9 ± 0.3 −6.1 ± 0.4	+1.2 +5.6 −6.8	−0.6 +5.3 −4.7
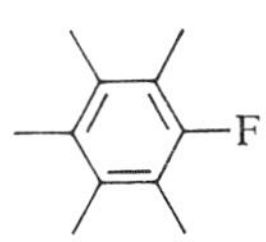	−1.9 ± 0.8 +5.1 ± 1.0 −3.2 ± 1.0		−3.0 +7.7 −4.6

Three points should be noted in connection with quadrupole moment determinations according to Eq. (II.1):

First, the error limits given in Table II.1 follow from the experimental uncertainties by standard error propagation and do not account for possible deficiencies of the rigid rotor model (see below).

Second, the g-value contribution [Eq. (II.1a)] and the susceptibility contribution [Eq. (II.1b)] usually have the same order of magnitude but opposite sign. Thus, part of the high accuracy with which the g- and χ-values may be determined is lost. In the case of the aa-component in ethylene oxide, for instance, the two contributions are:

$$-\frac{|e|h}{16\pi^2 M_p}\left(\frac{2g_{aa}}{A}-\frac{g_{bb}}{B}-\frac{g_{cc}}{C}\right)=+13.04\times 10^{-26}\text{ esu cm}^2 \qquad \text{(II.5a)}$$

$$-\frac{2mc^2}{|e|}(2\chi_{aa}-\chi_{bb}-\chi_{cc})=-10.45\times 10^{-26}\text{ esu cm}^2 \qquad \text{(II.5b)}$$

Third, the quadrupole moments are referred to the principal inertia axis system, *i.e.*, to the center of mass. Since the quadrupole moments are origin dependent for molecules which have nonzero electric dipole moments, the appropriate transformation to a common origin must be performed, if quadrupole moments of different molecules or if quadrupole moments determined by different methods are to be compared. For instance, if the reference system is shifted parallel to the a-axis by an amount Δa, $(a'_\nu = a_\nu + \Delta a,\ a'_\varepsilon = a_\varepsilon + \Delta a,\ b'_\nu = b_\nu, \ldots)$ the new primed quadrupole moments are:

$$\begin{aligned} Q_{aa}' &= Q_{aa} + 4\,\Delta a\,\mu_a \\ Q_{bb}' &= Q_{bb} - 2\,\Delta a\,\mu_a \\ Q_{cc}' &= Q_{cc} - 2\,\Delta a\,\mu_a \end{aligned}$$

with the electric dipole moment μ_a defined as

$$\mu_a = |e|\{\sum_{\nu}^{\text{nuclei}} Z_\nu a_\nu - \langle 0|\sum_{\varepsilon}^{\text{electrons}} a_\varepsilon|0\rangle\}\,.$$

From these equations it is obvious that it is possible, at least in principle, to determine the electric dipole moment of a molecule from the rotational Zeeman effect data of two isotopic species (see Fig. II.1). If the isotopic substitution causes a parallel shift of the principle axis system: $a'_\nu = a_\nu + \Delta a$, $b'_\nu = b_\nu + \Delta b$, $c'_\nu = c_\nu + \Delta c$, etc., where Δa is the a-coordinate of the center of mass of the daughter molecule referred to the principal axes of the parent molecule, the corresponding equations are given by [1,11]

$$-\frac{|e|h}{16M_p\pi^2}\left(\frac{g'_{aa}}{A'}-\frac{g_{aa}}{A}\right)=(\Delta b\,\mu_b+\Delta c\,\mu_c) \qquad \text{(II.6a)}$$

$$-\frac{|e|h}{16M_p\pi^2}\left(\frac{g'_{bb}}{B'}-\frac{g_{bb}}{B}\right)=(\Delta c\,\mu_c+\Delta a\,\mu_a) \qquad \text{(II.6b)}$$

$$-\frac{|e|h}{16M_p\pi^2}\left(\frac{g'_{cc}}{C'}-\frac{g_{cc}}{C}\right)=(\Delta a\,\mu_a+\Delta b\,\mu_b) \qquad \text{(II.6c)}$$

Using the following data for formaldehyde[1,12] where $|\mu_a| = 2.339(15)$ D

"parent molecule"

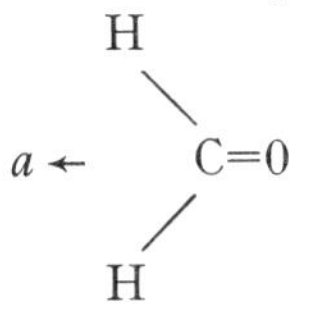

"daughter molecule"

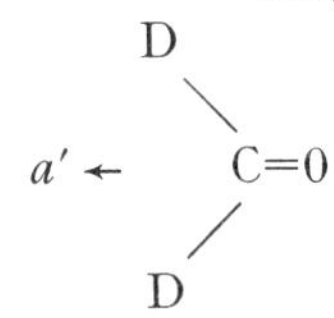

$\Delta a = +\,0.0737$ Å

$$\begin{array}{llll}
g_{aa} = -2.9024(6) & A = 282{,}106.0 \text{ MHz} & g'_{aa} = -1.445(2) & A' = 141{,}732 \text{ MHz} \\
g_{bb} = -0.2245(1) & B = 38{,}835.69 & g'_{bb} = -0.1917(5) & B' = 32{,}368.6 \\
g_{cc} = -0.0994(1) & C = 34{,}003.28 & g'_{cc} = -0.0788(4) & C' = 26{,}2725
\end{array}$$

leads to the following values for the electric dipole moment:

$$\mu_a = 1.2(3) \text{ D}$$
$$\mu_a = 2.4(3) \text{ D}$$

These results indicate that the oxygen atom is negative as expected. However, the control equation,

$$-\frac{|e|h}{16M_p\pi^2}\left(\frac{g'_{aa}}{A'} - \frac{g_{aa}}{A}\right) = -0.110(17) \text{ Å D} ,$$

is considerably different from zero. This large discrepancy indicates that a determination of the sign of the electric dipole moment from rotational Zeeman effect data is in many cases a marginal experiment. Part of the problem certainly lies in the neglect of vibrations.

We will now turn to a critical assessment of the range of validity of Eq. (II.1). We first briefly discuss the neglection of the $\frac{2}{I^{(n)}_{aa}}\left(\frac{L_a L_a}{\Delta}\right)$ terms as compared to unity in Eqs. (I.2) and (I.8). For molecules in which accurate information on the geometry of the nuclear frame is available from microwave spectra of different isotopic species or from electron scattering, this correction may be calculated from the experimental g-values. It has values on the order of 10^{-4} to 10^{-3} as might be guessed already from the expressions for the rotational constants in Eq. (I.8) where it accounts for the small electronic contribution to the molecular moment of inertia. The subsequent correction on the quadrupole moments is usually below 0.01×10^{-26} esu cm^2, which is small compared to the experimental uncertainties. Far more important is the neglect of vibrations. For molecules with only nondegenerate vibrational motions[13,14] ground state vibrational expectation values over the theoretical expressions in Eqs. (I.8), (I.2), and (I.4) will give reasonable approximations for the experimentally observed rotational constants, g-values and susceptibilities[15]. This is shown in Appendix III, where a simplified treatment of vibrational effects is given. Assuming that the experimentally determined rotational constants, g-values, and susceptibility anisotropies indeed correspond to the vibrational expectation values of the theoretical expressions given in Eqs. (I.2), (I.4), and (I.8) then leads to the conclusion that Eq. (II.1) essentially gives the vibrational average of the molecular electric quadrupole moment, although the rotational constants and the g-values enter as individually averaged quantities into this equation. This is seen by expanding the appropriate quantities about their equilibrium values as:

$$A(q) = A^{(e)} + \tilde{A}(q); \; g_{aa} = g_{aa}^{(e)} + \tilde{g}_{aa}(q); \; \chi_{aa}(q) = \chi_{aa}^{(e)} + \tilde{\chi}_{aa}(q) \qquad \text{(II.7)}$$

where q stands for all vibrational coordinates. Rewriting the vibrational average $\langle v|g_{aa}(q)|v\rangle / \langle v|A(q)|v\rangle$, which enters into Eq. (II.1):

$$\frac{\langle v|g_{aa}(q|v\rangle}{\langle v|A(q)|v\rangle} = \frac{g_{aa}^{(e)}}{A^{(e)}}\left(1 + \frac{\langle v|\tilde{g}_{aa}|v\rangle}{g_{aa}^{(e)}} - \frac{\langle v|\tilde{A}|v\rangle}{A^{(e)}} - \frac{\langle v|\tilde{g}_{aa}|v\rangle\langle v|\tilde{A}|v\rangle}{g_{aa}^{(e)}\,A^{(e)}} + \ldots\right) \qquad \text{(II.8)}$$

and comparing with:

$$\left\langle v\left|\frac{g_{aa}(q)}{A(q)}\right|v\right\rangle = \frac{g_{aa}^{(e)}}{A^{(e)}}\left(1 + \frac{\langle v|\tilde{g}_{aa}|v\rangle}{g_{aa}^{(e)}} - \frac{\langle v|\tilde{A}|v\rangle}{A^{(e)}} - \frac{\langle v|\tilde{g}_{aa}\cdot\tilde{A}|v\rangle}{g_{aa}^{(e)}\cdot A^{(e)}} + \ldots\right) \qquad \text{(II.9)}$$

shows that the quadrupole moments determined from Eq. (II.1) using ground state vibrational averages for g_{aa}, A, and so on will differ from the ground state vibrational average of Q_{aa} only in expressions which are small on second order. Under the assumption that over the zero point vibration instantaneous values for the rotational constants, etc., do not deviate more than 10% from their equilibrium values [16] we therefore conclude that Q-values determined from Eq. (II.1) will agree within about 5% with the vibrational expectation value (apart from standard error propagation due to experimental uncertainties).

C. Magnetic Susceptibilities

As we have seen in Chapter I, Fig. I.4, comparatively big magnetic susceptibility anisotropies are observed in unsaturated rings. These anisotropies, which may be attributed to a higher degree of delocalization of the electronic configuration about the ring, may be used as one among several other physical criteria for a quantitative definition of aromaticity. In this contribution, however, we will not enter into a detailed discussion on ring currents and aromaticity [17] but merely present some experimental results. Combining the magnetic susceptibility anisotropies from rotational Zeeman effect measurements with bulk susceptibility data [18], $\chi_{\text{bulk}} = (\chi_{aa} + \chi_{bb} + \chi_{cc})/3$, leads to the individual diagonal elements of the magnetic susceptibility tensor (see Fig. II.1). In view of the above described difference in the magnetic properties of unsaturated rings and open chain molecules, it was tempting to try to refine the criteria by breaking down the observed susceptibilities into a local (atomic) contribution and a nonlocal (molecular ring current) contribution. In 1973 Schmalz, Norris, and Flygare [19] published an extensive list of atom- and bond-susceptibilities, empirically determined from measured anisotropies and bulk values of molecules which were assumed to contain well localized orbitals only. In their paper they have also indicated possible applications, for instance, in conformational analysis, in liquid crystal investigations, or in connection with chemical shielding in NMR-spectroscopy. The underlying basic idea of localized atomic susceptibilities may be considered as an extension of Pascal's well known additivity scheme for atomic average susceptibilities [20,21]. In the following we report an extension and a slight modification of the values of this original list. This extension originates from the inclusion of a series of fluorine containing compounds [22]. Only atom susceptibilities are given, since the bond susceptibilities, though intuitively perhaps more attractive, are highly correlated in the presently available set of experimental data. This new list is given in Table II.2. We first demonstrate the use of the localized atom susceptibilities taking H_2O as a simple example. Within the additivity scheme, the field induced local component of the magnetic moment:

$$\boldsymbol{\mu}_{\text{ind,local}} = \underset{\sim}{\chi}_{\text{local}} \cdot \boldsymbol{H}$$

is regarded as a sum of three induced local moments, two hydrogen contributions and one contribution from an ether oxygen:

$$\boldsymbol{\mu}_{\text{ind,local}} = \underset{\sim}{\chi}_{\text{local(H}\diagup)} \cdot \boldsymbol{H} + \underset{\sim}{\chi}_{\text{local}(\diagup\text{O}\diagdown)} \cdot \boldsymbol{H} + \underset{\sim}{\chi}_{\text{local}(\diagdown\text{H})} \cdot \boldsymbol{H} \qquad \text{(II.10)}$$

or

$$\underset{\sim}{\chi}_{\text{local}}(\text{H}_2\text{O}) = \underset{\sim}{\chi}_{\text{local(H}\diagup)} + \underset{\sim}{\chi}_{\text{local}(\diagup\text{O}\diagdown)} + \underset{\sim}{\chi}_{\text{local}(\diagdown\text{H})} \qquad \text{(II.11)}$$

When using Eq. (II.11), care must be taken to rotate the constituent atomic susceptibility tensors, with their "natural principal axis system" oriented at the bond axes, into the principal inertia axis system of the molecule. With the components of the susceptibility tensor transforming like the corresponding coordinate products [compare Eq. (I.4)], the appropriate transformation is given by:

$$\chi_{aa} = \sum_{A}^{\text{atoms}} (\cos^2 ax^{(A)}\, \chi_{xx}^{(A)} + \cos^2 ay^{(A)}\, \chi_{yy}^{(A)} + \cos^2 az^{(A)}\, \chi_{zz}^{(A)}) \qquad \text{(II.12)}$$

with similar expressions for the *b*- and *c*-components. In Eq. (II.12), $\cos ax^{(A)}$, $\cos ay^{(A)}$, and $\cos az^{(A)}$ are the direction cosines between the molecular *a*-axis and the *x*-, *y*-, and *z*-principal axis of the atom susceptibility tensor of atom A respectively. The assumed orientations of the atomic tensors with respect to the bonds are in Table II.2. In the case of water (compare Fig. II.2 and Table II.2), the three equations corresponding to Eq. (II.12) become:

$$\begin{aligned}\chi_{aa} &= 2 \cdot \left(\cos^2 (37.6^0)\, \chi_{xx(\text{H-})} + \cos^2 (52.4^\circ)\, \chi_{yy(\text{H-})}\right) + \chi_{yy}\left(\text{O}\langle\right) \\ &= -\, 13.4 \times 10^{-6}\ \text{erg}/(\text{G}^2\ \text{mole})\end{aligned}$$

$$\chi_{bb} = 2\left(\cos^2(52.4^0)\,\chi_{xx(\mathrm{H-})} + \cos^2(37.6^0)\,\chi_{yy(\mathrm{H-})}\right) + \chi_{xx}\left(\mathrm{O}\langle\right)$$
$$= -12.2 \times 10^{-6}\ \mathrm{erg/(G^2\ mole)}$$

$$\chi_{cc} = 2\,\chi_{zz(\mathrm{H-})} + \chi_{zz}\left(\mathrm{O}\langle\right) = -12.4 \times 10^{-6}\ \mathrm{erg/(G^2\ mole)}$$

Table II.2. Table of atom susceptibilities in units of 10^{-6} erg/(G^2 mole). The values were fitted to the *aa*-, *bb*-, and *cc*-components of the molecules listed in Table III.3. The uncertainties are standard deviations from the least squares fit described in the text. For H- and methylcarbon, cylindrical symmetry about the *c*-axis was assumed. For practical reasons C_{sp3} (H_3C-) and C_{sp3} ($H_2C\langle$) were fitted separately. The values for nitrogen and S$\langle$ were calculated from the molecular susceptibilities of ammonia [Reijnders, J. M. H., Verhoeven, J., Dymanus, A.: Symp. Mol. Struct. Spectr. Ohio, R 13 (1971)] formamide [Tigelaar, H. C., Flygare, W. H.: J. Am. Chem. Soc. *94*, 343 (1972)], and dimethylsulfide [Hamer, E., Sutter, D. H., Dreizler, H.: Z. Naturforsch. *27a*, 1159 (1972)] together with the local susceptibilities listed in the first part of the table. To a first approximation one would expect all three components of C_{sp3} to be identical. The differences, although within the standard deviations of the fit, indicate limits of the simple additivity scheme. However, in general, reasonably good agreement has been found for nonaromatic molecules if experimental values are compared to values calculated from the structure and the local susceptibilities in this table (see Table II.3)

(axes: y ↑, x →)	χ_{xx}	χ_{yy}	χ_{zz}
H—	—1.17(53)	—2.08(38)	—2.08(38)
F—	—8.22(116)	—6.87(111)	—5.42(52)
C_{sp3}(H_3C—)	—9.92(181)	—8.27(149)	—8.27(149)
C_{sp3} ($H_2C\langle$)	—7.45(118)	—7.19(125)	—8.26(107)
C_{sp2} ($\rangle$C=)	—3.64(51)	—3.75(60)	—7.33(55)
O_{ether} (O$\langle$)	—8.73(101)	—10.35(104)	—8.23(95)
$O_{carbonyl}$(O=)	+1.90(120)	—1.29(120)	—5.70(107)
N (NH$_3$)[1]	—11.47	—13.27	—13.27
N_{sp2} ($\rangle$N—)	—13.82	—10.35	— 6.13
S$\langle$	—17.03	—17.07	—15.62

[1]) χ-bulk(NH_3) = —18.10^{-6} erg/(G^2 mole) from Ref. [77].

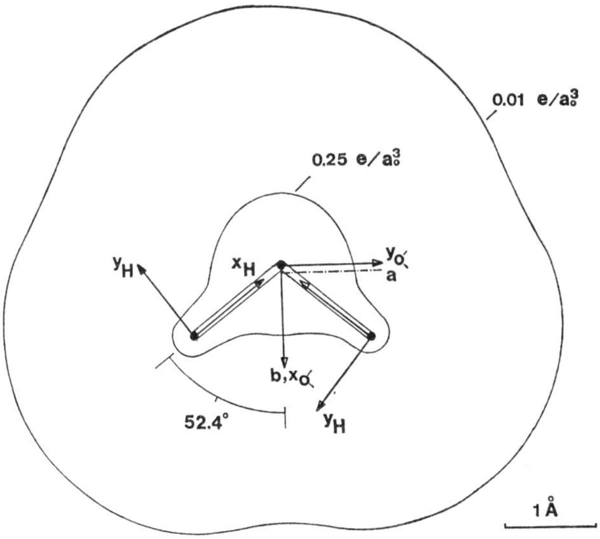

Fig. II.2. Water is used to illustrate the atomic coordinate systems used in the additivity scheme for atom susceptibilities. Also shown are contour lines for the overall electron density measured in units of electrons per cubic atomic unit ($a_0 = 0.529$ Å). The contour lines have been taken from Streitwieser, Andrew, Jr., Owens, Peter H.: Orbital and electron density diagrams. New York: McMillan and Co. 1973

A comparison with the experimentally determined values [23] $\chi_{aa} = -12.85$, $\chi_{bb} = -13.07$, $\chi_{cc} = -13.09$ [all values in 10^{-6} erg/(G^{2}mole)] shows that the local scheme reproduces rather well the observed susceptibilities.

For the empirical determination of the local atom susceptibilities, the above method is reversed. First, a sufficiently large set of molecules is selected for which accurate susceptibility anisotropies, bulk susceptibilities, and an accurate structure are available and for which the concept of well localized orbitals may be assumed to hold with a sufficiently high degree of approximation. Then, assuming that the observed susceptibilities are indeed only local in character, according to Eq. (II.12) each molecule yields a set of three linear equations for the unknown atom susceptibilities, and the latter are determined by a least square procedure. The values listed in Table II.2 were fitted to the molecules given in Table II.3, where the quality of the fit is also shown.

As is seen from Table II.2, the standard deviations of the fitted atom susceptibilities are rather large. This may have different reasons. First of all, part of the uncertainties are simply due to the experimental uncertainties of the input data since only 54 equations are used to fit 19 constants. In addition, nonlocal contributions may be present even in this selected set of molecules. Further, there may be differences between gas phase and liquid phase susceptibilities in those systems with strong intermolecular associations such as hydrogen bonding (bulk susceptibilities are usually measured in the liquid phase). Finally, it is possible that different substituents, for instance, at an ether oxygen, could cause

Table II.3. Diagonal values of molecular susceptibilities referred to the principal inertial axis system in units of 10^{-6} erg/(G^2 mole). In each case the first reference is for the susceptibilities, the second for the structure. The "local values" $\chi_{aa(\text{loc})}$, $\chi_{bb(\text{loc})}$, and $\chi_{cc(\text{loc})}$ have been calculated according to Eq. (II.12), using the known structure and the atom susceptibilities listed in Table II.2

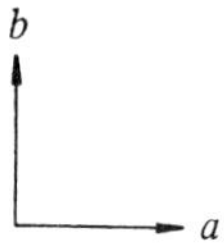		χ_{aa} (exp) χ_{aa} (loc)	χ_{bb} (exp) χ_{bb} (loc)	χ_{cc} (exp) χ_{cc} (loc)
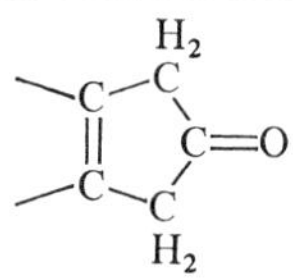	[1,2]	—34.3(45) —34.8	—40.0(40) —37.9	—53.9(60) —54.3
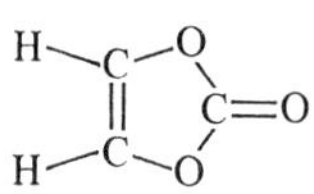	[1,3]	—35.3(40) —32.9	—30.5(40) —33.4	—47.4(50) —48.3
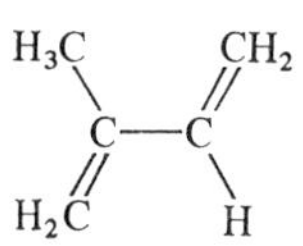	[1,4]	—35.4(40) —37.0	—34.5(40) —38.1	—52.8(50) —53.0
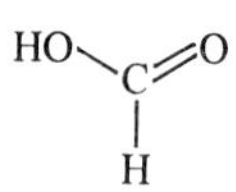	[1,5]	—18.8(15) —15.9	—16.8(15) —16.5	—24.2(20) —25.4
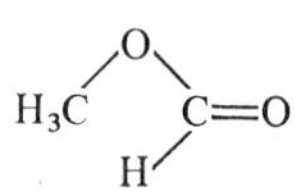	[1,6]	—28.3(5) —30.3	—30.9(6) —28.0	—36.7(8) —36.6
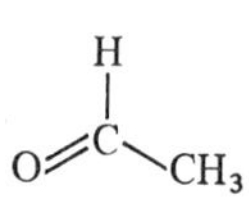	[1,7]	—20.0(9) —20.3	—19.5(9) —18.9	—28.6(15) —28.4
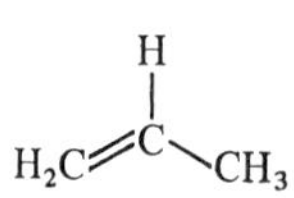	[1,8]	—30.9(10) —28.3	—26.2(10) —25.4	—34.9(11) —34.3
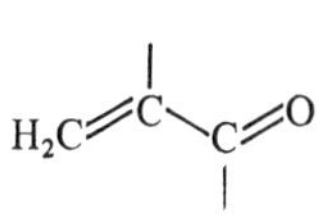	[1,9]	—16.0(30) —17.3	—18.3(30) —17.3	—37.7(40) —36.0
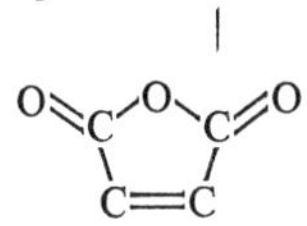	[1,10]	—25.7(15) —25.5	—28.2(15) —28.5	—53.5(17) —53.1

Table II.3 (continued)

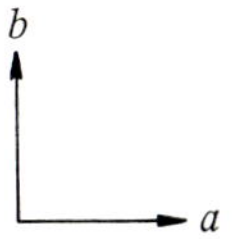		χ_{aa} (exp) χ_{aa} (loc)	χ_{bb} (exp) χ_{bb} (loc)	χ_{cc} (exp) χ_{cc} (loc)
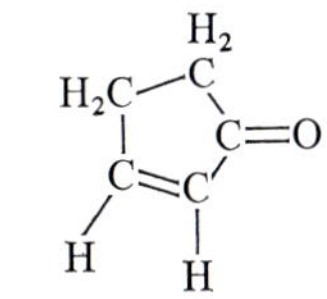	[1,11]	—33.1(40) —34.8	—39.6(40) —37.7	—55.4(60) —54.6
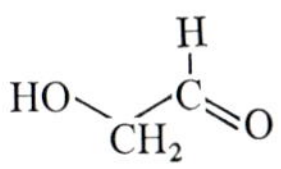	[1,12]	—27.6(40) —27.8	—23.7(40) —26.9	—38.6(45) —36.7
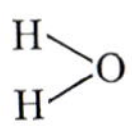	[22,13]	—12.85 —13.4	—13.07 —12.2	—13.09 —12.4
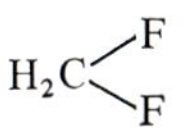	[14,15]	—25.3 —27.9	—23.7 —25.7	—23.0 —20.9
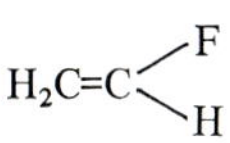	[16,17]	—22.0 —20.8	—18.5 —18.8	—24.6 —26.3
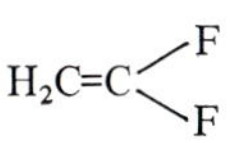	[18,19]	—29.3 —25.7	—26.0 —25.8	—30.4 —29.7
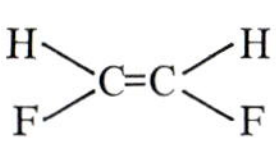	[18,19]	—25.3 —25.5	—27.7 —26.0	—28.6 —29.7
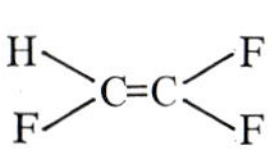	[16,20]	—33.6 —32.1	—29.6 —31.2	—33.4 —33.0
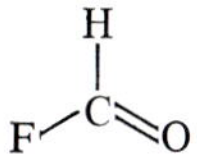	[16,21]	—11.7 —12.6	—11.7 —12.6	—17.7 —20.5

[1] Schmalz, T. G., Norris, C. L., Flygare, W. H.: J. Am. Chem. Soc. *95*, 7961 (1973).
[2] Bevan, J. W., Legon, A. C.: J. Chem. Soc. Faraday Trans. II *69*, 902 (1973).
[3] White, W. F., Boggs, J. E.: J. Chem. Phys. *54*, 4714 (1971). — Norris, C. L., Benson, R. C., Beak, P. A., Flygare, W. H.: J. Am. Chem. Soc. *95*, 2766 (1973). — See Ref. [4].
[4] Vilkov, L. V., Sadova, N. I.: J. Struct. Chem. *8*, 398 (1967).
[5] Kukolich, S. G., Flygare, W. H.: J. Am. Chem. Soc. *91*, 2433 (1969).
[6] Curl, R. F.: J. Chem. Phys. *30*, 1529 (1959).
[7] Tigina, T., Kimura, M.: Bull. Chem. Soc. Japan *42*, 2159 (1969).
[8] Lide, D. R., Christensen, O.: J. Chem. Phys. *35*, 1374 (1961).
[9] Cherniak, E. A., Costain, C. C.: J. Chem. Phys. *45*, 104 (1966).

10) Hildebrandt, R. L., Paixotol, E. M. A.: J. Mol. Struct. *12*, 31 (1972).
11) Determined from similar compounds.
12) Marstokk, K. M., Molendal, H.: J. Mol. Struct. *7*, 101 (1971).
13) Taft, H., Daily, B. P.: J. Chem. Phys. *51*, 1002 (1969).
14) Blickensderfer, R. P., Wang, J. H. S., Flygare, W. H.: J. Chem. Phys. *51*, 3196 (1969).
15) Lide, D. R.: J. Am. Chem. Soc. *74*, 3548 (1952).
16) Rock, S. C., Hankock, J. K., Flygare, W. H.: J. Chem. Phys. *54*, 3450 (1971).
17) Lide, D. R., Christensen, O.: Spectrochim. Acta *17*, 665 (1961).
18) Laurie, V. W., Pence, D. T.: J. Chem. Phys. *38*, 2693 (1963).
19) Laurie, V. W.: J. Chem. Phys. *34*, 291 (1961).
20) Bhanmik, A., Brooks, W. V. F., Pan, S. C.: J. Mol. Struct. *16*, 29 (1973).
21) Favero, P., Mirri, A. M., Baker, J. G.: Nuovo Cimento *17*, 740 (1960).
22) Verhoeven, J., Dymanus, A.: J. Chem. Phys. *52*, 3222 (1970).

a tilt of the x-principal axis of the atomic susceptibility tensor with respect to the line bisecting the bond angle, a tilt which is sufficiently large to partly spoil the quality of the fit. At present, none of these possibilities can be excluded and more experimental data is needed to improve the empirical basis for the concept of local susceptibilities.

Keeping the above restrictions in mind, we will now apply the concept of local susceptibilities to furane as an example for a ring with delocalized electrons. From the structure [24] and the local susceptibilities given in Table II.2, the following values are obtained and compared to the experimental results [25] [all in units of 10^{-6} erg/(G^2 mole)]

$$\chi_{aa}^{\text{local}} = -30.1 \qquad \chi_{aa}^{\text{exp}} = -30.35 \pm 1.6$$

$$\chi_{bb}^{\text{local}} = -31.5 \qquad \chi_{bb}^{\text{exp}} = -33.45 \pm 1.6$$

$$\chi_{cc}^{\text{local}} = -45.9 \qquad \chi_{cc}^{\text{exp}} = -70.5 \pm 1.6$$

While the in-plane χ_{aa}- and χ_{bb}-components are in close agreement with the values predicted from the additivity rules, the out-of-plane component χ_{cc} is considerably more negative, indicating a strong nonlocal, diamagnetic ring current contribution. This result is typical for rings with delocalized electrons. Schmalz, Norris, and Flygare [19] have also shown, on the basis of theoretical arguments, that only the out-of-plane magnetic susceptibility tensor element is related to the electron delocalization in ring systems. The comparison between local and experimental out-of-plane susceptibilities in several ring compounds is shown in Table II.4. The paper by Schmalz *et al.*[19], as well as earlier references from the same group as cited [26], suggests that the presence of a nonlocal magnetic susceptibility anisotropy might serve as a measure of electron delocalization in ring compounds and hence, if aromaticity is defined in terms of electron delocalization, of aromatic character. By this criterion, a compound is judged to have delocalized electrons not because it has a large out-of-plane magnetic susceptibility but because it has a more negative susceptibility than that which would be predicted from a localized model [26,27]. The advantages of the use of nonlocal molecular magnetic susceptibilities for evaluation of aromaticity lie in the reliability of

Table II.4. Local magnetic susceptibilities (from Table II.2 and the known structures) and the experimental values. Also shown is the component of out-of-plane susceptibility due to non-local or strain effects. The numbers are in units of 10^{-6} erg/(G^2 mole)

		χ_{aa}^{local} (χ_{aa}^{exp})	χ_{bb}^{local} (χ_{bb}^{exp})	χ_{cc}^{local} (χ_{cc}^{exp})	$\chi_{cc}^{exp} - \chi_{cc}^{local}$
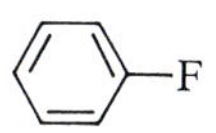	[1]	—38.9 (—35.3)	—36.7 (—31.7)	—59.8 (—91.7)	—31.9
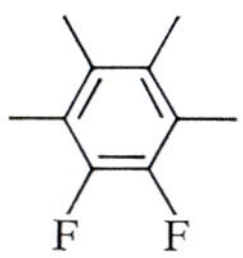	[2]	—43.4 (—46.9)	—44.0 (—44.7)	—63.2 (—100.1)	—36.9
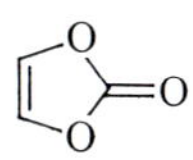	[3]	—32.9 (—34.1)	—33.4 (—34.8)	—48.3 (—48.2)	—
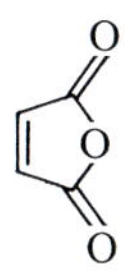	[4]	—25.5 (—25.7)	—28.5 (—28.2)	—53.1 (—53.5)	—
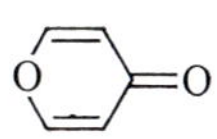	[4]	—32.3 (—34.9)	—36.1 (—35.6)	—58.9 (—58.2)	—
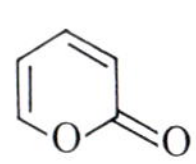	[4]	—33.1 (—37.7)	—35.4 (—34.0)	—58.9 (—63.6)	—4.7
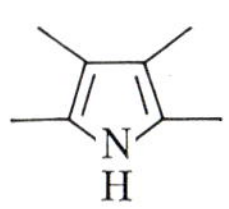	[5]	—36.5 (—31.9)	—33.5 (—37.0)	—45.9 (—76.8)	—30.9
	[6]	—38.6 (—40.9)	—38.1 (—40.5)	—53.3 (—90.8)	—43.4
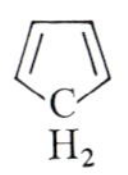	[7]	—32.4 (—34.5)	—32.5 (—32.1)	—48.8 (—67.5)	—18.7
	[8]	—30.1 (—30.4)	—31.5 (—33.5)	—45.9 (—70.5)	—24.6
	[9]	—32.5 (—30.9)	—31.4 (—30.2)	—56.5 (—67.7)	—11.2

Footnote Table II.4.:

1) Experimental bulk value from Gierke, T. D.: Ph. D. Thesis, University of Illinois, 1974, and anisotropies from Ref. [1].
2) Experimental bulk value from Gierke, T. D.: Anisotropies from Table A.2.
3) Benson, C., Flygare, W. H.: J. Chem. Phys. *58*, 2366 (1973).
4) Experimental bulk values from A. Burnham and anisotropies from Benson, R. C., Norris, C. L., Flygare, W. H., Beak, P.: J. Am. Chem. Soc. *93*, 5591 (1971).
5) Local value of 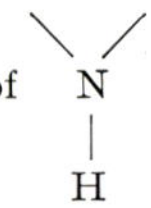 from 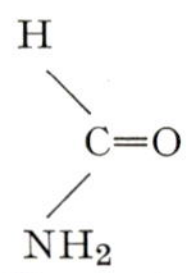 (Ref. [2]) and remaining experimental numbers from Sutter, D. H., Flygare, W. H.: J. Am. Chem. Soc. *91*, 6895 (1969).
6) Local value of  from CH_3—S—CH_3 (Ref. [49]) and the remaining experimental numbers from Sutter, D. H., Flygare, W. H.: J. Am. Chem. Soc. *91*, 4063 (1969).
7) Experimental numbers from Benson, R. C., Flygare, W. H.: J. Am. Chem. Soc. *92*, 7523 (1970).
8) Experimental numbers from Ref. [25].
9) Experimental numbers from Benson, R. C., Flygare, W. H.: J. Chem. Phys. *58*, 2366 (1973).

the assignments of magnetic susceptibilities to the hypothetical localized models and in the theoretical relationship of nonlocal contributions to electron delocalization. These relations have now been established [19] and discussion of the magnitude of the nonlocal contributions to the susceptibility have been documented [19,27]. Within limits, the π-electron theories have also been helpful in understanding the nature of electron delocalization in ring currents [28]. However, these calculations are limited to those systems in which the σ and π electrons are uncoupled and quantitative agreement between the calculated π-electron contribution to $\Delta\chi$ and the experimental nonlocal contribution discussed above is not expected [29]. These early workers [28] certainly understood these limitations. However, a new effort [30] at computing $\Delta\chi$ by the π-electron model has ignored these early warnings and the work of Schmalz, *et al.* by attempting to use the π-electron model as a quantitative measure of aromaticity. These new more elaborately evaluated results [30] are no less ambiguous than the earlier more simple models [28]; both sets of calculations are plagued by the extent of the π—σ electron correlation [29]. One prominent discrepancy is that the recent π-electron calculation [30] gives a paramagnetic delocalization in fulvene and the nonlocal value from the magnetic susceptibility results is diamagnetic. We conclude that the nonlocalized magnetic susceptibility anisotropies give a better view of the electron delocalization in five- and six-member rings than the π-electron models.

While the application of the local scheme in five- and six-membered aromatic systems quite generally yields strong negative ("diamagnetic") nonlocal contributions to the out-of-plane component of the susceptibility, $\chi_\perp$, positive ("paramagnetic") nonlocal contributions are obtained for the ten planar four-membered rings measured so far [30,31]. The three-membered rings in turn show again negative ("diamagnetic") nonlocal contributions [32]. Of course, it is an open question whether or not the concept of localized susceptibilities empirically fitted to the susceptibilities of open chain nonstrained molecules may be also applied to small strained rings. However, the observation of general trends, especially of the "paramagnetism" in the nonlocal contribution in four-membered rings certainly calls for a quantitative explanation.

We further note that there may be strong insertion effects on the nonlocal contribution to the susceptibilities. For instance, carbonyl groups, substituted into an aromatic ring, tend to quench the nonlocal contributions [33]. A different example, showing the effects of fluorine substitutions, is given in Fig. II.3, where the directly measured out-of-plane minus average in-plane susceptibility anisotropies for fluorine substituted benzenes and pyridines [34,22] are plotted versus the number of fluorine atoms. Also shown in this figure are values determined independently from Cotton-Mouton effect measurements [35]. Apart from the fact that the curve connecting the Cotton-Mouton data runs about 10 units below the curves for the rotational Zeeman effect data [36], two points should be mentioned. First, a marked influence of the position at which the second fluorine atom is substituted may be observed in 1,2-difluorobenzene and 1,3-difluorobenzene. It

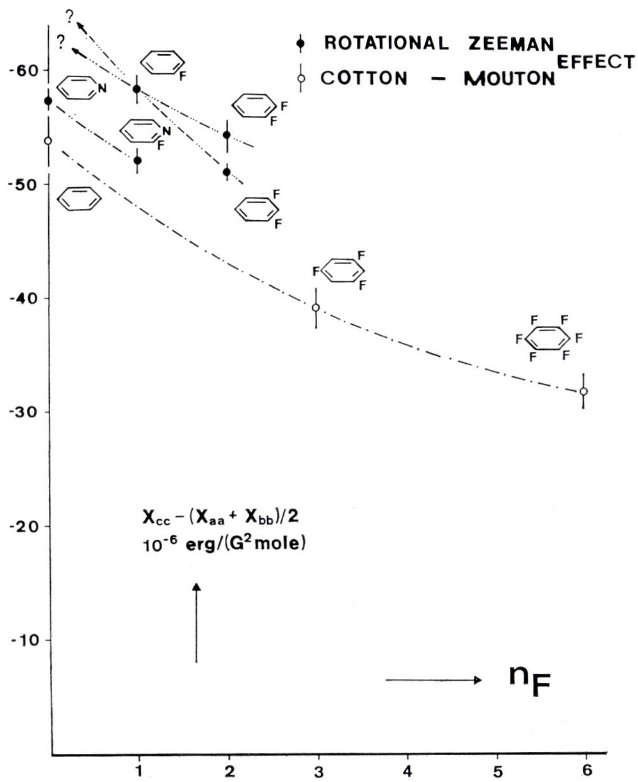

Fig. II.3. Effect of fluorine substitutions on the out-of-plane minus average in-plane component of the magnetic susceptibility are shown for several aromatic rings. The difference between the Cotton-Mouton and the rotational Zeeman effect data is probably due to the neglect of the field dependence of the *electric* polarizability in the analysis of the Cotton-Mouton data. Note that the difference in the results for 1,2- and 1,3-difluorobenzene indicates that the ring current quenching effects of substituents strongly depend on their position appears to be correlated to the alternation in the overall INDO-electron density (including σ-electrons), while a correlation to the π-electron density alone is not obvious. However, more experimental data on F-substituted pyridines, thiophenes, etc., is needed to substantiate this preliminary finding. Second, the extrapolated out-of-plane minus average in-plane susceptibility of benzene is considerably more negative than the value obtained from crystal measurements. This extrapolated value $(\chi_{\perp} - \chi_{||})_{\text{benzene}} \approx -62.5 \times 10^{-6}$ erg/(G^2 mole) would become even more negative if the 1,3,5-trifluorobenzene value from the Cotton-Mouton effect analysis would be used without further correction and if the series 1,3,5-trifluorobenzene — 1,3-difluorobenzene — fluorobenzene would be used for the extrapolation.

At the end of this Section we may venture to predict susceptibility anisotropies and g-values for a molecule which has not yet been measured, acetone. Assuming well localized orbitals, the atom values of Table II.2 and the known structure [37] make it possible to predict the following local values for the susceptibilities:

$$\begin{aligned} \chi_{aa}^{\text{pred}} &= -35.4 \\ \chi_{bb}^{\text{pred}} &= -29.7 \quad 10^{-6}\ \text{erg}/(\text{G}^2\ \text{mole}) \\ \chi_{cc}^{\text{pred}} &= -39.6 \end{aligned} \tag{II.12}$$

The missing g-values may be predicted indirectly by combining INDO-values or atom dipole values for the diamagnetic susceptibilities and the above values for the overall susceptibilities to calculate the paramagnetic susceptibilities and the perturbation sums entering into the theoretical expressions for the g-values [compare Eq. (I.2) and (I.4)]. The INDO-values for the diamagnetic susceptibilities, calculated with the known structure are given by

$$\begin{aligned} \chi_{aa,\,\text{dia}}^{(\text{INDO})} &= -168.2 \times 10^{-6}\ \text{erg}/(\text{G}^2\ \text{mole})\ ; \\ \chi_{bb,\,\text{dia}}^{(\text{INDO})} &= -205.2 \times 10^{-6}\ \text{erg}/(\text{G}^2\ \text{mole})\ ; \\ \chi_{cc,\,\text{dia}}^{(\text{INDO})} &= -308.3 \times 10^{-6}\ \text{erg}/(\text{G}^2\ \text{mole})\ . \end{aligned} \tag{II.13}$$

Combining (II.12) and (II.13) then leads to predicted values for the paramagnetic susceptibilities and g-values of acetone as:

$$\begin{aligned} \chi_{aa}^{\text{para}} &= +132.8 & g_{aa} &= -0.031 \\ \chi_{bb}^{\text{para}} &= +175.5 \times 10^{-6}\ \text{erg}/(\text{G}^2\ \text{mole}) & g_{bb} &= -0.034 \\ \chi_{cc}^{\text{para}} &= +268.6 & g_{cc} &= -0.004 \end{aligned}$$

In view of our present experience with the additivity scheme for atom susceptibilities, we would expect the predicted susceptibilities [Eq. (II.12)] to be correct within $\pm 3.10^{-6}$ erg/(G^2 mole) and the g-values within ± 0.02.

III. Instrumentation and Analysis of the Spectra

A. Instrumentation

The early experimental work on the rotational Zeeman effect was carried out using rather small magnets and comparatively low fields [38–41]. As a consequence, the volume of the absorption cell was small and the studies were restricted to a few light molecules with strong absorption lines and comparatively large g-values. Because of low field strengths, the second order Zeeman effect was not observed until 1967. In the following we will briefly describe two types of spectrometers which have been especially designed for rotational Zeeman effect experiments. Both types use conventional Stark-effect modulated microwave spectrometers of the Hughes-Wilson type [42] with phase stabilized backward wave oscillators or Klystrons as radiation sources. They differ, however, in the construction of the absorption cell and in the magnetic system. According to the universities where they were used first, we will call them "Urbana-type" spectrometers and "Berlin-type" spectrometers, respectively.

In the Urbana-type spectrometer, first introduced by W. H. Flygare *et al.* in 1968 [43], emphasis is put on a large absorption volume and on ease of operation. The absorption cell, a conventional rectangular waveguide cell, is located between the pole faces of a powerful electromagnet which is capable of delivering a field of up to approximately 30 kG over a gap length of 1.8 m (Urbana) to 2.5 m (Kiel). A schematic of the latter system is shown in Fig. III.1. A high sensitivity is obtained in these systems and rotational Zeeman effect studies can be carried out on practically all molecules that have been studied so far by microwave spectroscopy. In the design of the Berlin-type spectrometer developed by R. Honerjäger *et al.*[44] primarily for use in high temperature work on nonvolatile diatomics, more emphasis was put on a high magnetic field strength. This is important if molecules with small rotational g-values are examined and if more accurate susceptibility anisotropies in small molecules are desired. The absorption cell, a coaxial cell especially designed for high temperature work, is located within a super-conducting solenoid. With a comparably small usable field volume, sensitivity problems may restrict the range of application of this system to smaller molecules with sufficiently strong absorption lines. In general, by the use of high field super-conduction magnets, rotational Zeeman effect data obtained by microwave spectroscopy can reach the accuracy of data obtained by molecular beam spectrometers [45].

In the following the principles of operation and some performance data will be given for the spectrometer operated at Kiel University which at present is the most advanced instrument of the Urbana-type. For details of the electronics, the reader is referred to the literature [46,47].

In a typical experiment, monochromatic microwave radiation with a bandwidth less than 10 Hz [48] at a frequency of 10 GHz is slowly swept over the interesting frequency range of the rotational spectrum. Highly accurate frequency markers are superimposed automatically on the recorded spectrum. After having passed an attenuator, a crystal mixer which is part of the frequency stabilization system, and a second attenuator, the microwave radiation enters the absorption cell through a mylar window. The temperature of the absorption cell is con-

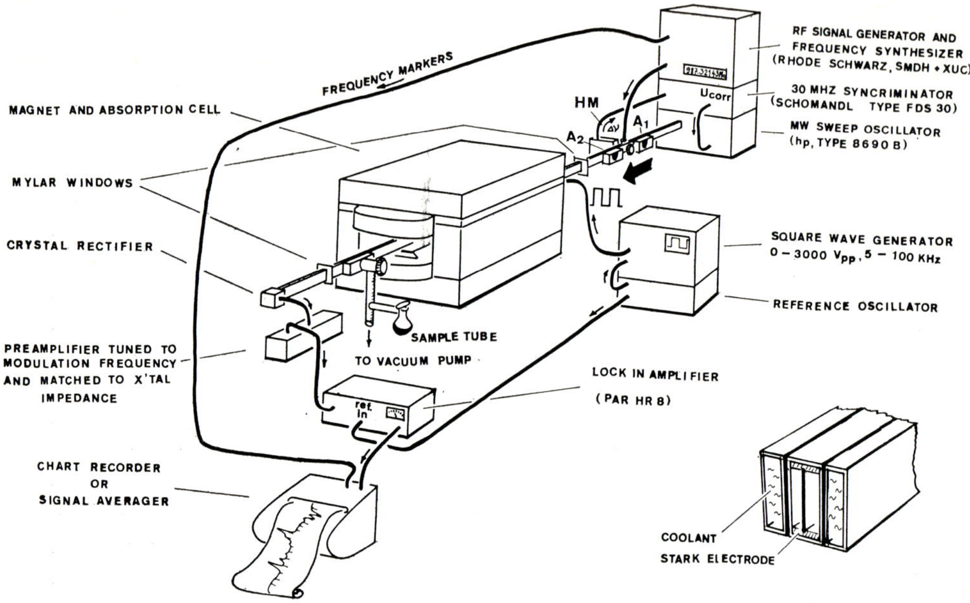

Fig. III.1. A typical setup of a microwave spectrograph used for rotational Zeeman effect studies is shown in this diagram. The monochromatic microwave radiation enters the absorption cell through a mica window, after having passed two attenuators, A_1 and A_2, and a harmonic mixer, HM, which is part of the frequency control system. Square wave Stark effect modulation (see text) is used in order to increase the sensitivity. A cross section of the waveguide absorption cell is shown in the insert at lower right. The absorption cell may be cooled (or heated) by a liquid circulating in two brass tubes soldered to the cell walls. The Stark voltage is applied to a central plate (Stark electrode) which is insulated from the walls by two teflon strips. Typical sample pressures are in the order of 10 m Torr. After having passed the absorption cell, the microwave radiation is rectified in a crystal detector. If the microwave frequency is swept slowly over the absorption line, a small ac-signal is superimposed on the rectified current due to the Stark effect modulated absorption. Narrow band preamplification followed by a phase sensitive amplifier makes it possible to detect rotational transitions with absorption coefficients as low as $\alpha = 10^{-10}$ cm^{-1}. The use of a signal averager may considerably reduce the measuring time

trolled by a cooling or heating liquid (methanol or water respectively) circulating in brass tubes which are attached to the cell on two sides. The temperature may be varied between $-80°$ and $+90$ °C. The cell contains the sample as a gas at a low pressure (typically 1 to 10 mtorr). This corresponds to a mean free path between collisions on the order of millimeters and to an average time of flight between collisions on the order of 10^{-5} seconds. Such low pressures are essential, if high resolution is to be achieved, since the observed linewidths are mainly due to collision broadening. If the frequency of the incident microwave radiation, ν_{MW}, corresponds to a rotational transition frequency and if the transition is electric dipole allowed, the radiation power or intensity at the detector crystal decreases slightly. However, due to the small absorption coefficient and the small differences in the Boltzmann populations of the lower and the upper rotational levels, only an extremely small fraction of the incident power is absorbed by the sample. The small changes in radiation power are measured in a crystal rectifier detector which, unfortunately, also generates a noise spectrum which is inversely proportional to the frequency. The effects of crystal noise can be reduced by the use of Stark modulation. For this purpose a central plate is introduced into the absorption cell which is insulated from the walls by teflon strips. If a square wave electric voltage is applied to the central plate, a periodic electric field is generated between the plate and the waveguide walls and causes a periodic Stark-effect splitting of the energy levels and absorption frequencies. Now most of the noise can be rejected by the use of narrow band preamplifiers tuned to the modulation frequency and followed by a lock-in amplifier as the final stage of the amplifier chain. With such a system modulated absorption signals of transitions with absorption coefficients as low as 10^{-10} cm^{-1} may still be detected. The additional use of computer controlled signal averaging may considerably reduce measuring time if transitions with such small absorption coefficients are examined.

The magnet, which was constructed by Bruker Physik, Karlsruhe–Forchheim, Germany, is shown in Fig. III.2. Magnetically soft steel with less than 0.06 weight % of carbon was used for the construction except for the cobalt iron caps of the tapered pole faces. Each pole is surrounded by 200 turns of 0.9 cm square by 0.4 cm bore copper conductor carrying up to 240 amps current. The power to the two coils is provided by a highly stable power supply also manufactured by Bruker. A typical field versus current plot is shown in Fig. III.3. Fig. III.4 shows the long

Fig. III.2. Electromagnet used for the rotational Zeeman effect measurements at Kiel. The upper yoke may be lifted by hydraulic jacks in order to insert spacers on top of the side yokes for different gap widths. Bearings in the side yokes are to allow for lateral access of the gaussmeter probe tip. The power connections ($I_{max} = 240$ Amps, each coil) are visible at the right front side. The overall length of the gap is 250 cm and the maximum field at a gap width of 6 cm is close to 21 kG, and at a gap of 0.6 cm the field is 31 kG

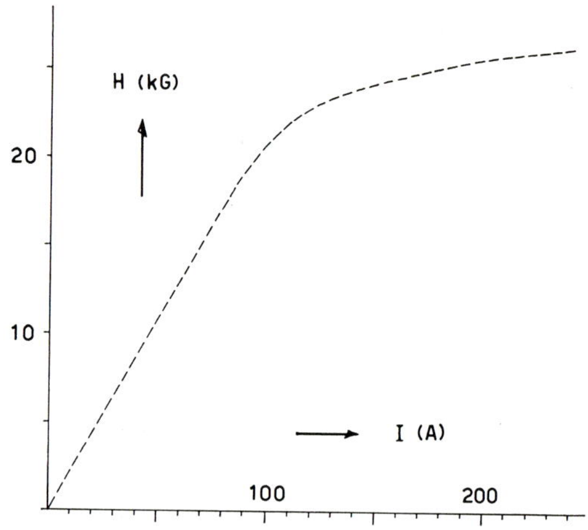

Fig. III.3. Field versus current plot of the Kiel magnet for a gap width of 3 cm

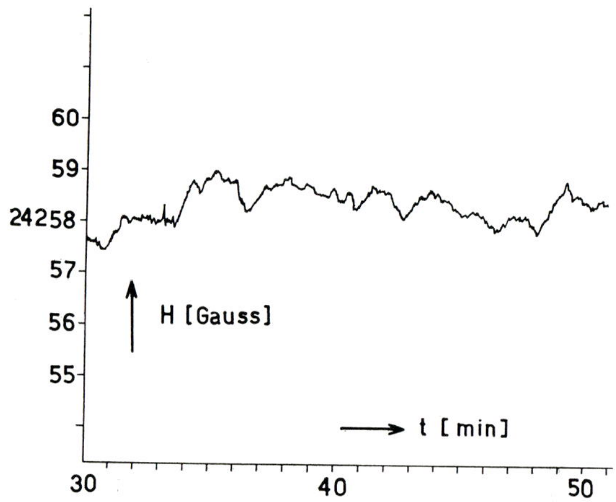

Fig. III.4. Long term stability of the magnetic field of the Kiel system

term stability of the field as measured by a Rawson Lush rotating coil Gaussmeter type 920 M. Although the requirements as to the homogeneity of the magnetic field are far less stringent than in NMR spectroscopy, a knowledge of the field distribution is necessary in order to avoid small systematic errors in the analysis of the data. Figs. III.5 and III.6 show typical field profiles. While the inhomogeneity perpendicular to the long axis of the magnet is sufficiently small to be negligible, the drop of the field toward the edges of the magnet has to be accounted for. This is done partly by reducing the effective length of the absorption cell by reducing the length of the Stark-electrode as indicated in Fig. III.6. The remaining inhomogeneity is accounted for numerically as described later in this Chapter.

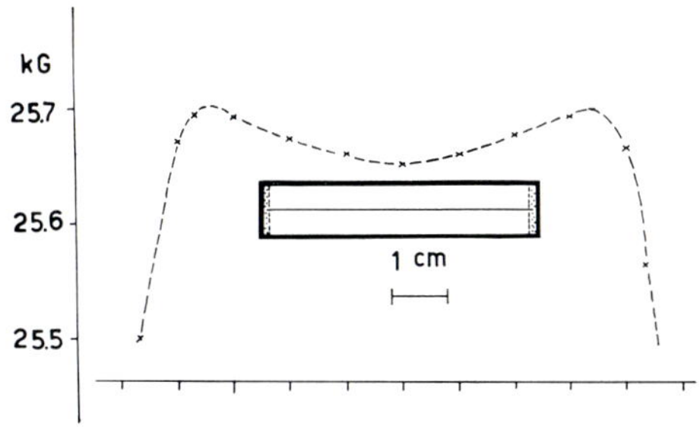

Fig. III.5. Magnetic field profile perpendicular to the long axis of the gap (gap width 3 cm, 210 Amps). Also shown is the position and the cross section of an oversized X-band absorption cell typically used for the recordings with $\Delta M=0$ selection rules

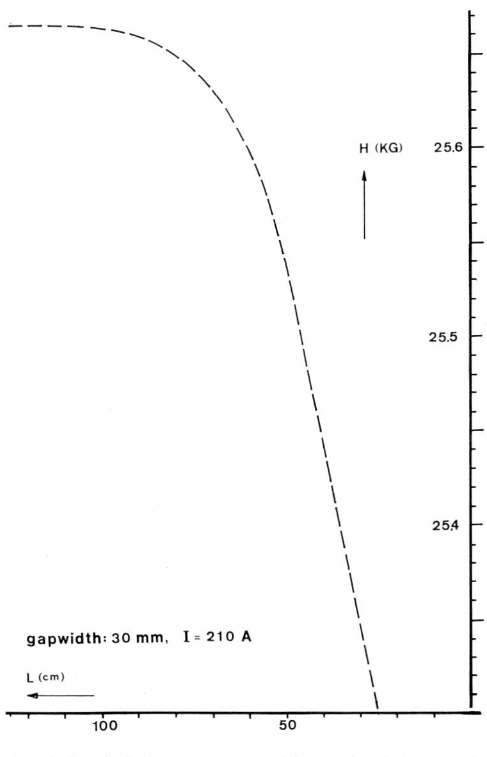

Fig. III.6. The longitudinal drop of the magnetic field towards the ends of the magnet limits the usable gap volume. In the $\Delta M=0$ absorption cells used in connection with the 3 cm gap, the Stark spectrum which determines the effectively absorption cell starts 35 cm inside the magnet. The remaining inhomogeneity is accounted for in the numerical analysis

Since the electric field vector of the incident microwave radiation, $\boldsymbol{E}_{MW}$, is linearly polarized perpendicular to the broad face of the waveguide absorption cell (TE_{10}-mode of propagation), the M-selection rules may be selected by the orientation of the waveguide cell with respect to the magnetic field. This may be

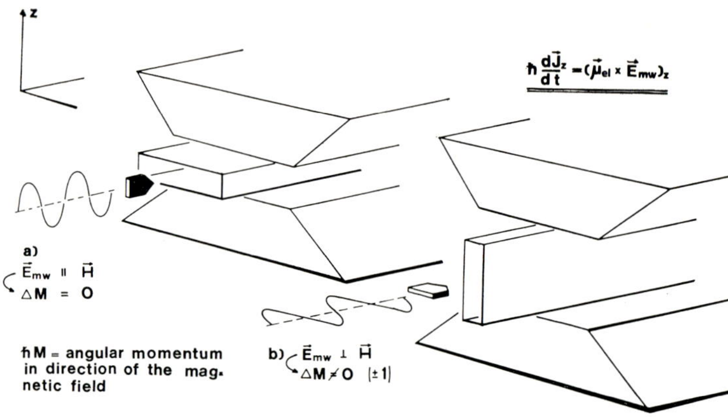

Fig. III.7. The selection rules $\Delta M = 0$ or $\Delta M = \pm 1$ may be selected by changing the orientation of the rectangular waveguide cell within the gap. In case a) the electrical vector of the incident microwave radiation being perpendicular to the broad face of the waveguide will produce no torque in the direction of the magnetic field which serves as the quantization axis. This geometry leads to the $\Delta M = 0$ selection rule. In case b) the same classical argument suggests $\Delta M \neq 0$ (quantum mechanics leads to $\Delta M = \pm 1$)

understood already with a classical picture as illustrated in Fig. III.7. The linearly polarized microwave field exerts a rapidly oscillating torque on the molecular electric dipole moment, $\boldsymbol{\mu}_{el}$, and the classical equation of motion for the angular momentum vector, $\hbar \boldsymbol{J}$, becomes:

$$\hbar \frac{d\boldsymbol{J}}{dt} = \boldsymbol{\mu}_{el} \times \boldsymbol{E}_{MW} .$$

Thus, the vector describing the time derivative of the angular momentum is perpendicular to the electric field vector of the incident microwave radiation and there will be no change of the angular momentum component in the direction of $\boldsymbol{E}_{MW}$. With the magnetic field providing the axis of M-quantization, we can therefore predict the following selection rules depending on whether $\boldsymbol{E}_{MW}$ is parallel or perpendicular to the magnetic field, $\boldsymbol{H}$.

a) $\Delta M = 0$ if $\boldsymbol{E}_{MW}$ is parallel to $\boldsymbol{H}$, *i.e.*, if the magnetic field is perpendicular to the broad face of the waveguide absorption cell.

b) $\Delta M \neq 0$ (quantum mechanics leads to $\Delta M = \pm 1$) if $\boldsymbol{E}_{MW}$ is perpendicular to $\boldsymbol{H}$, *i.e.*, if the magnetic field is parallel to the broad side of the waveguide cell.

At this point we have to note a restriction. The above selection rules are only valid as long as the translational Stark-effect due to the Lorentz forces may be neglected which is not always true for symmetric top molecules as is discussed in detail later in this Chapter.

In the solenoid-type spectrometer operated in the H_{10} mode of propagation, the vector of the microwave radiation field is always perpendicular to the axis of the waveguide while the magnetic field of the solenoid points in the direction of the waveguide axis. Thus, $\boldsymbol{E}_{\text{MW}}$ and $\boldsymbol{H}$ are perpendicular and $\Delta M = \pm 1$ transitions are generally observed (again this statement holds only as long as the translational Stark-effect is negligible). If, however, circularly polarized microwave radiation is used in a solenoid-type spectrometer, only $\Delta M = +1$ or $\Delta M = -1$ selection rules are observed, depending on the sense of rotation of the electric field vector of the incident microwave.

In cases of doubt about the sign of the rotational g-values, which cannot be determined from the experiment if both $\Delta M = +1$ and $\Delta M = -1$ transitions are observed simultaneously (compare Section B of this Chapter), the use of circularly polarized microwave radiation is necessary if an unambiguous experimental choice of the sign of the g-value is required.

B. The Analysis of Rotational Zeeman Effect Spectra in Asymmetric Top Molecules

In the following we will demonstrate how the effective Hamiltonian, Eq. (I.7), which will be discussed in more detail in the final Chapter, is used in practical spectroscopy. For this purpose we will discuss in detail the analysis of the Zeeman multiplets of an asymmetric top molecule with subsequent shorter sections on symmetric top molecules, linear molecules and molecules containing quadrupole nuclei.

Consider the rotational Zeeman effect of ethyleneoxide as an example for an asymmetric top molecule. The first investigation of the rotational Zeeman effect of ethyleneoxide has been carried out by the authors and W. Hüttner in 1968. In the meantime, the Zeeman splittings were remeasured with improved accuracy in an attempt to determine the sign of the electric dipole moment from the change of the g-values and rotational constants upon isotopic substitution (compare Chapter II). All numerical values will be taken from this later work [49].

Quite generally, the analysis of the rotational Zeeman effect of asymmetric top molecules may be broken up into a sequence of three steps:

1. The zero field Hamiltonian matrix of the asymmetric top is set up within the basis of eigenfunctions of the limiting symmetric top, $\phi_{J,K,M}(\phi, \theta, \chi)$, and the rotational eigenvalues and asymmetric top wavefunctions, $\psi_{J,\tau,M}(\phi, \theta, \chi)$ are determined in the course of a numerical diagonalization.

2. The asymmetric top matrix elements of the complete effective rotational Hamiltonian are then calculated and the eigenvalues are determined by a perturbation treatment. In view of the smallness of the Zeeman contributions, a first order perturbation treatment in the asymmetric top basis is sufficient in most cases so that only the diagonal elements of $\mathscr{H}_{\text{Zeeman}} = \mathscr{H}_{\text{g}} + \mathscr{H}_{\chi} + \mathscr{H}_{\text{TS}}$ need to be calculated explicitly. Symmetry considerations and order of magnitude estimates will prove helpful in this context.

3. The molecular g-values and susceptibility anisotropies are fitted to the observed multiplet splittings by a least squares procedure and the results are used to calculate derived molecular properties such as the molecular electric quadrupole moments.

Although the determination of the zero field eigenvalues and wavefunctions is described in many standard texts on rotational spectroscopy, we will briefly recall the principles and give some results for later reference. From Eq. (IV. 59a) and IV.59b), the zero field Hamiltonian of an asymmetric top molecule is given by (from now on quantum mechanical operators will be denoted by underlining):

$$\begin{aligned}\underline{\mathscr{H}}_{\text{rot}} &= \sum_{\gamma} \frac{\hbar^2}{2\,I^{(\text{n})}_{\gamma\gamma}} \left(1 + \frac{2}{I^{(\text{n})}_{\gamma\gamma}} \left(\frac{\underline{L}_\gamma \underline{L}_\gamma}{\Delta}\right)\right) \underline{J}_\gamma{}^2 \\ &+ \sum_{\substack{\gamma \\ \gamma \neq \gamma'}} \sum_{\gamma'} \frac{\hbar^2}{I^{(\text{n})}_{\gamma\gamma}\, I^{(\text{n})}_{\gamma'\gamma'}} \left\{\left(\frac{\underline{L}_\gamma \underline{L}_{\gamma'}}{\Delta}\right) \underline{J}_\gamma \underline{J}_{\gamma'} + \left(\frac{\underline{L}_{\gamma'} \underline{L}_\gamma}{\Delta}\right) \underline{J}_{\gamma'} \underline{J}_\gamma\right\}\end{aligned} \tag{III.1}$$

This is usually written in a more compact form as

$$\underline{\mathscr{H}}_{\text{rot}} = h(A\underline{J}_a^2 + B\underline{J}_b^2 + C\underline{J}_c^2) \tag{III.2}$$

[compare Eqs. (IV.67) and (IV.71)], where A, B, and C are the measured rotational constants, and the second sum in Eq. (III.1) may give rise to a small tilt in the molecular principle axis system as compared to the principal axis system of the nuclear frame. In ethylene oxide the experimental rotational constants are $A = 25{,}483.66 \pm 0.16$ MHz, $B = 22{,}121.13 \pm 0.18$ MHz, and $C = 14{,}097.95 \pm 0.16$ MHz, and the second sum in Eq. (III.1) vanishes due to the symmetry of the nuclear frame.

We will prove this explicitly as an example for how symmetries of the nuclear frame may be used to show that certain off-diagonal elements of the moment of inertia tensor, the $\underset{\sim}{g}$-tensor, and $\underset{\sim}{\chi}$-tensor must be zero.

According to Fig. III.8, ethylene oxide contains two mirror planes perpendicular to one another, the a,b-plane (m_{ab}) and the b,c-plane (m_{bc}) and the line of intersection of the two mirror planes generates a twofold axis, C_{2b}. Due to this symmetry of the nuclear Coulomb potential, the electronic Hamiltonian, Eq. (IV.55a), is symmetric with respect to the operations:

$$\begin{array}{llll} m_{ab}\colon & a_\varepsilon \longrightarrow a_\varepsilon & m_{bc}\colon \; a_\varepsilon \longrightarrow -a_\varepsilon & C_{2b}\colon \; a_\varepsilon \longrightarrow -a_\varepsilon \\ & b_\varepsilon \longrightarrow b_\varepsilon & \quad\;\; b_\varepsilon \longrightarrow b_\varepsilon & \quad\;\; b_\varepsilon \longrightarrow b_\varepsilon \\ & c_\varepsilon \longrightarrow -c_\varepsilon & \quad\;\; c_\varepsilon \longrightarrow c_\varepsilon & \quad\;\; c_\varepsilon \longrightarrow -c_\varepsilon \end{array}$$

where all electronic coordinates change simultaneously and the electronic wavefunctions may be classified according to their symmetry species with respect to the C_{2v} group (see the character table in Table III.1). All representations of C_{2v} are one dimensional and apart from accidental degeneracies, the electronic wavefunctions are nondegenerate.

As an example we will now show that the coefficient which is connected with the product $J_a J_b$ is zero, *i.e.*,

$$\frac{\hbar^2}{I^{(\text{n})}_{aa}\, I^{(\text{n})}_{bb}} \sum_{n}^{\substack{\text{excited}\\\text{states}}} \frac{\langle 0|\underline{L}_a|n\rangle\langle n|\underline{L}_b|0\rangle}{E_0 - E_n} \overset{!}{=} 0\,. \tag{III.3}$$

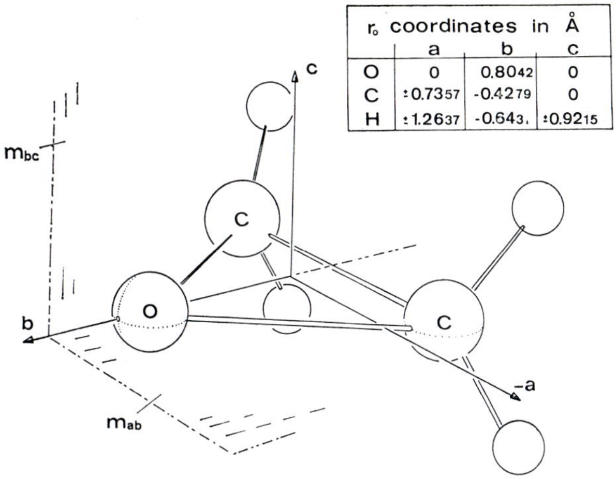

r_0 coordinates in Å	a	b	c
O	0	0.8042	0
C	±0.7357	-0.4279	0
H	±1.2637	-0.643,	±0.9215

Fig. III.8. The equilibrium configuration of the nuclear frame of ethyleneoxide contains two perpendicular symmetry planes, m_{ab} and m_{bc}. They generate a twofold symmetry axis which coincides with the b-axis, the axis of intermediate moment of inertia. The experimentally determined rotational constants of the most abundant isotopic species are: $A = 25483.66 \pm 0.16$ MHz, $B = 22121.13 \pm 0.18$ MHz, and $C = 14097.95 \pm 0.16$ MHz and the electric dipole moment has a value of 1.88 ± 0.01 Debye (negative end at oxygen). The reference for the structure is Cunningham, G. L., Boyd, A. W., Myers, R. J., Gwinn, W. D., Le Van, W. I.: J. Chem. Phys. *19*, 676 (1951)

Table III.1. Character table for the C_{2v} group

Symmetry Species (representation)	Symmetry Operation E	C_{2b}	m_{ab}	m_{bc}
A_1	1	1	1	1
A_2	1	1	—1	—1
B_1	1	—1	1	—1
B_2	1	—1	—1	1

We will show this by proving that from symmetry all matrix products in the numerator of Eq. (III.3) must vanish for a rigid placement of nuclei. For this purpose we first note that the angular momentum operators $\underline{L}_a$ and $\underline{L}_b$ transform according to the B_2 and A_2 species respectively in Table III.1. Now for a nonvanishing matrix element, the integrand must belong to the unit representation A_1. From Table III.1, it is therefore immediately clear that $\underline{L}_a$ may have nonvanishing matrix elements only if $\langle m|$ and $|n\rangle$ transform according to A_1 and B_2 (or vice versa) or according to B_1 and A_2 (or vice versa), while $\underline{L}_b$ has nonvanishing matrix elements only if $\langle n|$ and $|m\rangle$ transform according to B_2 and B_1 (or vice versa) or according to A_2 and A_1 (or vice versa). Thus, since these conditions for non zero matrixelements exclude each other, one factor in the product $\langle m|\underline{L}_a|n\rangle\langle n|\underline{L}_b|m\rangle$ is necessarily zero which completes the proof.

We now turn to the calculation of the asymmetric top eigenvalues and wavefunctions from the limiting symmetric top energy levels and wavefunctions. Since the A and B rotational constants are nearly equal in ethyleneoxide, the limiting symmetric top is an oblate top with the c-axis the axis of K quantization. Thus, we choose a representation $\phi_{JKM}(\phi, \theta, \chi)$ in which J^2, J_c, and J_z are diagonal. From the Appendix II the nonvanishing matrix elements for the squares of the angular momentum operators may then be written as:

$$\langle J,K,M\,|\underline{J}_a^2|\,J,K,M\rangle = \frac{1}{2}\,[J(J+1) - K^2] = \langle J,K,M\,|\underline{J}_b^2|\,J,K,M\rangle \tag{III.4a}$$

$$\begin{aligned}\langle J,K,M\,|\underline{J}_a^2|\,J,K\pm 2,M\rangle &= -\frac{1}{4}\Big([J(J+1) - K(K\pm 1)]\,[J(J+1) \\ &\quad - (K\pm 1)(K\pm 2)]\Big)^{1/2} \\ &= -\langle J,K,M\,|\underline{J}_b^2|\,J,K\pm 2,M\rangle\end{aligned} \tag{III.4b}$$

$$\langle J,K,M\,|\underline{J}_c^2|\,J,K,M\rangle = K^2\,. \tag{III.4c}$$

The resulting matrix elements are diagonal in the quantum numbers J and M and may be arranged as is shown in Fig. III.9. The infinite Hamiltonian matrix is factored into submatrices corresponding to the different J-values and each J-submatrix in itself is factorized into $2J+1$ identical submatrices corresponding to the $2J+1$ different values of the projection quantum number M. This M-degeneracy will be lifted later upon application of the exterior magnetic field. Due to the favorable structure of the Hamiltonian matrix each submatrix of rank $2J+1$ may be diagonalized individually in order to obtain the corresponding eigenvalues and wavefunctions. For numerical methods compare references.[50,51)]

In Table III.2 this procedure is illustrated for one M-submatrix with $J=1$, where the diagonalization is still almost trivial.

In the course of the diagonalization, the corresponding unitary transformation matrix $\langle K|U_J|\tau\rangle$ which diagonalizes the original Hamiltonian matrix according to Eq.I.5), is also determined.

$$(E_{J,\tau,M}) = (\langle K|\underline{U}_J|\tau\rangle)^{-1}\cdot(\langle J,K,M|\underline{\mathscr{H}}|J,K',M\rangle)\cdot(\langle K|\underline{U}_J|\tau\rangle) \tag{III.5}$$

The index J is used to indicate the submatrix of rank $2J+1$. The column vectors of the matrix $(\langle K|U_J|\tau\rangle)$ which are the eigenvectors of the original Hamiltonian matrix $(\langle J,K,M|\mathscr{H}|J,K',M\rangle)$ lead to the asymmetric top wavefunctions:

$$\psi_{J,\tau,M}(\phi,\theta,\chi) = \sum_{K=-J}^{+J} \langle K|U_J|\tau\rangle \phi_{J,K,M}(\phi,\theta,\chi) \tag{III.6}$$

(compare Table III.2). For later reference, the lowest rotational energy levels and the asymmetric top expectation values of the squares of the angular momentum

Table III.2. The $\langle J,K,M|\mathscr{H}_0|J.K',M\rangle$ submatrix of the asymmetric top Hamiltonian, $\mathscr{H}_0 = A\underline{J}_a^2 + B\underline{J}_b^2 + C\underline{J}_c^2$; matrix of the diagonalizing unitary transformation, $\langle K|\underline{U}_J|\tau\rangle$ ($\mathscr{H}_{0,\text{diag}} = U^{-1}\cdot\mathscr{H}_0\cdot U$, where $U^{-1} = U^t$); and asymmetric top eigenvalues, $E_{J,\tau,M}$, and eigenfunctions, $\psi_{J,\tau,M}(\phi,\theta,\chi)$ for $J=1$. (Compare to Fig. III.3). As original basis functions, $\phi_{J,K,M}(\phi,\theta,\chi)$, the eigenfunctions of the angular momentum operators J^2, J_c, and J_Z are used, *i.e.*,

$$\underline{J}^2\phi_{J,K,M}(\phi,\theta,\chi) = J(J+1)\phi_{J,K,M}(\phi,\theta,\chi)$$
$$\underline{J}_c\phi_{J,K,M}(\phi,\theta,\chi) = K\,\phi_{J,K,M}(\phi,\theta,\chi)$$
$$\underline{J}_Z\phi_{J,K,M}(\phi,\theta,\chi) = M\,\phi_{J,K,M}(\phi,\theta,\chi)$$

$\langle 1,K,M|\mathscr{H}_0|1,K',M\rangle$:

$K \backslash K'$	-1	0	1
-1	$\frac{A+B}{2}+C$	0	$-\frac{A-B}{2}$
0	0	$A+B$	0
1	$-\frac{A-B}{2}$	0	$\frac{A+B}{2}+C$

$\langle 1,K,M|U|1,\tau,M\rangle$:

$K \backslash \tau$	$+1$	0	-1
-1	0	$1/\sqrt{2}$	$1/\sqrt{2}$
0	1	0	0
$+1$	0	$-1/\sqrt{2}$	$1/\sqrt{2}$

τ	$E_{J\tau M}$	$\psi_{J\tau M} = \sum_{K=-J}^{+J} \langle J,K,M|U|J,\tau,M\rangle\,\phi_{JKM}$
1	$A+B$	$1\sqrt{2}\,(\phi_{1,1,M} - \phi_{1,-1,M})$
0	$A+C$	$\phi_{1,0,M}$
-1	$B+C$	$1\sqrt{2}\,(\phi_{1,1,M} + \phi_{1,-1,M})$

operators are listed in Table III.3. The latter may be calculated by several methods which include

a) a continuous fraction method described by Kivelson and Wilson [52],

b) starting with the symmetric top matrix elements and obtaining $\langle K|U_J|\tau\rangle$ by direct transformation,

$$\langle J,\tau,M|\underline{J}_\gamma^2|J,\tau,M\rangle = \sum_{K'}\sum_{K}\langle\tau|\underline{U}_J|K'\rangle\langle J,K',M|\underline{J}_\gamma^2|J,K,M\rangle\langle K|\underline{U}_J|\tau\rangle$$

c) taking the derivatives of the rotational energy levels with respect to the rotational constants,[53]

$$\langle J,\tau,M|\underline{J}_a^2|J,\tau,M\rangle = \frac{\partial E_{J,\tau,M}(A,B,C)}{\partial A},\ \text{etc.}$$

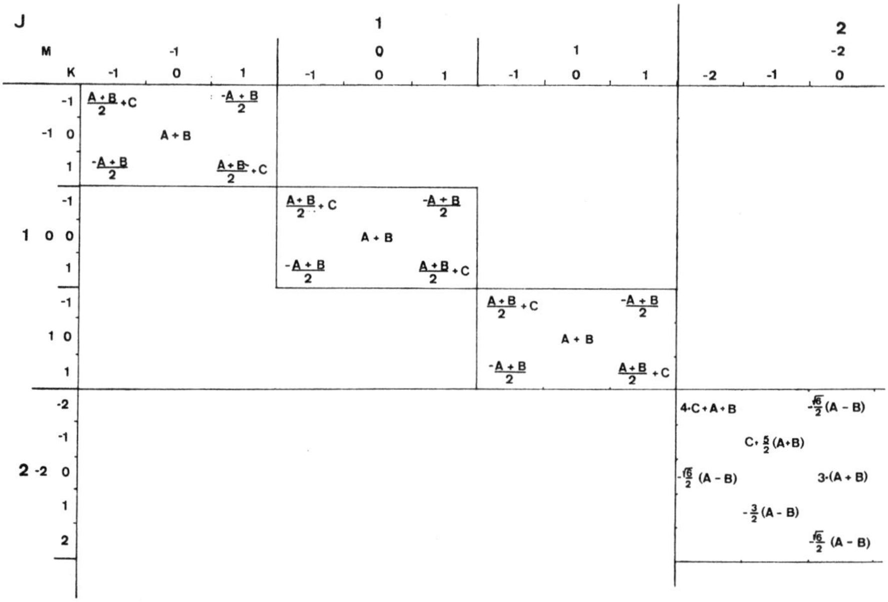

Fig. III.9. Low J section of the Hamiltonian matrix of an asymmetric top molecule such as for instance ethyleneoxide in the absence of exterior fields. The matrix is set up in the eigenfunction basis of the limiting oblate symmetric top, $\phi_{J,K,M}(\phi,\theta,\chi)$. $(\underline{J}^2\phi_{J,K,M}(\phi,\theta,\chi) = J(J+1)\phi_{J,K,M}(\phi,\theta,\chi); \underline{J}_c\phi_{J,K,M}(\phi,\theta,\chi) = K\phi_{J,K,M}(\phi,\theta,\chi); \underline{J}_Z\phi_{J,K,M}(\phi,\theta,\chi) = M\phi_{J,K,M}(\phi,\theta,\chi))$

The values given in Table III.3 were calculated with method c). For this purpose the derivatives were approximated by quotients of differences. In each case it is sufficient to calculate only one expectation value, say $\langle|\underline{J}_b^2|\rangle$ according to one of the three methods described above. The other two follow from

$$E_{J,\tau,M} = h(A\langle|\underline{J}_a^2|\rangle + B\langle|\underline{J}_b^2|\rangle + C\langle|\underline{J}_c^2|\rangle)$$

and

$$J(J+1) = \langle|\underline{J}_a^2|\rangle + \langle|\underline{J}_b^2|\rangle + \langle|\underline{J}_c^2|\rangle$$

as:

$$\langle J,\tau,M\,|\underline{J}_a^2|\,J,\tau,M\rangle = \frac{E_{J,\tau,M}(A,B,C)/\mathrm{h} - (B-C)\langle|\underline{J}_b^2|\rangle - C\,J(J+1)}{A-C} \qquad \text{(III.7)}$$

$$\langle J,\tau,M\,|\underline{J}_c^2|\,J,\tau,M\rangle = \frac{E_{J,\tau,M}(A,B,C)/\mathrm{h} - (B-A)\langle|\underline{J}_b^2|\rangle - A\,J(J+1)}{C-A} \qquad \text{(III.8)}$$

In Table III.3 we have introduced the (J,K_-K_+) notation for the rotational states instead of the (J,τ) notation which was used up until now in this paper. K_- and K_+ are the K-quantum numbers of the limiting prolate and oblate symmetric tops respectively as is illustrated in Fig. III.10.

For future reference, we recall the symmetry properties of the asymmetric top wavefunctions with respect to 180° rotations, $\underline{C}_{2a}$, $\underline{C}_{2b}$, and $\underline{C}_{2c}$ about the

Table III.3. Lowest rotational energy levels and expectation values for the squares of the angular momentum components (measured in units of $\hbar$) for ethylene oxide

$A = 25.48366$ GHz, $B = 22.12113$ GHz, $C = 14.09795$ GHz

J	K_-	K_+	$E_{JK_-K_+}$ (GHz)	$\langle J_a^2\rangle$	$\langle J_b^2\rangle$	$\langle J_c^2\rangle$
0	0	0	0.000000	0.000000	0.000000	0.000000
1	0	1	36.219080	0.000000	1.000000	1.000000
1	1	1	39.581610	1.000000	0.000000	1.000000
1	1	0	47.604790	1.000000	1.000000	0.000000
2	0	2	103.141600	0.544381	1.540004	3.915614
2	1	2	103.996590	1.000000	1.000000	4.000000
2	1	1	128.066130	1.000000	4.000000	1.000000
2	2	1	138.153720	4.000000	1.000000	1.000000
2	2	0	143.669360	3.455619	2.459996	0.084386
3	0	3	197.666763	1.142443	1.929887	8.927669
3	1	3	197.808033	1.253575	1.789788	8.956637
3	1	2	242.846700	2.011663	6.327551	3.660785
3	2	2	246.810960	4.000000	4.000000	4.000000
3	2	1	266.458677	2.857557	8.070113	1.072331
3	3	1	286.492587	8.746425	2.210212	1.043363
3	3	0	289.593000	7.988337	3.672449	0.339215
4	0	4	320.081820	1.614838	2.459965	15.925197
4	1	4	320.101061	1.634810	2.434021	15.931169
4	1	3	385.908874	3.863186	7.473939	8.662875
4	2	3	386.858635	4.590379	6.560356	8.849266
4	2	2	427.492972	3.582437	13.055344	3.362218
4	3	2	437.905838	8.365190	7.565979	4.068831
4	3	1	452.329825	6.136814	12.526061	1.337125
4	4	1	485.005365	15.409621	3.439644	1.150734
4	4	0	486.480008	14.802724	4.484691	0.712585

Table III.4. Transformation properties of the asymmetric top wavefunctions $\psi_{J,K_-K_+,M}(\phi,\theta,\chi)$ under the 180° rotations of the four group:

$$\underline{C}_{2\gamma}\,\psi_{J,K_-K_+,M}(\phi,\theta,\chi) = f_{\gamma,K_-K_+}\;\psi_{J,K_-K_+,M}(\phi,\theta,\chi) \quad (\gamma = a,b,c)$$

with $f_{\gamma,K_-K_+} = +1$ or -1 as given in the table

K_-K_+	Symmetry species	$\underline{C}_{2a}$ (I_{aa} = least moment of inertia)	$\underline{C}_{2b}$ (I_{bb} = intermediate moment of inertia)	$\underline{C}_{2c}$ (I_{cc} = greatest moment of inertia)
ee	A	1	1	1
eo	B_a	1	-1	-1
oo	B_b	-1	1	-1
oe	B_c	-1	-1	1

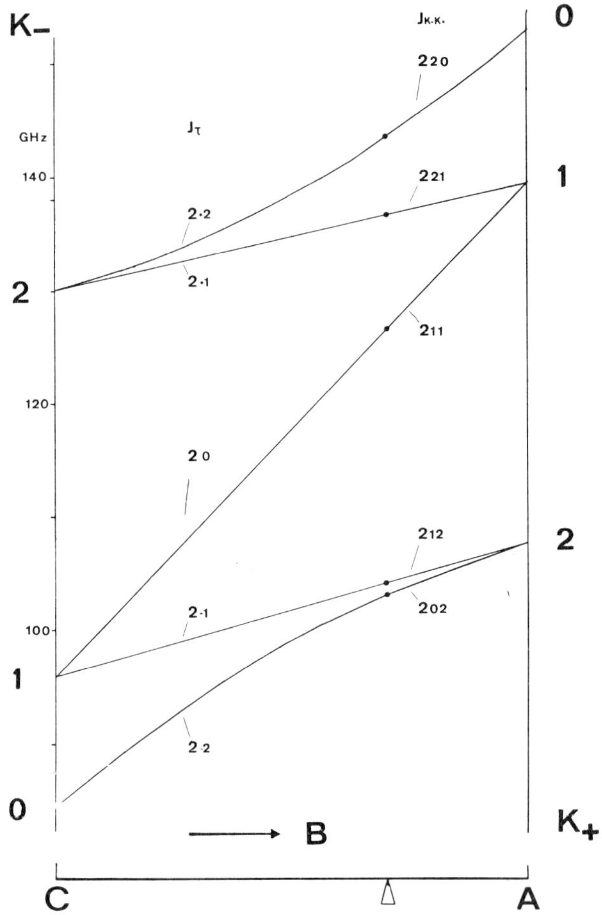

Fig. III.10. The $J = 2$ rotational levels of an asymmetric top molecule as a function of the rotational constant B. $A = 25.48366$ GHz and $C = 14.09795$ GHz have been fixed to their values for ethyleneoxide. As soon as the moment of inertia tensor becomes asymmetric, the K-degeneracy of the limiting prolate (left) and oblate (right) symmetric tops is lifted. The actual B value for ethyleneoxide is marked by a dagger. Both conventions of labelling the rotational levels, the J_τ designation and the $J_{K_-K_+}$ designation are shown

principal axes of the moment of inertia tensor. Since the zero field Hamiltonian is symmetric with respect to these rotations, the asymmetric wavefunctions may be classified according to their symmetry properties with respect to these operations. Together with the identity operation $\underline{C}_{2a}$, $\underline{C}_{2b}$ and $\underline{C}_{2c}$ form a group which is isomorphous to Klein's four group. It may be shown [54] that the symmetry of the asymmetric top wavefunctions depends only on the eveness or oddness of the quantum numbers K_- and K_+ as is summarized in Table III.4.

If one compares measured rotational frequencies with values calculated from the rotational constants and the expectation values of the squares of the angular momentum operators listed in Table III.3, small discrepancies will become obvious. These are due to minor centrifugal distortions. For instance, the frequency of the $1_{01} \to 1_{10}$ rotational transition ($1_{-1} \to 1_{+1}$ in the J,τ-notation), which should be equal to $(A-C)$ in the rigid rotor approximation, was measured as $11{,}385.908 \pm 0.005$ MHz. From this experimental value the frequency of the $2_{12} \to 2_{21}$ rotational transition shown in Fig. I.4, which should be $3 \cdot (A-C)$ from Table III.3, is predicted as 34,157.724 MHz. The observed value is $34{,}156.990 \pm 0.005$ MHz or 734 kHz, "too low". This corresponds to an increase of 0.0002% in the difference of the rotational constants. Such an increase is not unexpected since a change of the molecular structure due to centrifugal forces usually leads to increased moments of inertia and smaller effective rotational constants for states with higher J- and K-values. The rotational state dependence of the g- and χ-values is expected to have the same order of magnitude, *i.e.*, three orders of magnitude below the experimental uncertainties of the microwave spectroscopical determination described here.

If the exterior magnetic field is switched on, the more complicated effective Hamiltonian [Eq. (I.7) or Eq. (IV.59)] applies:

$$\mathscr{H}_{\rm eff} = \underbrace{\sum_{\gamma} \frac{\hbar^2}{2I^{(\rm n)}_{\gamma\gamma}} \left(1 + \left(\frac{2}{I^{(\rm n)}_{\gamma\gamma}} \left(\frac{\underline{L}_\gamma \underline{L}_\gamma}{\Delta}\right)\right)\right) \underline{J}_\gamma^2 + \sum_{\substack{\gamma,\gamma' \\ \gamma \neq \gamma'}} \frac{\hbar^2}{I^{(\rm n)}_{\gamma\gamma} I^{(\rm n)}_{\gamma'\gamma'}} \left\{ \left(\frac{\underline{L}_\gamma \underline{L}_{\gamma'}}{\Delta}\right) \underline{J}_\gamma \underline{J}_{\gamma'} + \left(\frac{\underline{L}_{\gamma'} \underline{L}_\gamma}{\Delta}\right) \underline{J}_{\gamma'} \underline{J}_\gamma \right\}}_{\mathscr{H}_{\rm rot}}$$

$$\underbrace{- H_z \frac{|e|\hbar}{2M_{\rm p}c} M_{\rm p} \left[\sum_{\gamma,\gamma'} \frac{S_{\gamma\gamma'}}{2I^{(\rm n)}_{\gamma\gamma}} \left\{ \underline{J}_\gamma \underline{\cos\gamma' Z} + \underline{\cos\gamma' Z}\, \underline{J}_\gamma \right\} + \frac{1}{m} \sum_{\gamma,\gamma'} \frac{1}{I^{(\rm n)}_{\gamma\gamma}} \left\{ \left(\frac{\underline{L}_\gamma \underline{L}_{\gamma'}}{\Delta}\right) \underline{J}_\gamma \underline{\cos\gamma' Z} + \left(\frac{\underline{L}_{\gamma'} \underline{L}_\gamma}{\Delta}\right) \underline{\cos\gamma' Z}\, \underline{J}_\gamma \right\} \right]}_{\mathscr{H}_{\rm g}} \tag{III.9}$$

$$\underbrace{+ \frac{1}{2} H_z^2 \frac{e^2}{4mc^2} \left[\sum_{\gamma,\gamma'} \langle 0 | \underline{s_{\gamma\gamma'}} | 0 \rangle \underline{\cos\gamma Z}\, \underline{\cos\gamma' Z} + \frac{2}{m} \sum_{\gamma,\gamma'} \left\{ \left(\frac{\underline{L}_\gamma \underline{L}_{\gamma'}}{\Delta}\right) \underline{\cos\gamma Z}\, \underline{\cos\gamma' Z} + (1 - \delta_{\gamma\gamma'}) \left(\frac{\underline{L}_{\gamma'} \underline{L}_\gamma}{\Delta}\right) \underline{\cos\gamma' Z}\, \underline{\cos\gamma Z} \right\} \right]}_{\mathscr{H}_\chi}$$

$$\underbrace{- E_{\rm TS} \sum_\gamma \langle 0 | \underline{\mu_\gamma} | 0 \rangle \underline{\cos\gamma Y}\,.}_{\mathscr{H}_{\rm TS}}$$

For ethylene oxide this Hamiltonian is considerably simplified due to the C_{2v}-symmetry of the nuclear frame. Using arguments similar to those which have been used to show that there are no nonzero off-diagonal elements in the molecular moment of inertia tensor, it may be shown that the $\underset{\sim}{\boldsymbol{g}}$- and $\underset{\sim}{\boldsymbol{\chi}}$-tensors must be diagonal, too. Thus, for ethylene oxide the effective rotational Hamiltonian is reduced to:

$$\underline{\mathscr{H}}_{\text{eff}} = \frac{\hbar^2}{2I_{aa}} \underline{J}_a^2 + \frac{\hbar^2}{2I_{bb}} \underline{J}_b^2 + \frac{\hbar^2}{2I_{cc}} \underline{J}_c^2 \qquad (\underline{\mathscr{H}}_{\text{rot}})$$

$$- \mu_{\text{N}} H_z \sum_{\gamma} g_{\gamma\gamma}(\underline{J}_\gamma \, \underline{\cos \gamma Z} + \underline{\cos \gamma Z} \, \underline{J}_\gamma)/2 \qquad (\underline{\mathscr{H}}_{\text{g}})$$

$$- \frac{1}{2} H_z^2 \sum_{\gamma} \chi_{\gamma\gamma} \, \underline{\cos^2 \gamma Z} \qquad (\underline{\mathscr{H}}_{\chi})$$

$$- \mu_{\text{b}} E_{\text{TS}} \, \underline{\cos bY} \qquad (\mathscr{H}_{\text{TS}})$$

with $g_{\gamma\gamma}$ and $\chi_{\gamma\gamma}$ from Eqs. (IV.65) and (IV.66) respectively and with the molecular electric dipole moment in the direction of the twofold b axis.

Due to the four group symmetry of the asymmetric top wavefunctions discussed earlier, only the rotational operators $\underline{J}_\gamma^2$, $\underline{J}_\gamma \, \underline{\cos \gamma Z}$, $\underline{\cos \gamma Z} \, \underline{J}_\gamma$, and $\underline{\cos^2 \gamma Z}$, which all transform according to the unit representation A (compare Table III.4), have nonvanishing diagonal elements. $\underline{\cos bY}$, which enters into the expression for the translational Zeeman effect, $\underline{\mathscr{H}}_{\text{TS}}$, changes its sign under $\underline{C}_{2a}$ and $\underline{C}_{2b}$ and thus transforms according to the B_b representation of the four group. It has zero expectation values but nonvanishing off-diagonal elements connecting B_b and A or B_a and B_c rotational states. These lead to small second order corrections as will be discussed later in this section.

After the neglect of the translational Zeeman effect, $\underline{\mathscr{H}}_{\text{TS}}$, the remaining Hamiltonian, which commutes with the space fixed Z-component of the angular momentum operator, is diagonal in M (compare too Table 2 in Appendix II). All expectation values may be factored into M and J dependent algebraic expressions and a term which depends on the molecular g- or χ-values and on the expectation values of the squares of the angular momentum components, the latter having been determined already in the course of the determination of the asymmetric top energy levels (see Table III.3). We will first treat the g-tensor contributions.

From the close relationship between the direction cosine matrix elements diagonal in J, $\langle J,K,M|\underline{\cos \gamma Z}|J,K,M\rangle$, and the matrix elements of the components of the angular momentum, $\langle J,K,M|\underline{J}_\gamma|J,K',M\rangle$, we have:

$$\langle J,K,M|\underline{\cos \gamma Z}|J,K',M\rangle = \frac{M}{J(J+1)} \langle J,K\|\underline{\cos \gamma Z}\|J',K\rangle$$

$$= \frac{M}{J(J+1)} \langle J,K,M|\underline{J}_\gamma|J,K',M\rangle$$

(see Table 2 in Appendix II). This relation directly translates into the asymmetric top basis since the unitary transformations $\langle K|\underline{U}_J|\tau\rangle$ are diagonal in M and act only on the K-dependent part of the matrix elements:

$$\langle J,\tau,M|\underline{\cos \gamma Z}|J,\tau',M\rangle = \frac{M}{J(J+1)} \langle J,\tau,M|\underline{J}_\gamma|J,\tau',M\rangle .$$

From this relation, we immediately obtain the matrix elements for the g-tensor contribution as:

$$\langle J,\tau,M \,|\underline{\mathscr{H}}_{g}|\, J,\tau,M \rangle = -\mu_{N} H_z \frac{M}{J(J+1)} \sum_{\gamma} g_{\gamma\gamma} \sum_{\tau'} \langle J,\tau,M \,|\underline{J}_{\gamma}|\, J,\tau',M \rangle \cdot \langle J,\tau',M \,|\underline{J}_{\gamma}|\, J,\tau,M \rangle \quad \text{(III.11)}$$

$$= -\mu_{N} H_z \frac{M}{J(J+1)} \sum_{\gamma} g_{\gamma\gamma} \langle J,\tau,M \,|\underline{J}_{\gamma}^{2}|\, J,\tau,M \rangle$$

In a similar, although considerably more tedious way, it is possible to show that the diagonal elements of the susceptibility contribution to the Zeeman energy may be written as:

$$\langle J,\tau,M \,|\underline{\mathscr{H}}_{\chi}|\, J,\tau,M \rangle = -\frac{1}{2} H_z^2 \chi - \frac{H_z^2}{2} \sum_{\gamma} \chi_{\gamma\gamma} \left\langle J,\tau,M \,\middle|\, \cos^2 \gamma Z - \frac{1}{3} \,\middle|\, J,\tau,M \right\rangle$$

$$= -\frac{1}{2} H_z^2 \chi - H_z^2 \frac{3M^2 - J(J+1)}{(2J-1)(2J+3)\,J(J+1)} \cdot \sum_{\gamma} (\chi_{\gamma\gamma} - \chi) \langle J,\tau,M \,|\underline{J}_{\gamma}^{2}|\, J,\tau,M \rangle \quad \text{(III.12)}$$

with $\qquad \chi = (\chi_{aa} + \chi_{bb} + \chi_{cc})/3.$

In Eq. (III.12) the isotropic contribution, $-\frac{1}{2}\chi H_z^2$, has been separated. It leads to a constant field dependent shift of all rotational energy levels and cancels in the energy differences of pure rotational transitions. A proof of Eq. (III.12) which makes use of the fact that $\left(\cos^2 \gamma Z - \frac{1}{3}\right)$ has the same transformation properties [55] under the full rotation group as the spherical harmonic $Y_{2,0}$ or $3Z_{\gamma}^2 - (X_{\gamma}^2 + Y_{\gamma}^2 + Z_{\gamma}^2)$, where X_{γ}, Y_{γ}, and Z_{γ} are the space fixed coordinates of the gyrating unit vector, $\boldsymbol{e}_{\gamma}$, is given in a basic paper by Hüttner and Flygare [56].

Eq. (III.12) may be simplified even further since the sum of the three susceptibility anisotropies, $(\chi_{aa} - \chi)$, $(\chi_{bb} - \chi)$, and $(\chi_{cc} - \chi)$, is zero.

This makes it possible to eliminate, for instance, $(\chi_{cc} - \chi)$ and leads to the final expression for the diagonal elements of the effective rotational Hamiltonian.

$$\langle J,\tau,M \,|\underline{\mathscr{H}}_{\text{eff}}|\, J,\tau,M \rangle = \sum_{\gamma} \frac{\hbar^2}{2I_{\gamma\gamma}} \langle J,\tau,M \,|\underline{J}_{\gamma}^{2}|\, J,\tau,M \rangle \quad \text{(III.13)}$$

$$- \mu_{N} H_z \frac{M}{J(J+1)} \sum_{\gamma} g_{\gamma\gamma} \langle J,\tau,M \,|\underline{J}_{\gamma}^{2}|\, J,\tau,M \rangle$$

$$- \frac{1}{2} \chi H_z^2$$

$$- H_z^2 \frac{3M^2 - J(J+1)}{(2J-1)(2J+3)\,J(J+1)} \{(\chi_{aa} - \chi) \langle |\underline{J}_{a}^{2} - \underline{J}_{c}^{2}| \rangle + (\chi_{bb} - \chi) \langle |\underline{J}_{b}^{2} - \underline{J}_{c}^{2}| \rangle \}.$$

This expression also holds in the more general case of a completely asymmetric molecule where the off diagonal elements in the $\underset{\sim}{g}$- and $\underset{\sim}{\chi}$-tensors are not zero, since these elements are connected with rotational operators which do not transform according to the unit representation A under the four group ($\underline{E}$, $\underline{C}_{2a}$, $\underline{C}_{2b}$, $\underline{C}_{2c}$) and therefore only lead to off-diagonal elements in the asymmetric top basis.

We will now turn to the small corrections to the rotational energies which stem from matrix elements which are off-diagonal in the asymmetric top basis. For this discussion we will use second order perturbation theory together with group theoretical arguments. By using order of magnitude considerations, we will be able to show that the off-diagonal elements may be neglected in most cases.

The J and M selection rules for the nonvanishing matrix elements, $\langle J,\tau,M|\mathscr{H}_{\text{eff}}|J',\tau',M'\rangle$, may be obtained from the transformation properties of the wavefunctions and rotational operators under the full rotation group. We will not discuss this point in detail and simply refer to the table of matrix elements given in Appendix II. In general, there may be nonvanishing off-diagonal elements with $J'=J$, $J\pm1$, $J\pm2$ and with $M'=M$, $M\pm1$. The operators $\mathscr{H}_{\text{rot}}$, $\mathscr{H}_{\text{g}}$, and $\mathscr{H}_{\chi}$ all commute with $\underline{J}_z$ and are diagonal in the quantum number M, while the contribution of the translational Zeeman effect $\mathscr{H}_{\text{TS}}$ is off-diagonal in M with $M'=M\pm1$.

The K_-K_+-selection rules for the nonvanishing matrix elements may be obtained from the transformation properties of the wavefunctions and the rotational operators under the four group, *i.e.*, under $\underline{E}$, $\underline{C}_{2a}$, $\underline{C}_{2b}$, $\underline{C}_{2c}$ (compare Table III.4). With the rotational operators $\underline{\cos^2 \gamma Z}$ and $(\underline{J}_\gamma\, \underline{\cos \gamma Z} + \underline{\cos \gamma Z}\, \underline{J}_\gamma)$ transforming according to the unit representation, A, the leading and trailing wavefunctions in the matrix element, $\langle J,K_-K_+,M|$ and $|J,K_-K_+,M\rangle$ respectively, must belong to the same species of the four group in order to give an A-species integrand and thus to allow for a nonvanishing matrix element of $\mathscr{H}_{\text{g}}$ or $\mathscr{H}_{\chi}$. In Fig. III.11 the corresponding chessboard pattern of possibly nonzero matrix elements is shown for ethylene oxide with the letter Z denoting matrix elements of the operators $\mathscr{H}_{\text{g}}$ and $\mathscr{H}_{\chi}$. The letter S denotes matrix elements due to the translational Zeeman effect $\mathscr{H}_{\text{TS}}$. Their positions are determined by the transformation properties of the rotational operator $\underline{\cos bZ}$ transforming according to the B_b-species of the four group. Thus, for nonzero matrix elements leading and trailing wavefunctions must transform according to B_a,B_c or B_c,B_a or to A,B_b or B_b,A respectively and the translational Zeeman effect operator, $\mathscr{H}_{\text{TS}}$, will connect states with $K_-K_+ \longleftrightarrow K_-'K_+'$ equal to $eo \longleftrightarrow oe$ or $ee \longleftrightarrow oo$.

For printing purposes, only the reduced Hamiltonian matrix, $\langle J,K_-K_+ ||\mathscr{H}_{\text{eff}}|| J',K_-'K_+'\rangle$, is shown in Fig. III.11, and the reader should keep in mind that the matrix elements indicated by the letter Z are diagonal in M while the matrix elements indicated with the letter S are off diagonal in M with $M'=M\pm1$.

We now turn to the order of magnitude considerations that will eventually lead to the neglect of the off-diagonal matrix elements in most asymmetric top molecules. From second order perturbation theory, the contribution of the off diagonal elements to the rotational energies is given by

$$E^{(2)}_{J,K_-K_+,M} = \sum_{\substack{J',K_-'K_+',M' \\ (J',K_-'K_+')\neq(J,K_-K_+)}} \frac{|\langle J,K_-K_+,M|\mathscr{H}_{\text{g}}+\mathscr{H}_{\chi}+\mathscr{H}_{\text{TS}}|J',K_-'K_+',M'\rangle|^2}{E^{(0)}_{J,K_-K_+} - E^{(0)}_{J',K_-'K_+'}} \,.$$

						z				z		s		z		s	
3_{30}	Bc				z		z		s		s		z		s		289.6
3_{31}	Bb			z		s		z		s		z		s		286.5	
3_{21}	Ba		z				s		z		z		s		266.5		s
3_{22}	A					z		s		z		s		246.8		s	
3_{12}	Bc				z		z		s		s		242.8		s		z
3_{13}	Bb			z		s		z		s		197.8		s		z	
3_{03}	Ba		z				s		z		197.7		s		z		s
2_{20}	A	z		s		z		s		143.7		s		z		s	
2_{21}	Ba		z		s		s		138.2		z		s		z		s
2_{11}	Bb			z		s		128.1		s		z		s		z	
2_{12}	Bc		s		z		104.0		s		s		z		s		z
2_{02}	A	z		s		103.1		s		z		s		z		s	
1_{10}	Bc		s		47.6		z		s				z				z
1_{11}	Bb	s		39.6		s		z		s		z				z	
1_{01}	Ba		36.2		s		s		z		z				z		
0_{00}	A	z		s		z				z							
		A	Ba	Bb	Bc	A	Bc	Bb	Ba	A	Ba	Bb	Bc	A	Ba	Bb	Bc
		0_{00}	1_{01}	1_{11}	1_{10}	2_{02}	2_{12}	2_{11}	2_{21}	2_{20}	3_{03}	3_{13}	3_{12}	3_{22}	3_{21}	3_{31}	3_{30}

Fig. III.11. Pattern of nonvanishing elements of the reduced Hamiltonian matrix $\langle J,K_-K_+ \|\mathscr{H}_{\text{eff}}\| J',K'_-K'_+\rangle$ for ethyleneoxide. The symmetry species of the rotational states under the four group is indicated by the symbols A, B_a, B_b, and B_c (compare Table III.1). s indicates matrix elements due to the translational Zeeman effect, $\mathscr{H}_{\text{TS}}$, which is equivalent to a velocity dependent Stark effect in an apparent electric field $\boldsymbol{E}_{\text{Lorenz}} = \frac{1}{c}(\boldsymbol{V}_0 \times \boldsymbol{H})$. These matrix elements are off-diagonal in the quantum number M, with the selection rule $\Delta M = \pm 1$. z indicates matrix elements due to the first- and second-order rotational Zeeman effect $\mathscr{H}_{\text{g}}$ and $\mathscr{H}_{\chi}$. These matrix elements are diagonal in M. Except for $J = 0$, the diagonal elements of $\mathscr{H}_{\text{g}}$ and $\mathscr{H}_{\chi}$ are not indicated for printing purposes. Rounded values for the rotational energies are given in frequency units (GHz). At field strengths close to 25 kG the magnitude of the Zeeman matrix elements are in the order of MHz

For the order of magnitude consideration we replace the matrix elements of $\mathscr{H}_{\text{g}}$ in the numerator of the perturbation sum by an average value, $\mu_{\text{N}} H_z \bar{g}$, where $\bar{g}$, a "typical g-value", is taken as 0.1 (compare the listing of g-values in Appendix I). Together with $\mu_{\text{N}} = 0.76$ MHz/kGauss and at a field strength of 25 kGauss, this leads to an estimate of 1.9 MHz for the matrix elements of $\langle\, |\mathscr{H}_{\text{g}}|\, \rangle$. The susceptibility contribution is of the same order of magnitude. With rotational energy differences in the denominator on the order of 40 GHz or more (see Fig. III.11), the second order contributions of the off-diagonal elements, $\langle\, |\mathscr{H}_{\text{g}} + \mathscr{H}_{\chi}|\, \rangle$, are seen to be less than 0.1 kHz, which is far below the experimental uncertainties.

In order to show that the translational Zeeman effect is negligible, we take an average translational velocity, $\boldsymbol{V}_0 = \sqrt{kT/M}$, which at $T = -60$ °C leads to $V_0 = 201$ m/sec for ethylene oxide. At 25 kGauss this corresponds to a cross field E_{TS} of 5 V/cm. With $\mu_{\text{b}} = 1.88$ Debye for ethylene oxide, this leads to an estimate of the Stark effect energy, $\langle\, |\mathscr{H}_{\text{TS}}|\, \rangle \cong |\mu_{\text{b}} E_{\text{TS}}| = 4.7$ MHz. We now take the most critical case, the Stark effect perturbation of the 1_{01} and 1_{10} rotational levels respectively, where the rotational energy difference in the denominator is smallest (11.4 GHz), and we get a second order perturbation in the order of 2 kHz which may come into the range of the experimental accuracy of a careful investigation of the Zeeman splittings. However, in the case of ethylene oxide a more detailed analysis in which the actual values of the off-diagonal matrix elements are taken into account shows that their contributions are below 1 kHz for all lines involving J values, $J \leq 4$, and the effect may thus be neglected.

From the above discussion we conclude that off-diagonal elements of $\langle\ |\mathscr{H}_{\text{eff}}|\ \rangle$ may be neglected in the case of ethylene oxide and that the energy expression given in Eq. (III.13) with appropriate g-values and susceptibility anisotropies should make it possible to calculate Zeeman splittings of low J transitions within the experimental accuracies. However, since we have seen that the second order contributions lead to corrections which may reach into the kHz range, we note that one has to check for each molecule whether or not neglection of off diagonal elements is permissible.

Now we return to Eq. (III.13) which, as we have discussed, gives the rotational energy levels within the experimental uncertainties for many asymmetric top molecules including ethylene oxide. Since, according to this equation, the Zeeman perturbation of the levels is linear with respect to the g-values and susceptibility anisotropies, the same must be true for the Zeeman splittings of the rotational transition frequencies. Thus, from each measured Zeeman satellite with a frequency shift $\Delta\nu(H)$ with respect to the zero field frequency:

$$\Delta\nu_{J,K_-K_+,M\to J',K'_-K'_+,M'} = \nu_{J,K_-K_+,M\to J',K'_-K'_+,M'}(H) - \nu_{J,K_-K_+,M\to J',K'_-K'_+,M'}(0)$$

an equation for g_{aa}, g_{bb}, g_{cc}, $\chi_{aa}-\chi$, and $\chi_{bb}-\chi$ is obtained in which the coefficients may be calculated from the measured magnetic field strength and the appropriate $\langle \| J\gamma^2 \| \rangle$ values (see Table III.3). While the rotational quantum numbers J,K_-K_+ and J', $K'_-K'_+$ may be assumed to be known from a previous assignment of the microwave spectrum, the M-quantum numbers must still be assigned. In principle this might be achieved by trial and error; however, their assignment is facilitated by the typical relative intensities within each Zeeman multiplet. These intensities are proportional to the squares of the electric dipole transition matrix elements, $\langle J,K_-K_+,M|\underline{\boldsymbol{\mu}_{\text{el}}\cdot\boldsymbol{E}_{\text{MW}}}|J',K'_-K'_+,M'\rangle$, which may be calculated from the factorized direction cosine matrix elements given in the Appendix. Specializing for ethylene oxide ($\mu_a=\mu_b=0$, $\mu_b=1.880$ D) for linearly polarized microwave radiation with the electric field vector either parallel ($E_X=0$, $E_Y=0$, $E_Z=E_{\text{MW}}\cos\omega_{\text{MW}}t$) or perpendicular with respect to the static magnetic field, $\boldsymbol{H}$ ($E_X=0$, $E_Y=E_{\text{MW}}\cos\omega_{\text{MW}}t$, $E_Z=0$; compare Fig. III.7), the transition matrix elements take the form:

$$\langle J,K_-K_+,M|\underline{\cos bZ}|J',K'_-K_+,M\rangle\, \mu_{\text{el},b}E_{\text{MW}}\,; \quad (\boldsymbol{E}_{\text{MW}} \text{ parallel to } \boldsymbol{H})$$

or

$$\langle J,K_-K_+,M|\underline{\cos bY}|J',K'_-K'_+,M'\rangle\, \mu_{\text{el},b}E_{\text{MW}}\,; \quad (\boldsymbol{E}_{\text{MW}} \text{ perpendicular to } \boldsymbol{H}).$$

Since for an individual rotational transition $J_{K_-K_+} \to J'_{K'_-K'_+}$ the transition matrix elements of the different satellites differ only in their reduced matrix elements $\langle J,M\,||\underline{\cos\gamma Z}||\,J',M'\rangle$ or $\langle J,M\,||\underline{\cos\gamma Y}||\,J',M'\rangle$ respectively, the relative intensities within a multiplet are given by

$|\langle J,M\,||\underline{\cos\gamma Z}||\,J',M\rangle|^2$ (with $J'=J-1$, J, or $J+1$) leading to the $\Delta M=0$ selection rule for parallel fields,

or by

$|\langle J,M\,||\underline{\cos\gamma Y}||\,J',M\pm 1\rangle|^2$ (with $J'=J-1$, J, $J+1$) leading to the $\Delta M=\pm 1$ selection rule for perpendicular fields.

The corresponding relative intensities which are independent of the rotational constants A, B, and C are listed in Table III.5.

In the course of a typical rotational Zeeman effect investigation of an asymmetric top molecule 40 to 100 Zeeman satellites of different rotational transitions are recorded with both $\Delta M=0$ and $\Delta M=\pm 1$ selection rules. According to Eq. (III.13), this corresponds to a set of 40 to 100 linear equations from which the g-values and susceptibility anisotropies are calculated by a least squares procedure. As an illustration, Fig. III.12 shows recordings of the $2_{12}\rightarrow 2_{21}$ rotational transition of ethylene oxide in exterior magnetic fields close to 25 kG. The $+2.259$

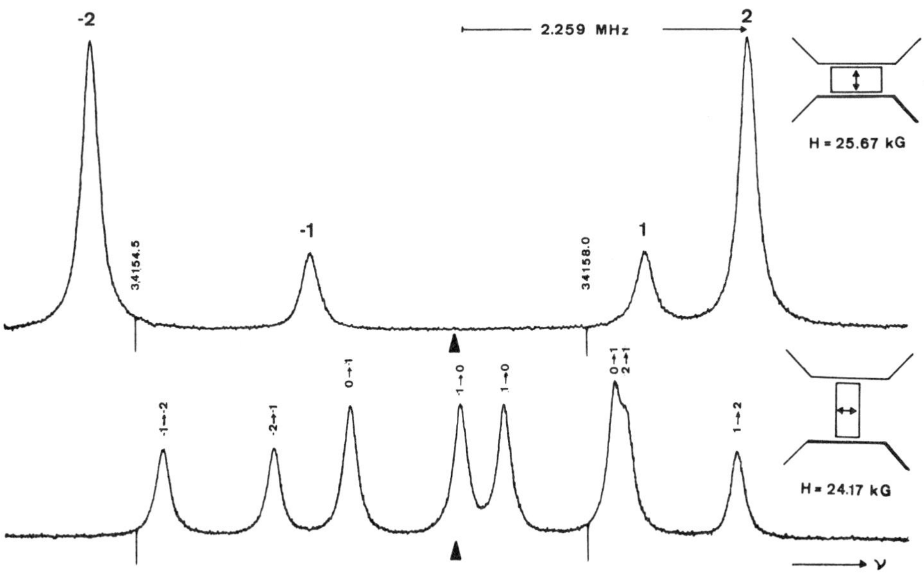

Fig. III.12. Zeeman multiplets of the 2_{12}—2_{21} rotational transition of ethyleneoxide measured with $\Delta M=0$ (upper trace) and $\Delta M=\pm 1$ (lower trace) selection rule. The zero field transition frequency is marked by a dagger. The Stark lobes are pushed out of the frequency range shown in the figure by application of a sufficiently high square wave voltage to the Stark electrode

MHz shift of the $M=2\rightarrow M'=2$ satellite of the $\Delta M=0$ recording, for instance, leads to the equation:

$$-19.5675\,\text{MHz}\,g_{aa}+19.5675\,\text{MHz}\,g_{cc}-0.015722\,\frac{\text{MHz G}^2\text{ mole}}{10^{-6}\text{ erg}}$$
$$\times\,[2\,\chi_{aa}-\chi_{bb}-\chi_{cc}]-0.007862\,\frac{\text{MHz G}^2\text{ mole}}{10^{-6}\text{ erg}}\,[2\,\chi_{bb}-\chi_{cc}-\chi_{aa}]=2.259\,\text{MHz}.$$

The coefficients follow according to Eq. (III.13) from the center gap field strength and the $\langle\|J\gamma^2\|\rangle$ values given in Table III.3. The final set of g-values and susceptibility anisotropies of ethylene oxide is given in the upper part of Table III.6; these results were obtained from a least squares fit to the Zeeman

Table III.5. The relative intensities within a Zeeman multiplet of an asymmetric top molecule which depend solely on the quantum numbers J and M of the lower and the upper rotational states

	$J\rightarrow J-1$	$J\rightarrow J$	$J\rightarrow J+1$
$\boldsymbol{E}_{\text{MW}}$ and $\boldsymbol{H}_{\text{static}}$ parallel, $\Delta M=0$	(J^2-M^2)	M^2	$(J+1)^2-M^2$
$\boldsymbol{E}_{\text{MW}}$ and $\boldsymbol{H}_{\text{static}}$ perpendicular, $M\rightarrow M+1$	$(J-M)\,(J-M-1)$	$J(J+1)-M(M+1)$	$(J+M+1)\,(J+M+2)$
$M\rightarrow M-1$	$(J+M)\,(J+M-1)$	$J(J+1)-M(M-1)$	$(J-M+1)\,(J-M+2)$

splittings of the $0_{00}\rightarrow 1_{10}$, $2_{12}\rightarrow 2_{21}$, $2_{11}\rightarrow 2_{20}$, $3_{21}\rightarrow 3_{30}$, and $4_{31}\rightarrow 4_{40}$ rotational transitions all measured at fields close to 25 kG and with both $\Delta M=0$ and $\Delta M=\pm 1$ selection rules.

For the fitting procedure, the inhomogeneity of the magnetic field was taken into account. For this purpose the field distribution (see Figs. III.5 and III.6) was approximated by a step function, *i.e.*, the absorption volume was divided into 21 sections of equal length where the field strength in each section was approximated by its value in the center of the section. The experimental spectrum was then approximated as a superposition of the 21 spectra corresponding to the different segments and the g-values and susceptibility anisotropies were optimized to give the best least squares fit of the superposition spectra to the observed peak frequencies of the recordings. The effect of the longitudinal inhomogeneity of the magnetic field on the peak frequency of the 2—2 Zeeman satellite of the $2_{12}-2_{21}$ transition is shown in Fig. III.13. If calculated from the superposition spectra, the peak of this satellite is shifted about 7 kHz to lower frequencies as compared to the center field calculation, which is in good agreement with the experimental value.

It should be noted that as long as the sign of the M-values is not determined from an experiment using circularly polarized microwave radiation, only the relative signs of the g-values are determined by the experiment (compare Fig.

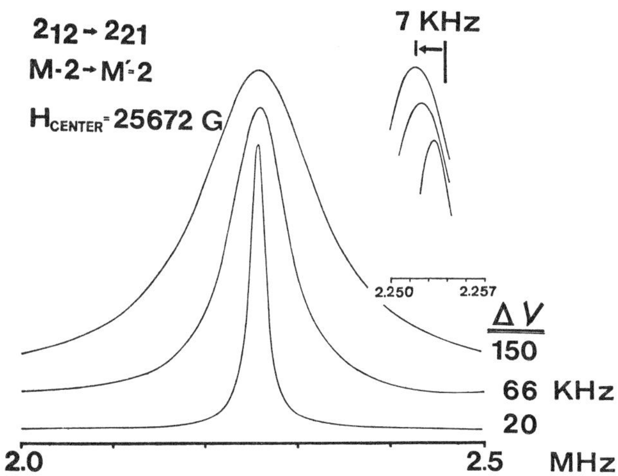

Fig. III.13. The effect of the longitudinal inhomogeneity of the magnetic field is illustrated in this calculated line profile of the $M=2$—$M'=2$ satellite of the 2_{12}—2_{21} rotational transition of ethyleneoxide. As described in the text, the line profile was calculated as a superposition of 21 Lorentzians corresponding to the different sections of the absorption cell. Half-widths of 20, 66, and 150 KHz (full width at half height) were used for the calculations which were based on a relative field distribution corresponding to the one shown in Fig. III.6. The insert gives an enlarged view of the center region. At a half-width of 150 KHz the peak frequency is shifted 6.8 kHz to lower frequencies (downfield) as compared to the position expected from a center field calculation

III.7). This follows from Eq. (III.13) where a simultaneous change of sign of M and the g-values leads to the same energy levels. We will come back to this point in the discussion of the rotational Zeeman effect in the presence of quadrupole nuclei where the sign of the g-values may be determined unambiguously if the sign and magnitude of the nuclear g-values are known. In ethylene oxide the choice of the correct sign may be based upon the quadrupole moments. The set with g_{aa} positive can be excluded because they would lead to unreasonably large molecular quadrupole moments. Furthermore a negative value for $\langle o|\sum c_\varepsilon^2|o\rangle$ would be obtained.

C. The Rotational Zeeman Effect of Symmetric Top Molecules

A symmetric top molecule has at least one rotational symmetry axis C_n with $n \geq 3$. CH_3F and CH_3—C≡C—H may serve as examples where the moment of inertia tensor, the molecular g-tensor, and the tensor of the magnetic susceptibility are simultaneously diagonal with two identical diagonal elements. This may be demonstrated by symmetry arguments similar to those used in the case of ethylene oxide and we will not elaborate on this further. In CH_3F and CH_3—C≡C—H, the a-axis, the axis of the least moment of intertia, is the symmetry axis and we have $I_{bb}=I_{cc}\neq I_{aa}$, $g_{bb}=g_{cc}\neq g_{aa}$, $\chi_{bb}=\chi_{cc}\neq \chi_{aa}$. We also introduce the notation $I_{||}$, $g_{||}$, and $\chi_{||}$ for the tensor elements in the direction of the figure axis and $I_\perp$, $g_\perp$, and $\chi_\perp$ for the components perpendicular to the figure axis. For example, in CH_3F, $g_{aa}=g_{||}$, and $g_{bb}=g_{cc}=g_\perp$, etc.

Table III.6. Molecular Zeeman parameters for ethylene oxide. For the rotational constants and the geometry of the nuclear frame, see Fig. III.8. Quoted uncertainties of g-values and susceptibility anisotropies are standard deviations from the least squares fit described in the text. They do not account for errors introduced through the neglect of vibrations. Uncertainties of derived quantities follow from standard error propagation

Molecular g-values	g_{aa}	-0.09692 ± 0.00004
	g_{bb}	$+0.01848 \pm 0.00005$
	g_{cc}	$+0.03361 \pm 0.00007$
Magnetic susceptibility anisotropies in units of 10^{-6} erg/(G^2 mole)	$2\chi_{aa} - \chi_{bb} - \chi_{cc}$	18.46 ± 0.07
	$2\chi_{bb} - \chi_{cc} - \chi_{aa}$	-0.05 ± 0.10
	$2\chi_{cc} - \chi_{aa} - \chi_{bb}$	-18.41 ± 0.12
Molecular quadrupole moments in units of 10^{-26} esu cm^2	Q_{aa}	2.60 ± 0.05
	Q_{bb}	-3.69 ± 0.07
	Q_{cc}	$+1.10 \pm 0.11$
Second moments of the nuclear charge distribution[1]) in units of 10^{-16} cm^2	$\sum_n Z_n a_n^2$	12.81 ± 0.02
	$\sum_n Z_n b_n^2$	8.99 ± 0.05
	$\sum_n Z_n c_n^2$	3.41 ± 0.02
Paramagnetic susceptibilities in units of 10^{-6} erg/(G^2 mole)	χ^p_{aa}	60.7 ± 0.3
	χ^p_{bb}	67.1 ± 0.2
	χ^p_{cc}	87.4 ± 0.2
Bulk magnetic susceptibility[2]) in units of 10^{-6} erg/(G^2 mole)	$\chi_M = \frac{1}{3}(\chi_{aa} + \chi_{bb} + \chi_{cc})$	-30.7 ± 1.5
Diamagnetic susceptibilities in units of 10^{-6} erg/(G^2 mole)	χ^d_{aa}	-85.3 ± 1.3
	χ^d_{bb}	-97.8 ± 1.5
	χ^d_{cc}	-124.3 ± 1.6
Second moments of the electronic charge distribution in Å^2	$\langle \lvert \sum_\varepsilon a_\varepsilon^2 \rvert \rangle$	16.1 ± 0.6
	$\langle \lvert \sum_\varepsilon b_\varepsilon^2 \rvert \rangle$	13.2 ± 0.6
	$\langle \lvert \sum_\varepsilon c_\varepsilon^2 \rvert \rangle$	6.9 ± 0.6

[1]) From the nuclear coordinates given in Fig. III.8.
[2]) From Lacher, J. R., Pollack, J. W., Park, J. D.: J. Chem. Phys. *20*, 1047 (1957). An uncertainty of 5% was assumed for χ_M in order to account for possible differences between gas and liquid phases.

In a certain respect the analysis of the rotational Zeeman effect may appear simpler in the case of symmetric top molecules as compared to the asymmetric top case, since the angular momentum eigenfunctions $\phi_{J,K,M}(\phi,\theta,\chi)$ with K referring to the component of the angular momentum in direction of the figure axis, are also the eigenfunctions of the zero field Hamiltonian. Thus, the diagonal elements of the effective rotational Hamiltonian may be written down immediately:

$$\begin{aligned}\langle J,K,M|\mathscr{H}_{\text{eff}}|J,K,M\rangle = {} & \frac{\hbar^2}{2}\left(\frac{1}{I_{\|}}-\frac{1}{I_{\perp}}\right)K^2+\frac{\hbar^2}{2I_{\perp}}J(J+1)\\ & -\mu_N H_z \frac{M}{J(J+1)}\left[(g_{\|}-g_{\perp})K^2+g_{\perp}J(J+1)\right]\\ & -\frac{1}{2}H_z^2\chi \\ & -H_z^2\frac{3M^2-J(J+1)}{3(2J-1)(2J+3)}\left[\frac{3K^2}{J(J+1)}-1\right](\chi_{\|}-\chi_{\perp})\end{aligned} \quad \text{(III.14)}$$

[Compare Eq. (III.13) with $g_{aa}=g_{\|}$, $\langle\|\underline{J}_a^2\|\rangle = K^2$, $\langle\|\underline{J}_b^2+\underline{J}_c^2\|\rangle = J(J+1)-K^2$, etc.]. This indeed simplifies the analysis in the case of $K=0$ rotational transitions. However, in the case of $K\neq 0$ rotational transitions which must be investigated to obtain $g_{\|}$, the analysis is considerably complicated by the strong off-diagonal perturbations due to the translational Zeeman effect. Since up until now, $\mathscr{H}_{\text{TS}}$ has not been accounted for in the microwave spectroscopical determinations of $g_{\|}$, we will discuss this point in some detail.

Replacing $(\boldsymbol{V}_0\times\boldsymbol{H})/c$ in $\mathscr{H}_{\text{TS}}$ by the virtual electric field $\boldsymbol{E}_{\text{TS}}$ and choosing the space fixed coordinate system so that its unit vector $\boldsymbol{e}_Y$ points in the direction of $\boldsymbol{E}_{\text{TL}}$ in order to avoid complex numbers in the subsequent numerical treatment, the nonvanishing matrix elements of $\mathscr{H}_{\text{TS}}$ which are diagonal in the rotational quantum numbers J and K are:

$$\langle J,K,M|\mathscr{H}_{\text{TS}}|J,K,M\pm 1\rangle = -\mu_{\text{el}}E_{\text{TS}}\frac{K}{2J(J+1)}\sqrt{J(J+1)-M(M\pm 1)} \quad \text{(III.15)}$$

(Compare Appendix II and note that $\boldsymbol{\mu}_{\text{el}}$ is aligned along the direction of the figure axis). The matrix elements which are off-diagonal in J and K may be neglected from order of magnitude considerations as discussed earlier in the case of the asymmetric top molecules.

Since at typical thermal velocities around 300 m/sec these matrix elements already reach the same order of magnitude as the diagonal Zeeman terms given in Eq. (III.14), they cause considerable mixing of states and not only shift the M-sublevels but also change the M-selection rules as shown in Figs. III.14 and III.16 for the $\Delta M=0$ arrangement of the spectrometer. Figure III.14 shows calculated Zeeman multiplets of the $J=1\rightarrow J=2$, $|K|=1$ rotational transition of CH_3—C≡C—H for molecules having different translational velocities perpendicular to the magnetic field. In this calculation the M-submatrices of the lower and upper rotational state were diagonalized numerically and the electric dipole transition matrix elements were subjected to the corresponding unitary transformation, *i.e.*, the M-submatrices were treated in a manner similar to the different J-submatrices in the discussion on asymmetric top molecules. Figure III.14 illustrates how the selection rules change from $\Delta M=0$ for molecules having low perpendicular velocities to $\Delta M=\pm 1$ for molecules moving with high perpendicular velocities. This change of selection rules may be understood qualitatively from the fact that at high perpendicular velocities, $\boldsymbol{E}_{\text{TS}}$ rather than $\boldsymbol{H}$ determines the axis of quantization. Since $\boldsymbol{E}_{\text{TS}}$ is perpendicular to the polarization of the incident microwave radiation in the spectrometer configuration shown, the change in

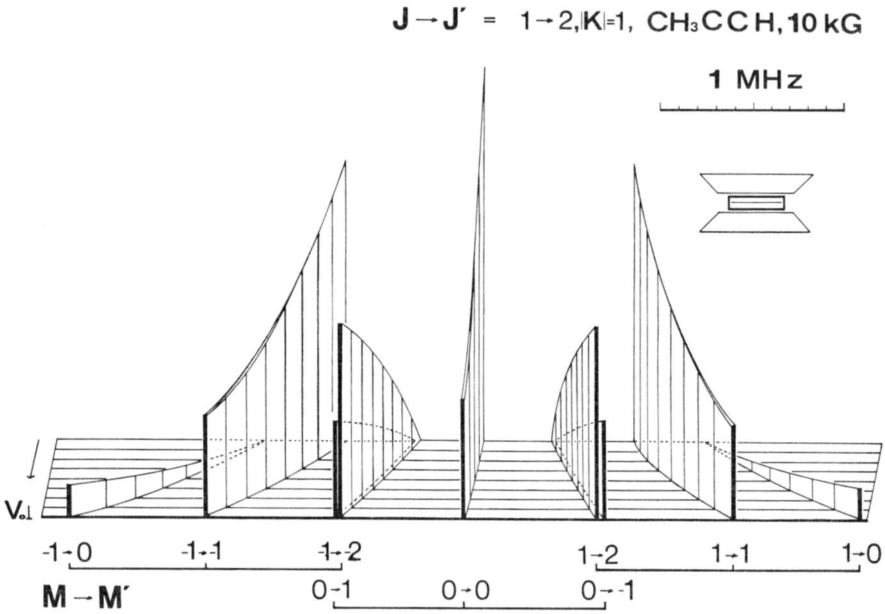

Fig. III.14. The effect of the translational Zeeman effect on the absorption spectra of symmetric top molecules moving at different velocities perpendicular to the exterior magnetic field is shown for the $J = 1 \rightarrow J = 2$, $K = 1$ rotational transition of methylacethylene ($\boldsymbol{V}_{0\perp} = 0$, 100, 200, ..., 800 m/sec). With increasing velocity the aligning force of the Lorentz cross field $\boldsymbol{E}_{\mathrm{TS}} = \frac{1}{c}\,(\boldsymbol{V}_0 \times \boldsymbol{H})$ acting on the molecular electric dipole moment begins to compete with the magnetic field acting on the rotational magnetic moment and the selection rules change from $\Delta M = 0$ to $\Delta M = \pm 1$ for the cell configuration shown here

selection rule follows from the arguments used in Chapter III. In this context we have continued to use M for labeling the energy levels also for nonzero $V_{0\perp}$ values, although it loses its meaning as a projection quantum number for the angular momentum.

If the different Zeeman multiplets shown in Fig. III.14 are weighted with the appropriate Maxwell-Boltzmann probability $P(V_{0\perp})$:

$$\mathrm{P}(V_{0\perp}) = \frac{M V_{0\perp}}{k\,T}\,\mathrm{e}^{-MV_{0\perp}^2/2kT} \qquad \text{(III.16)}$$

by integration from $V_{0\perp} = 0$ to $V_{0\perp} = \infty$, the averaged Zeeman pattern shown in Fig. III.15 is obtained.

Since methylacetylene has a comparatively small electric dipole moment, the change in the $\Delta M = \pm 1$ selection rule occurs at comparatively high perpendicular velocities with correspondingly small Boltzmann populations. Thus, the most intense peaks of the velocity averaged spectrum still arise from the ex-

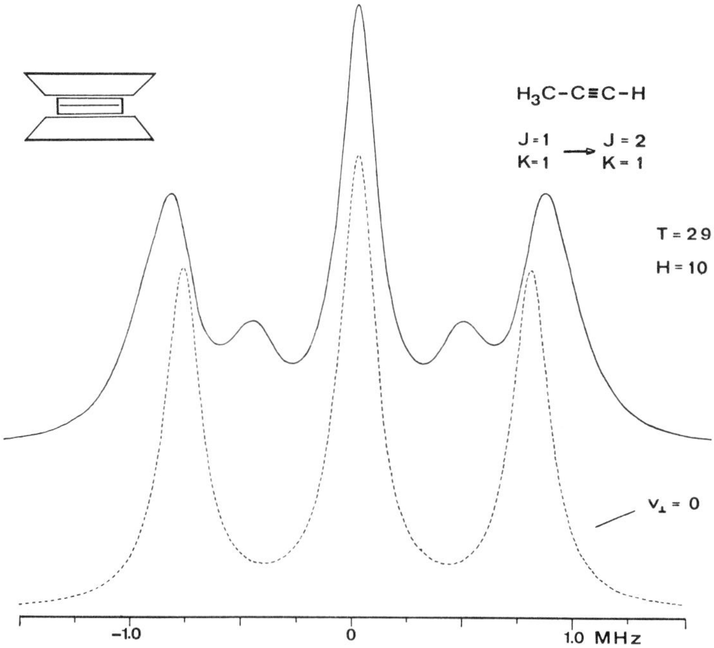

Fig. III.15. Velocity averaged Zeeman multiplet of the $J=1$—$J=2$, $K=1$ rotational transition of $CH_3C{\equiv}C$—H at $H=10$ kG and $T=20$ °C. The line profile was calculated as a superposition of 10 multiplets such as shown in Fig. III.13, with $v_\perp = 100, 200, \ldots, 1100$ m/sec. Lorentzian line shapes with a typical half-width of 300 kHz were assumed for the satellites and each multiplet was weighted with its appropriate Boltzmann probability. For comparison the corresponding $\Delta M=0$ multiplet, in the absence of the translational Zeeman effect, $\mathscr{H}_{TS}$, is shown by the dotted curve

pected selection rules in the absence of the translational Zeeman effect. In lighter molecules with bigger electric dipole moments, this is no more the case as is demonstrated in Fig. III.16 using methylfluoride ($\mu_{el}=1.847$ Debye) as an example. In this case, the overall appearance of the Zeeman multiplets may be influenced considerably by the translational Zeeman term and the observed spectral pattern corresponds more to a Stark-effect pattern in an inhomogeneous electric field centered around $E_{TS,max}=(H/c)\sqrt{kT/M}$ with perturbations due to the $\boldsymbol{g}$- and $\boldsymbol{\chi}$-tensor contributions. Thus, we conclude that the $g_{||}$ values obtained from rotational Zeeman effect studies by microwave spectroscopy should be redetermined after inclusion of the translational Zeeman effect contribution into the analysis. The changes in $g_{||}$ may be well out of the quoted experimental uncertainties. Using published data, L. Engelbrecht [57] has recalculated $g_{||}$ for methylfluoride and obtained a value of $g_{||}=+0.245$ (as compared to $+0.265 \pm 0.008$ as reported by C. L. Norris *et al.*[58]). Since the temperature of the absorption cell has to be assumed, these calculations are only preliminary but they do give an idea of the order of magnitude of the corrections to the $g_{||}$ values which should be expected.

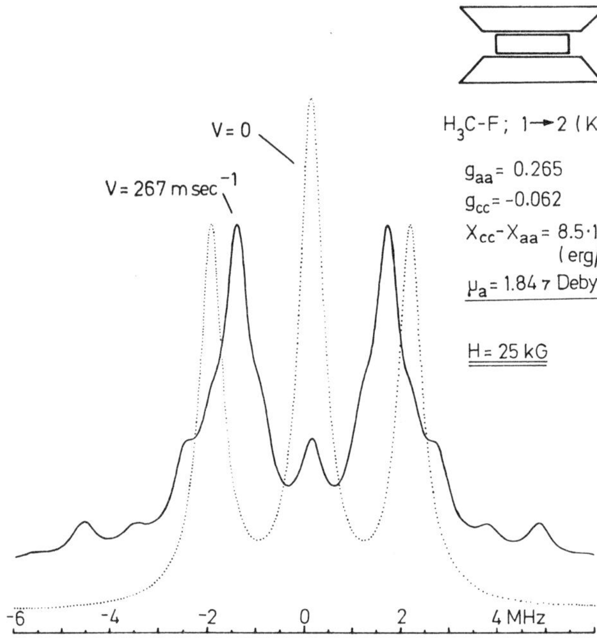

Fig. III.16. In light symmetric top molecules with reasonably large electric dipole moments such as for instance methylfluoride the change of the absorption spectrum due to the translational Zeeman effect occurs at comparatively low perpendicular velocities. The spectrum shown here corresponds to the absorption of a group of molecules moving at 267 m/sec (maximum of the Maxwell-Boltzmann probability distribution) perpendicular to the magnetic field. The dotted line gives the spectrum calculated neglecting the translational Zeeman effect. The Lorentz cross field has caused considerable mixing of M_J substates resulting in considerable changes in the selection rules

D. The Rotational Zeeman Effect of Linear Molecules and Zeeman Effects Due to Intramolecular Rotational Motions

A linear molecule may be regarded as limiting case of a prolate symmetric top with $K = 0$ and $I_{aa} = 0$. The diagonal elements of the effective rotational Hamiltonian follow from Eq. (III.14) with $K = 0$.:

$$\begin{aligned}\langle J,M|\mathscr{H}_{\text{eff}}|J,M\rangle &= \frac{\hbar^2}{2I_\perp} J(J+1) - \mu_{\text{N}} H_z M g_\perp \\ &\quad - \frac{1}{2} H_z^2 \chi - \frac{H_z^2}{3}\left[\frac{3M^2 - J(J+1)}{(2J-1)(2J+3)}\right](\chi_\perp - \chi_{\parallel})\end{aligned} \tag{III.17}$$

Considerable work has been done on linear molecules as discussed earlier [1]. In excited bending vibrational states a vibrational Zeeman effect is observable which has been studied by Hüttner and Morgenstern [59] in the $J = 1 \rightarrow 2$ transition in OCS and l-type doubling transitions in HCN, both molecules being in their first excited bending vibrational states. Since these states are very sensitive with respect to small electric fields because of the close degeneracy of the l-type doublets, the translational Zeeman term causes similar complications as in the case of symmetric top molecules discussed in the previous section. Moss and Perry [60] have extended the theoretical treatment of this vibrational Zeeman effect to symmetric top molecules in excited states of degenerate vibrations. Although this effect should be observable, no experimental results have been published up until now.

A different Zeeman effect which is also due to an intramolecular circular motion may be observed in molecules with very low barrier internal rotations of methyl groups such as, for instance, CH_3NO_2[61,57] and CH_3BF_2[57] with sixfold barriers of approximately 6 [62] and 14 [63] cal/mole respectively. For these molecules, an additional g-value for the methyl-top internal rotation could be determined. For the underlying theory, see Refs. [57,61,64].

E. The Rotational Zeeman Effect in Molecules Containing Quadrupolar Nuclei

As mentioned earlier, the neglect of nuclear spin contributions is not possible in the presence of quadrupole nuclei such as, for instance, ^{14}N (nuclear spin $I=1$, nuclear quadrupole moment $Q=0.190\times10^{-34}$ esu cm^2, [65] nuclear magnetic g-value $g_I=0.4036$). For a detailed discussion of the nuclear quadrupole interactions in the absence of exterior fields the reader is referred to the review article by W. Zeil in Vol. 30 (1972) of this series as well as other references [5].

In the following we will limit the discussion to the case where only one quadrupole nucleus is present in the molecule. Ethylenimine will be used as an example in all numerical calculations. As mentioned above, the ^{14}N nucleus has a small positive prolate shaped quadrupole moment. If the potential well at the equilibrium position of the nucleus lacks spherical symmetry, the nucleus will tend to align itself with respect to the molecular frame as depicted in Fig. III.17.

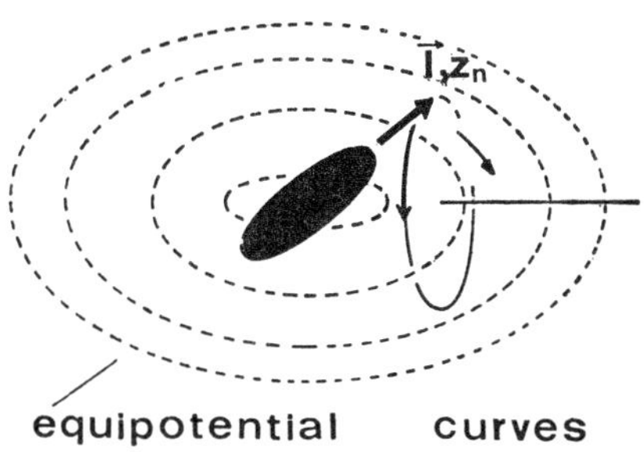

Fig. III.17. The origin of the quadrupole hyperfine interaction is depicted schematically for a nucleus with positive quadrupole moment $Q=\frac{1}{2}\int \varrho_n(3Z_n^2-r_n^2)\,d\tau_n$ (ϱ_n = nuclear charge density, Z_n-axis = symmetry axis = axis of nuclear spin). In the nonspherical intramolecular potential well a torque acts on the nucleus which tends to align its axis to the long axis of the well. This results in a perturbation of the overall rotation which may be observed as a splitting of the rotational absorption lines. If measured in frequency units, typical orientational energies are on the order of MHz

The corresponding potential energy contribution to the effective rotational Hamiltonian may be expressed by [5)]

$$\underline{\mathscr{H}}_{\mathrm{Q}} = -\frac{1}{6}\underline{Q} : \underline{\nabla E} \tag{III.18}$$

where the double dot designates the inner product of the nuclear quadrupole dyadic, Q, and the dyadic of the electric field gradient, ∇E, caused by the molecular charge distribution outside a small sphere around the nucleus in question. If expressed in matrix notation, Eq. (III.18) corresponds to the trace of the matrix product of the quadrupole tensor matrix and the electric field gradient matrix respectively.

In an exterior magnetic field, the torque on the magnetic moment of the spinning nucleus, $\boldsymbol{\mu}_{\mathrm{nucl}} \times \boldsymbol{H}$, competes with the aligning force of the intramolecular potential. The corresponding potential energy (nuclear Zeeman effect) is given by:

$$\underline{\mathscr{H}}_{\mathrm{NZ}} = -\boldsymbol{\mu}_{\mathrm{nucl}} \cdot \boldsymbol{H}\,. \tag{III.19}$$

The small modification due to electronic shielding of the exterior field is negligible in most cases of interest here. As will be demonstrated below, at field strengths close to 30 kG as are generally used in rotational Zeeman effect studies, the magnetic field can uncouple the ^{14}N nuclear spin from the overall rotation. Thus, we will use the uncoupled basis $|J,\tau,M_J,I,M_I\rangle$ in order to set up the matrix of the effective rotational Hamiltonian given by the combination of Eqs. (III.10), (III.18), and (III.19). The subscripts J and I are used to discriminate between projection quantum numbers of the rotational angular momentum, M_J, and the spin angular momentum, M_I.

Within the uncoupled basis the nonvanishing matrix elements of the nuclear Zeeman contribution are:

$$\langle J,\tau,M_J,I,M_I|\underline{\mathscr{H}}_{\mathrm{NZ}}|J,\tau,M_J,I,M_I\rangle = -\mu_{\mathrm{N}}\, g_I\, M_I H_z \tag{III.20}$$

After application of the Wigner-Eckart theorem [5,66)], the quadrupole hyperfine interaction, Eq. (III.18), may be shown to be equivalent to:

$$\begin{aligned}\langle J,\tau,M_J,I,M_I|\underline{\mathscr{H}}_{\mathrm{Q}}|J,\tau,M_J',I,M_I'\rangle &= \{\sum_{\gamma} eQV_{\gamma\gamma}\langle J,\tau,M_J=J\,|\underline{\cos^2\gamma Z}|\,J,\tau,M_J=J\rangle\}\\ &\cdot\frac{1}{2J(2J-1)\,I(2I-1)}\left\langle J,M_J,I,M_I\left\|3\,(\underline{\boldsymbol{J}}\cdot\underline{\boldsymbol{I}})^2+\frac{3}{2}(\underline{\boldsymbol{J}}\cdot\underline{\boldsymbol{I}})-\underline{\boldsymbol{J}}^2\cdot\underline{\boldsymbol{I}}^2\right\|J,M_J',I,M_I'\right\rangle\end{aligned} \tag{III.21}$$

where $V_{\gamma\gamma} = \left\langle 0 \left| \left(\frac{\partial^2 V_{\mathrm{coul}}}{\partial\gamma^2}\right)_{\boldsymbol{r}_Q} \right| 0 \right\rangle$, $\gamma = a,b,c$ are the electronic ground state averaged second derivatives of the Coulomb potential due to the charge distribution outside the nucleus. The matrix elements, which are off diagonal in J and τ, may be neglected except for the case of close degeneracies of rotational states. From the matrix elements of the space fixed components of the angular momentum $\underline{J}_X, \underline{J}_Y, \underline{J}_Z$ (see Appendix II), and the similar matrix elements for the spin operators $\underline{I}_X, \underline{I}_Y, \underline{I}_Z$, the nonvanishing matrix elements corresponding to Eq. (III.21) are given by:

$$\langle J,\tau,M_J,I,M_I|\mathscr{H}_Q|J,\tau,M_J,I,M_I\rangle = \frac{1}{2}\,C\;(J(J+1)-3\,M_J^2)\;(I(I+1)-3\,M_I^2). \tag{III.22}$$

$$\begin{aligned}\langle J,\tau,M_J \pm 1,I,M_I\mp 1|\mathscr{H}_Q|J,\tau,M_J,I,M_I\rangle = \frac{3}{4}\,C\;(2\,M_J\pm 1)\;(2\,M_I\mp 1)\\ ((J\mp M_J)\;(J\pm M_J+1)\;(I\pm M_I)(I\mp M_I+1))^{\frac{1}{2}}.\end{aligned} \tag{III.23}$$

$$\begin{aligned}\langle J,\tau,M_J \pm 2,I,M_I \mp 2\,|\mathscr{H}_Q|J,\tau,M_J,I,M_I\rangle = \frac{3}{4}\,C\;((J\mp M_J)\;(J\mp M_J-1)\\ (J\pm M_J+1)\;(J\pm M_J+2)\;(I\pm M_I)\;(I\pm M_I-1)\;(I\mp M_I+1)\;(I\mp M_I+2))^{\frac{1}{2}}\end{aligned}$$

$$C = \frac{1}{J(2J-1)\,I(2I-1)}\left\{\sum_{\gamma} eQV_{\gamma\gamma}\,\frac{\langle J_\tau\,||J_\gamma^2||\,J_\tau\rangle}{(J+1)\,(2J+3)}\right\} \tag{III.24}$$

$eQV_{\gamma\gamma}/h$, $(\gamma = a,b,c)$, usually abbreviated as $\chi^{\mathrm{N}}_{\gamma\gamma}$, are called the nuclear quadrupole coupling constants. They may be determined from the hyperfine splittings of the rotational transitions in the absence of the magnetic field. For ethylenimine they have the values $\chi^{\mathrm{N}}_{aa} = 0.685$ MHz, $\chi^{\mathrm{N}}_{bb} = 2.170$ MHz, $\chi^{\mathrm{N}}_{cc} = -2.855$ MHz [67]. The $M_J M_I$ sublevels of the 2_{11} rotational state in ethylenimine are shown in Fig. III.18 for a magnetic field of 5 kG. This diagram gives an idea of the relative importance of the different contributions: nuclear Zeeman effect, Eq. (III.20), first and second order rotational Zeeman effect, Eq. (III.13), and quadrupole hyperfine interaction, Eqs. (III.22), (III.23), and (III.24). In the case of ^{14}N nuclei, a field of 5 kG is already sufficient to separate effectively the diagonal matrix elements into three groups corresponding to the parallel, $M_I = 1$, perpendicular, $M_I = 0$, and antiparallel, $M_I = -1$, orientations of the nuclear spin, However, at such low fields the off diagonal matrix elements (which are entirely due to the quadrupole coupling) are only one order below the energy differences between connected diagonal elements and thus cause some mixing of the $M_I M_J$ substates. Although this complicates the analysis, this mixing may be used to determine the sign of the g-values if the nuclear g-value is known. This also is illustrated in Fig. III.18. As long as only the diagonal elements are considered, the energy level schemes are identical for both possible sets of g-values. However, the $M_I M_J$ assignment is different in both cases. Thus, for the two sets of g-values, the off-diagonal elements connect different sublevels causing different shifts, different mixing, and different transition probabilities. Fig. III.19 shows a comparison of the resulting Zeeman multiplets for the 2_{11}—2_{20} rotational transitions for ethylenimine. The experimentally observed spectrum corresponds to the calculated spectrum for the set with g_{aa} negative and thus confirms the original choice of sign which had been based on the fact that unreasonably large values for the molecular electric quadrupole moments [see Eq.(II.1)] would follow from the set with g_{aa} positive.

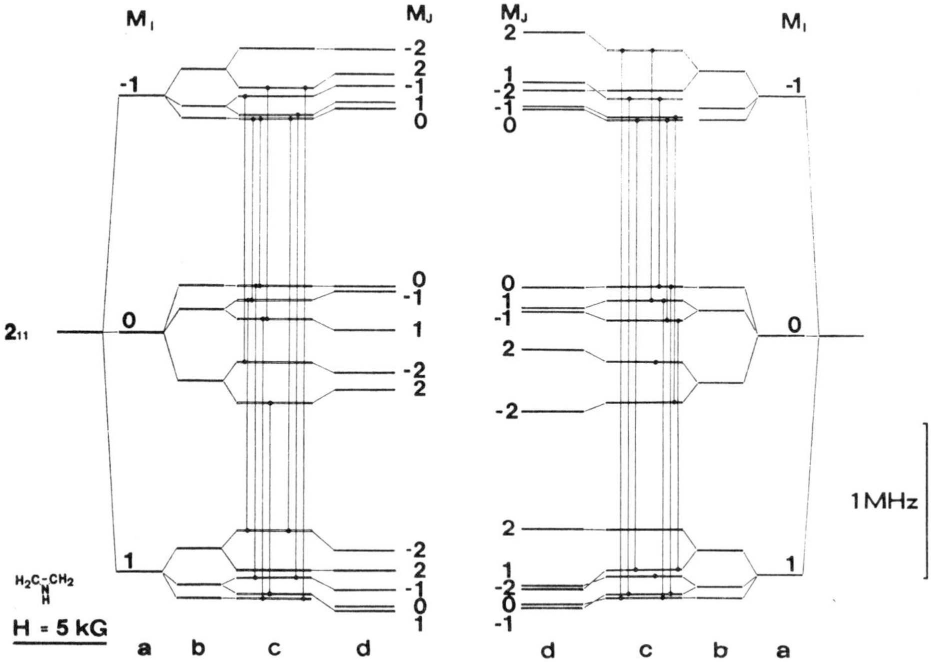

Fig. III.18. Accurate scale drawing of the M_IM_J-substructure of the 2_{11} rotational state of ethylenimine in an exterior magnetic field of 5 kG. The different contributions to the energies are introduced in a step by step fashion: (a) nuclear Zeeman effect [Eq. (III.20)]; (b) diagonal elements of the quadrupole hyperfine interaction [Eq. (III.23)]; (c) rotational Zeeman effect [Eq. (III.13b, c)]; the isotropic shift $-H^2\chi/2$ is neglected. Vertical arrows indicate connecting off-diagonal matrixelements due to quadrupole coupling; (d) final positions of the M_IM_J-sublevels after numerical diagonalization of the 15 × 15 submatrix corresponding to the 2_{11} rotational state. Note that the $(M_I, M_J) = (+1, +2)$ and $(-1, -2)$ states are not affected by the off-diagonal contributions, which makes these states especially suited for accurate evaluations of the magnetic constants.
The left pattern was calculated for the true set of g-values (g_{aa} negative). The right pattern follows for the set with reversed signs. Up until stage (c) both energy patterns are identical. However because of the different assignment of the M_J quantum numbers, the off-diagonal elements couple different levels and thus cause different mixing and different spectra for the two choices of sign. At high magnetic fields this quadrupole mixing is negligible and the two calculated spectra are indistinguishable

Fig. III.20 illustrates how the overall pattern of the Zeeman multiplet changes with increasing strength of the exterior magnetic field. The higher the field, the more the nuclear Zeeman effect separates the three groups of $M_I = 1$, $M_I = 0$, and $M_I = -1$ levels (compare to Fig. III.18) and the less important are the field independent off-diagonal elements which stem from the quadrupole hyperfine interaction. At 30 kG, the uncoupled basis already is a good approximation to the true eigenfunctions and the spectrum may be calculated using the uncoupled selection rules: $\Delta M_I = 0$ and $\Delta M_J = 0$ (or 1 depending on the orientation of the waveguide cell). The matrix elements off-diagonal in M_I and M_J can now be accounted for by second order perturbation theory.

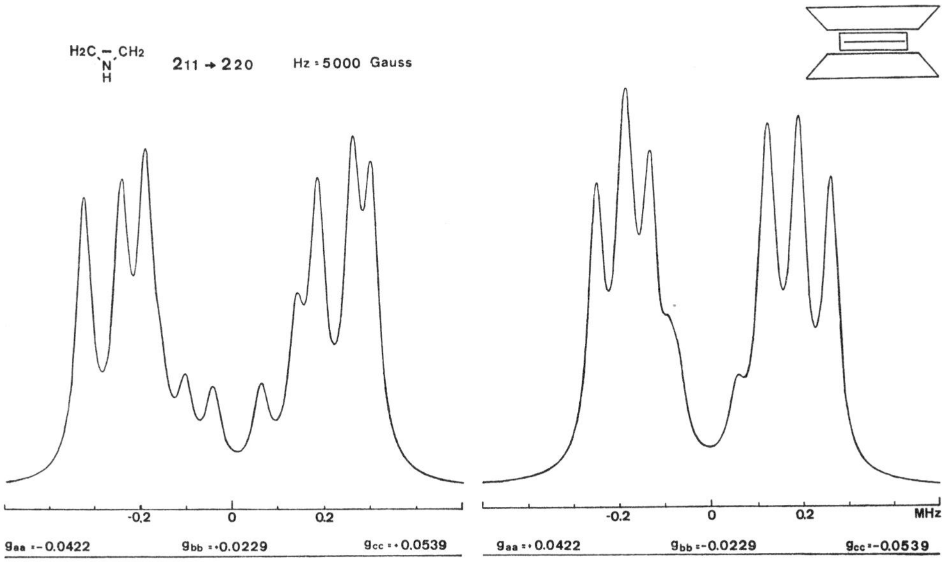

Fig. III.19. Calculated Zeeman spectra of the 2_{11}—2_{20} rotational transition show how the correct set of rotational g-values may be deduced from the Zeeman splittings at intermediate fields where the off-diagonal quadrupole hyperfine matrix elements cause different mixing of states and thus different Zeeman patterns for the two choices. (The observed spectra corresponds to the pattern on the left with g_{aa} negative). The calculated patterns were obtained by numerical diagonalization assuming Lorentzian lineshapes with half-widths of 40 kHz for the satellites

In molecules containing quadrupole nuclei with bigger coupling constants, the uncoupling of spin and overall rotation occurs at higher fields and one has to resort to numerical diagonalization for the analysis of the spectrum. In such a case it may be an advantage to fit the g-values and susceptibility anisotropies to components or frequency combinations which are not affected by the off-diagonal quadrupole interaction [68,69] (compare Fig. III.18). More recently, Suzuki and Guarnieri [70] have obtained very accurate g-values and susceptibility anisotropies for H—O—^{35}Cl, as well as also redetermining the nuclear g-value for the ^{35}Cl nucleus: $g_I = 0{,}5490(14)$, in close agreement with the generally accepted value of $g_I = 0.5479$ [78].

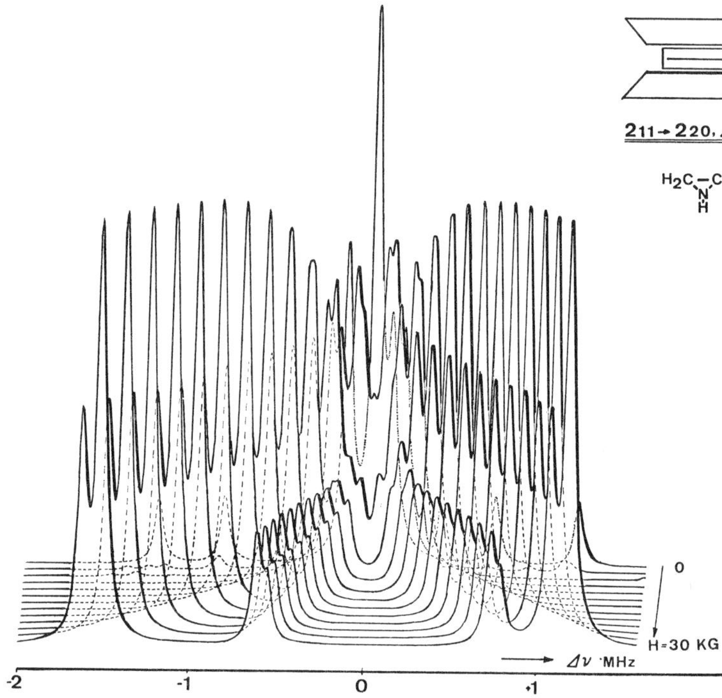

Fig. III.20. The change in the overall appearance of the Zeeman pattern with increasing magnetic field is shown using the $2_{11} \rightarrow 2_{20}$ rotational transition of ethylenimine as an example. At intermediate fields (5 kG) the quadrupole coupling causes sufficient mixing of the $M_I M_J$ sublevels to produce electric dipole transitions between almost all $M_I M_J$ sublevels of the two rotational states (there are several other low intensity satellites which fall out of the frequency range shown in the figure). At high fields the nuclear Zeeman effect more effectively uncouples the different M_I states. The $\Delta M_I = 0$ and $\Delta M_J = 0$ selection rules become predominant and each $M_J \rightarrow M_J$ transition for $M_I = \pm 1$ is accompanied by a satellite of half intensity corresponding to $M_I = 0$. The splitting between the $M_I = \pm 1$ and $M_I = 0$ satellites is essentially due to the differences in the diagonal elements of the quadrupole coupling Hamiltonian (Eq. III.23) and becomes independent of the magnetic field. Furthermore, at high fields the two sets of g-values can no longer be distinguished

IV. Derivation of the Effective Hamiltonian

A. The Classical Hamilton Function

In this Section we will derive the effective rotational Hamiltonian within the rigid nuclear frame approximation, *i.e.*, under the simplifying assumption that the nuclei may be considered as frozen at their equilibrium positions within the molecule. (For a discussion of vibrational effects compare Appendix III.) Furthermore all intramolecular magnetic interactions are neglected since they lead only to comparatively small shifts and splittings of the rotational absorption lines, which in most cases cannot be observed with the standard resolution of a microwave spectrograph.

The starting point of the theoretical treatment is the Lagrangian for an ensemble of charged particles in an exterior magnetic field $\boldsymbol{H}$. Under the neglect of all intramolecular magnetic interactions this Lagrangian, if referred to the space fixed laboratory coordinate system; reduces to [71]

$$\mathscr{L} = \underbrace{\sum_i \frac{m_i}{2} \boldsymbol{V}_i^2}_{\text{kinetic energy}} \quad \underbrace{+ \frac{1}{c} \sum_i Q_i \boldsymbol{V}_i \cdot \boldsymbol{A}_i}_{\text{velocity dependent potential (Zeeman term)}} \quad \underbrace{- \frac{1}{2} \sum_{\substack{i \\ i \neq j}} \sum_j \frac{Q_i Q_j}{|\boldsymbol{R}_i - \boldsymbol{R}_j|}}_{\text{Coulomb potential}} \, .$$

In this equation the sums are over all particles of the molecule (electrons and nuclei), and m_i, $\boldsymbol{v}_i$, Q_i, $\boldsymbol{A}_i$ and $\boldsymbol{R}_i$ are mass, velocity, electric charge, vector potential of the exterior magnetic field, and position vector of particle number i, respectively. c is the velocity of light (the Gaussian system of units will be used throughout).

Assuming the exterior field to be homogeneous in space and independent of time, the vector potential in the second sum of Eq. (IV.1) may be written as

$$\boldsymbol{A}_i = \frac{1}{2} (\boldsymbol{H} \times \boldsymbol{R}_i) \, . \tag{IV.2}$$

This leads to:

$$\mathscr{L} = \sum_i \frac{m_i}{2} \boldsymbol{V}_i^2 + \frac{1}{2c} \sum_i Q_i \boldsymbol{V}_i \cdot (\boldsymbol{H} \times \boldsymbol{R}_i) - \frac{1}{2} \sum_i \sum_j \frac{Q_i Q_j}{|\boldsymbol{R}_i - \boldsymbol{R}_j|} \, . \tag{IV.2'}$$

As the second step in our derivation we will introduce a molecule fixed coordinate system and we will rewrite the Lagrangian using the corresponding generalized coordinates.

Because of the assumption of a rigid nuclear frame, it is most convenient to identify the molecular system with the principal axis system of the nuclear moment of inertia tensor (compare Fig. IV.1).

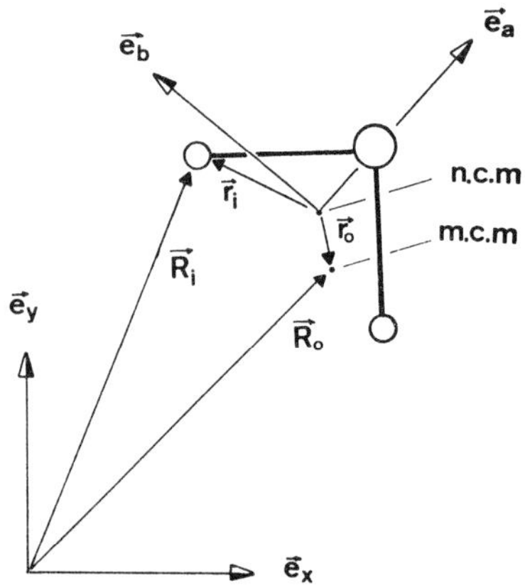

Fig. IV.1. Coordinate systems used in the derivation of the classical Hamiltonian. The difference between the nuclear center of mass (n.c.m.) and the molecular center of mass (m.c.m.) which is typically on the order of 10^{-3} to 10^{-4} Å is vastly exagerated for illustration. $\boldsymbol{e}_X$, $\boldsymbol{e}_Y$, ($\boldsymbol{e}_Z$) are the basis vectors of the space fixed coordinate system. $\boldsymbol{e}_a$, $\boldsymbol{e}_b$, ($\boldsymbol{e}_c$) are the rotating basis vectors of the principle moment of inertia tensor of the rigid nuclear frame

Now let $\boldsymbol{e}_a$, $\boldsymbol{e}_b$, and $\boldsymbol{e}_c$ be the basis (unit) vectors of the molecular coordinate system and $\boldsymbol{e}_X$, $\boldsymbol{e}_Y$ and $\boldsymbol{e}_Z$ the basis vectors (unit vectors) of the space fixed coordinate system (lab system) which leads to (compare Fig. IV.1):

$$\boldsymbol{R}_i - \boldsymbol{R}_0 - \boldsymbol{r}_0 + \boldsymbol{r}_i \tag{IV.3}$$

where

$$\boldsymbol{R}_0 = X_0\boldsymbol{e}_X + Y_0\boldsymbol{e}_Y + Z_0\boldsymbol{e}_Z = \sum_i \frac{m_i \boldsymbol{R}_i}{M} \tag{IV.4}$$

is the vector pointing from the origin of the space fixed coordinate system to the molecular center of mass ($M = \sum_i m_i =$ mass of the molecule),

$$\boldsymbol{r}_i = a_i\boldsymbol{e}_a + b_i\boldsymbol{e}_b + c_i\boldsymbol{e}_c \tag{IV.5}$$

is the vector pointing from the nuclear center of mass to the position of particle number i, and

$$\boldsymbol{r}_0 = a_0\boldsymbol{e}_a + b_0\boldsymbol{e}_b + c_0\boldsymbol{e}_c = \frac{m}{M} \sum_{\varepsilon}^{\text{electrons}} \boldsymbol{r}_\varepsilon \tag{IV.6}$$

is the vector pointing from the nuclear center of mass ($\sum_\nu^{\text{nuclei}} m_\nu \boldsymbol{r}_\nu = 0$) to the molecular center of mass.

Before rewriting the velocities, $\boldsymbol{V}_i$ in Eq. (IV.2′), a remark on the time dependence of the unit vectors defining the molecular coordinate system may be helpful. If the molecule rotates, these basis vectors $\boldsymbol{e}_a$, $\boldsymbol{e}_b$, and $\boldsymbol{e}_c$ will change their orientation in space and their time derivatives are given by

$$\frac{d\boldsymbol{e}_a}{dt} = \boldsymbol{\omega} \times \boldsymbol{e}_a$$
$$\frac{d\boldsymbol{e}_b}{dt} = \boldsymbol{\omega} \times \boldsymbol{e}_b \qquad \text{(IV.7)}$$
$$\frac{d\boldsymbol{e}_c}{dt} = \boldsymbol{\omega} \times \boldsymbol{e}_c ,$$

where $\boldsymbol{\omega}$ is the instantaneous angular velocity of the nuclear frame.

Keeping in mind that the basis vectors of the laboratory system are assumed to be independent of time we combine Eqs. (IV.3) through (IV.6) to give the following relation for the velocities $\boldsymbol{V}_i$:

$$\begin{aligned}\boldsymbol{V}_i = \frac{d\boldsymbol{R}_i}{dt} &= \dot{X}_0\boldsymbol{e}_x + \dot{Y}_0\boldsymbol{e}_y + \dot{Z}_0\boldsymbol{e}_z - \dot{a}_0\boldsymbol{e}_a - \dot{b}_0\boldsymbol{e}_b - \dot{c}_0\boldsymbol{e}_c - a_0(\boldsymbol{\omega} \times \boldsymbol{e}_a) \\ &- b_0(\boldsymbol{\omega} \times \boldsymbol{e}_b) - c_0(\boldsymbol{\omega} \times \boldsymbol{e}_c) + \dot{a}_i\boldsymbol{e}_a + \dot{b}_i\boldsymbol{e}_b + \dot{c}_i\boldsymbol{e}_c \qquad \text{(IV.8)} \\ &+ a_i(\boldsymbol{\omega} \times \boldsymbol{e}_a) + b_i(\boldsymbol{\omega} \times \boldsymbol{e}_b) + c_i(\boldsymbol{\omega} \times \boldsymbol{e}_c) ,\end{aligned}$$

or in more compact form:

$$\boldsymbol{V}_i = \boldsymbol{V}_0 - (\boldsymbol{v}_0 + (\boldsymbol{\omega} \times \boldsymbol{r}_0)) + (\boldsymbol{v}_i + (\boldsymbol{\omega} \times \boldsymbol{r}_i)) . \qquad \text{(IV.9)}$$

In (IV.9) the following abbreviations have been used:

$$\boldsymbol{V}_0 = \dot{X}_0\boldsymbol{e}_X + \dot{Y}_0\boldsymbol{e}_Y + \dot{Z}_0\boldsymbol{e}_Z \qquad \text{(IV.10)}$$

represents the velocity of the molecular center of mass with respect to the laboratory system,

$$\boldsymbol{v}_0 = \dot{a}_0\boldsymbol{e}_a + \dot{b}_0\boldsymbol{e}_b + \dot{c}_0\boldsymbol{e}_c = \frac{m}{M}\sum_{\varepsilon}^{\text{electrons}} (\dot{a}_\varepsilon\boldsymbol{e}_a + \dot{b}_\varepsilon\boldsymbol{e}_b + \dot{c}_\varepsilon\boldsymbol{e}_c) \qquad \text{(IV.11)}$$

is the velocity of the molecular center of mass with respect to the principal axis system of the nuclear moment of inertia tensor [compare Eq. (IV.6)], and

$$\boldsymbol{v}_i = \dot{a}_i\boldsymbol{e}_a + \dot{b}_i\boldsymbol{e}_b + \dot{c}_i\boldsymbol{e}_c \qquad \text{(IV.12)}$$

Is the velocity of the i-th particle with respect to the principle axis system of the nuclear moment of inertia tensor.

In Eqs. (IV.8) and (IV.9) the rigidity of the nuclear frame may be accounted for by the requiring that all of the relative velocities of the nuclei with respect to the nuclear frame be zero, *i.e.*, $\dot{a}_\nu = \dot{b}_\nu = \dot{c}_\nu = 0$ or $\dot{\boldsymbol{v}}_\nu = 0$ for $\nu = 1,2,\ldots,N_n$ ($N_n =$ number of nuclei in the molecule).

Substitution of Eqs. (IV.2), (IV.3) and (IV.9) into the Eq. (IV.1) leads to the Lagrangian (the sums over ε and ν are over electrons and nuclei respectively):

$$\begin{aligned}\mathscr{L} = {} & \frac{M}{2}\boldsymbol{V}_0^2 + \frac{M}{2}(\boldsymbol{v}_0 + (\boldsymbol{\omega}\times\boldsymbol{r}_0))^2 + \frac{m}{2}\sum_{\varepsilon}(\boldsymbol{v}_\varepsilon + (\boldsymbol{\omega}\times\boldsymbol{r}_\varepsilon))^2 \\ & + \sum_{\nu}\frac{m_\nu}{2}(\boldsymbol{\omega}\times\boldsymbol{r}_\nu)^2 - m\sum_{\varepsilon}(\boldsymbol{v}_0 + (\boldsymbol{\omega}\times\boldsymbol{r}_0))\cdot(\boldsymbol{v}_\varepsilon + (\boldsymbol{\omega}\times\boldsymbol{r}_\varepsilon)) \\ & - \frac{|e|}{2c}\sum_{\varepsilon}\{\boldsymbol{V}_0\cdot(\boldsymbol{H}\times\boldsymbol{r}_\varepsilon) - (\boldsymbol{v}_0 + (\boldsymbol{\omega}\times\boldsymbol{r}_0))\cdot(\boldsymbol{H}\times\boldsymbol{r}_\varepsilon) \\ & + (\boldsymbol{v}_\varepsilon + (\boldsymbol{\omega}\times\boldsymbol{r}_\varepsilon))\cdot(\boldsymbol{H}\times(\boldsymbol{R}_0 - \boldsymbol{r}_0)) + (\boldsymbol{v}_\varepsilon + (\boldsymbol{\omega}\times\boldsymbol{r}_\varepsilon))\cdot(\boldsymbol{H}\times\boldsymbol{r}_\varepsilon)\} \\ & + \frac{|e|}{2c}\sum_{\nu} Z_\nu\{\boldsymbol{V}_0\cdot(\boldsymbol{H}\times\boldsymbol{r}_\nu) - (\boldsymbol{v}_0 + (\boldsymbol{\omega}\times\boldsymbol{r}_0))\cdot(\boldsymbol{H}\times\boldsymbol{r}_\nu) \\ & + (\boldsymbol{\omega}\times\boldsymbol{r}_\nu)\cdot(\boldsymbol{H}\times(\boldsymbol{R}_0 - \boldsymbol{r}_0)) + (\boldsymbol{\omega}\times\boldsymbol{r}_\nu)\cdot(\boldsymbol{H}\times\boldsymbol{r}_\nu)\} \\ & - V_{\text{Coulomb}}\end{aligned} \tag{IV.13}$$

$$V_{\text{Coulomb}} = -\sum_{\nu}\sum_{\varepsilon}\frac{Z_\nu e^2}{|\boldsymbol{r}_\varepsilon - \boldsymbol{r}_\nu|} + \frac{1}{2}\sum_{\substack{\varepsilon \\ \varepsilon\neq\varepsilon'}}\sum_{\varepsilon'}\frac{e^2}{|\boldsymbol{r}_\varepsilon - \boldsymbol{r}_{\varepsilon'}|} + \frac{1}{2}\sum_{\substack{\nu \\ \nu\neq\nu'}}\sum_{\nu'}\frac{Z_\nu Z_{\nu'} e^2}{|\boldsymbol{r}_\nu - \boldsymbol{r}_{\nu'}|}.$$

The center of mass conditions for the nuclei, $\sum_\nu m_\nu \boldsymbol{r}_\nu = 0$, and for the molecule as a whole, $\sum_\nu m_\nu(\boldsymbol{r}_\nu - \boldsymbol{r}_0) + \sum_\varepsilon m(\boldsymbol{r}_\varepsilon - \boldsymbol{r}_0) = 0$, as well as the condition of electroneutrality, $\sum_\nu Z_\nu - N_e = 0$, where Z_ν is the atomic number of ν-th nucleus and N_e is the number of electrons in the molecule, have also been used in deriving Eq. (IV.13).

The Lagrangian as given in Eq. (IV.13) shows an explicit $\boldsymbol{R}_0$-dependence which may initially appear rather disturbing, since the magnetic field was assumed to be homogeneous in space and thus independent of the position of the molecule. One should keep in mind, however, that the Lagrangian has no direct physical meaning by itself and that it may be changed by subtracting the total derivative of a scalar function $F(q_1, q_2, \ldots, t)$ with respect to time t without changing the equations of motion *i.e.*

$$\frac{d}{dt}\left(\frac{\partial\mathscr{L}}{\partial\dot{q}_k}\right) = \frac{\partial\mathscr{L}}{\partial q_k}$$

where $q_1, q_2, \ldots$ stand for the generalized coordinates and may be identified with X_0, Y_0, etc., and

$$\frac{d}{dt}\left(\frac{\partial\left(\mathscr{L} - \frac{dF}{dt}\right)}{\partial\dot{q}_k}\right) = \frac{\partial\left(\mathscr{L} - \frac{dF}{dt}\right)}{\partial q_k}$$

with

$$\frac{dF}{dt} = \sum_k \frac{\partial F}{\partial q_k} \dot{q}_k + \frac{\partial F}{\partial t}$$

are equivalent.

It is not too difficult to guess such a scalar function F which indeed removes all explicit $\boldsymbol{R}_0$-dependence from the Lagrangian: $F = \frac{1}{2c} \boldsymbol{\mu}_{el} \cdot (\boldsymbol{H} \times (\boldsymbol{R}_0 - \boldsymbol{r}_0))$ where $\boldsymbol{\mu}_{el} = |e| \{ \sum_\nu Z_\nu \boldsymbol{r}_\nu - \sum_\varepsilon \boldsymbol{r}_\varepsilon \}$ is the instantaneous electric dipole moment of the molecular charge distribution. Details of the calculation will be given in Appendix IV.

After subtraction of

$$\frac{dF}{dt} = \frac{\partial F}{\partial X_0} \dot{X}_0 + \frac{\partial F}{\partial Y_0} \dot{Y}_0 + \frac{\partial F}{\partial Z_0} \dot{Z}_0 + \sum_\varepsilon \frac{\partial F}{\partial a_\varepsilon} \dot{a}_\varepsilon + \frac{\partial F}{\partial b_\varepsilon} \dot{b}_\varepsilon + \frac{\partial F}{\partial c_\varepsilon} \dot{c}_\varepsilon + \frac{\partial F}{\partial \phi_a} \omega_a$$
$$+ \frac{\partial F}{\partial \phi_b} \omega_b + \frac{\partial F}{\partial \phi_c} \omega_c$$

where the use of ϕ_a, ϕ_b, and ϕ_c is permissible as long as only infinitesimal rotations about the principal axes are considered, the final Lagrangian becomes:

$$\begin{aligned}
\mathscr{L} = {} & \frac{M}{2} \boldsymbol{V}_0^2 - \frac{M}{2} (\boldsymbol{v}_0 + (\boldsymbol{\omega} \times \boldsymbol{r}_0))^2 + \frac{m}{2} \sum_\varepsilon (\boldsymbol{v}_\varepsilon + (\boldsymbol{\omega} \times \boldsymbol{r}_\varepsilon))^2 + \sum_\nu \frac{m_\nu}{2} (\boldsymbol{\omega} \times \boldsymbol{r}_\nu)^2 \\
& + \sum_\varepsilon \sum_\nu \frac{Z_\nu e^2}{|\boldsymbol{r}_\nu - \boldsymbol{r}_\varepsilon|} - \frac{1}{2} \sum_{\varepsilon} \sum_{\substack{\varepsilon' \\ \varepsilon' \neq \varepsilon}} \frac{e^2}{|\boldsymbol{r}_\varepsilon - \boldsymbol{r}_{\varepsilon'}|} - \frac{1}{2} \sum_{\nu} \sum_{\substack{\nu' \\ \nu' \neq \nu}} \frac{Z_\nu Z_{\nu'} e^2}{|\boldsymbol{r}_\nu - \boldsymbol{r}_{\nu'}|} \\
& + \frac{|e|}{2c} \sum_\nu Z_\nu (\boldsymbol{\omega} \times \boldsymbol{r}_\nu) \cdot (\boldsymbol{H} \times \boldsymbol{r}_\nu) - \frac{|e|}{2c} \sum_\varepsilon (\boldsymbol{v}_\varepsilon + (\boldsymbol{\omega} \times \boldsymbol{r}_\varepsilon)) \cdot (\boldsymbol{H} \times \boldsymbol{r}_\varepsilon) \\
& + \frac{1}{c} \boldsymbol{\mu}_{el} \cdot ((\boldsymbol{V}_0 - (\boldsymbol{v}_0 + (\boldsymbol{\omega} \times \boldsymbol{r}_0))) \times \boldsymbol{H}) .
\end{aligned} \qquad \text{(IV.14)}$$

This Lagrangian should be thought of as dependent on $3N_e + 6$ generalized coordinates, q_k, and velocities, $\dot{q}_k$, respectively. These are the $3N_e$ coordinates a_ε, b_ε, c_ε which describe the relative positions of the N_e electrons with respect to the nuclear frame; three coordinates X_0, Y_0 and Z_0 which describe the position of the molecular center of mass as referred to the laboratory coordinate system, and three Eulerian angles ϕ, θ, and χ which describe the instantaneous orientation of the molecular coordinate system with respect to the space fixed X-, Y- and Z-axes. There are numerous ways of specifying Eulerian angles. Because of later reference we will follow the choice used by Wilson *et al.*[72] where ϕ and θ are the ordinary polar coordinates of the molecular c-axis $0 \leq \theta \leq \pi$; $0 \leq \phi \leq 2\pi$) and χ is the angle between the nodal line N and the positive b axis as is illustrated in Fig. IV.2. χ is positive for clockwise rotation about the c axis.

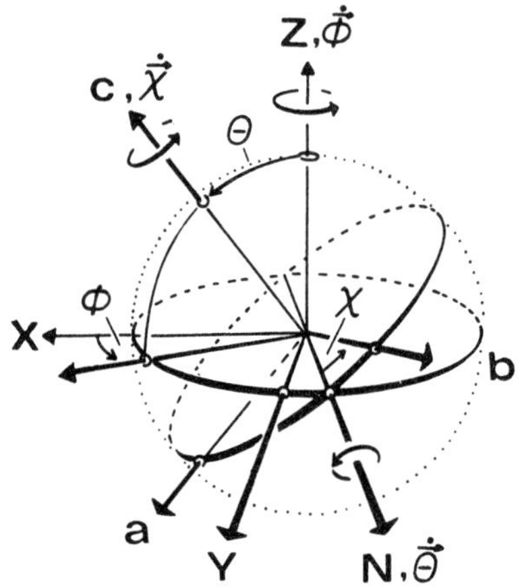

Fig. IV.2. Eulerian angles used to describe the instantaneous orientation of the nuclear frame with respect to the space fixed coordinate system

Using this choice, the direction cosines between the basis vectors of the two coordinate systems and the instantaneous angular velocities about the molecular axes — ω_a, ω_b, ω_c — are related to the Eulerian angles and their time derivatives $\dot{\phi}$, $\dot{\theta}$ and $\dot{\chi}$ through

$$\begin{pmatrix} \cos aX & \cos aY & \cos aZ \\ \cos bX & \cos bY & \cos bZ \\ \cos cX & \cos cY & \cos cZ \end{pmatrix} =$$

$$= \begin{pmatrix} \cos\theta\cos\phi\cos\chi - \sin\phi\sin\chi & \cos\theta\sin\phi\cos\chi + \cos\phi\sin\chi & -\sin\theta\cos\chi \\ -\cos\theta\cos\phi\sin\chi - \sin\phi\cos\chi & -\cos\theta\sin\phi\sin\chi + \cos\phi\cos\chi & \sin\theta\sin\chi \\ \sin\theta\cos\phi & \sin\theta\sin\phi & \cos\theta \end{pmatrix} \tag{IV.15}$$

$$\begin{pmatrix} \omega_a \\ \omega_b \\ \omega_c \end{pmatrix} = \begin{pmatrix} -\sin\theta\cos\chi & \sin\chi & 0 \\ \sin\theta\sin\chi & \cos\chi & 0 \\ \cos\theta & 0 & 1 \end{pmatrix} \cdot \begin{pmatrix} \dot{\phi} \\ \dot{\theta} \\ \dot{\chi} \end{pmatrix} \tag{IV.16}$$

$$\begin{pmatrix} \dot{\phi} \\ \dot{\theta} \\ \dot{\chi} \end{pmatrix} = \begin{pmatrix} -\dfrac{\cos\chi}{\sin\theta} & \dfrac{\sin\chi}{\sin\theta} & 0 \\ \sin\chi & \cos\chi & 0 \\ \cot\theta\cos\chi & -\cot\theta\sin\chi & 1 \end{pmatrix} \cdot \begin{pmatrix} \omega_a \\ \omega_b \\ \omega_c \end{pmatrix} \tag{IV.16'}$$

Is is convenient to use a more compact matrix notation in place of Eq. (IV.14). For this purpose we introduce a column matrix $\boldsymbol{\Omega}$ corresponding to a generalized velocity vector, $\boldsymbol{\Omega}$. In the fourth, fifth and sixth position of $\boldsymbol{\Omega}$ we still use the angular velocities ω_a, ω_b, ω_c and we will continue to use them as long as possible. For later reference the relation between $\boldsymbol{\Omega}$ and the corresponding column matrix of the time derivatives of the generalized coordinates which we will call $\underset{\sim}{\boldsymbol{\Omega}}_q$ is given at the right hand side of the following equation [compare Eq. (IV.16)].

$$\underset{\sim}{\boldsymbol{\Omega}} = \begin{pmatrix} \Omega_1 \\ \Omega_2 \\ \Omega_3 \\ \Omega_4 \\ \Omega_5 \\ \Omega_6 \\ \Omega_7 \\ \Omega_8 \\ \Omega_9 \\ \cdot \\ \cdot \\ \cdot \\ \Omega_{3\mathrm{N_e}+4} \\ \Omega_{3\mathrm{N_e}+5} \\ \Omega_{3\mathrm{N_e}+6} \end{pmatrix} = \begin{pmatrix} \dot{X}_0 \\ \dot{Y}_0 \\ \dot{Z}_0 \\ \omega_a \\ \omega_b \\ \omega_c \\ \dot{a}_1 \\ \dot{b}_1 \\ \dot{c}_1 \\ \cdot \\ \cdot \\ \cdot \\ \dot{a}_{\mathrm{N_e}} \\ \dot{b}_{\mathrm{N_e}} \\ \dot{c}_{\mathrm{N_e}} \end{pmatrix} = \underbrace{\begin{pmatrix} 1 &&&&&&&&&&&& \\ & 1 &&&&&&&&&&& \\ && 1 &&&&&&&&&& \\ &&& -\sin\theta\cos\chi & \sin\chi & 0 &&&&&&& \\ &&& \sin\theta\sin\chi & \cos\chi & 0 &&&&&&& \\ &&& \cos\theta & 0 & 1 &&&&&&& \\ &&&&&& 1 &&&&&& \\ &&&&&&& 1 &&&&& \\ &&&&&&&& 1 &&&& \\ &&&&&&&&& \ddots &&& \\ &&&&&&&&&& 1 && \\ &&&&&&&&&&& 1 & \\ &&&&&&&&&&&& 1 \end{pmatrix}}_{\underset{\sim}{\boldsymbol{T}}_q} \cdot \underbrace{\begin{pmatrix} \dot{X}_0 \\ \dot{Y}_0 \\ \dot{Z}_0 \\ \dot{\phi} \\ \dot{\theta} \\ \dot{\chi} \\ \dot{a}_1 \\ \dot{b}_1 \\ \dot{c}_1 \\ \cdot \\ \cdot \\ \cdot \\ \dot{a}_{\mathrm{N_e}} \\ \dot{b}_{\mathrm{N_e}} \\ \dot{c}_{\mathrm{N_e}} \end{pmatrix}}_{\underset{\sim}{\boldsymbol{\Omega}}_q} \tag{IV.17}$$

With the $\underset{\sim}{\boldsymbol{\Omega}}$ defined in Eq. (IV.17) $\mathscr{L}_{\text{kinetic}}$ [the first four terms of Eq. (IV.14)] which depends quadratically on the velocities, may be written as:

$$\mathscr{L}_{\text{kinetic}} = \frac{1}{2}\underset{\sim}{\boldsymbol{\Omega}}^t \cdot \underset{\sim}{\boldsymbol{\theta}} \cdot \underset{\sim}{\boldsymbol{\Omega}} = \frac{1}{2}\underset{\sim}{\boldsymbol{\Omega}}_q^t \cdot \underset{\sim}{\boldsymbol{T}}_q^t \cdot \underset{\sim}{\boldsymbol{\theta}} \cdot \underset{\sim}{\boldsymbol{T}}_q \cdot \underset{\sim}{\boldsymbol{\Omega}}_q = \frac{1}{2}\underset{\sim}{\boldsymbol{\Omega}}_q^t \cdot \underset{\sim}{\boldsymbol{\theta}}_q \cdot \underset{\sim}{\boldsymbol{\Omega}}_q\,, \tag{IV.18}$$

where super t denotes the transposed matrices.

$\underset{\sim}{\boldsymbol{\theta}}$ in Eq. (IV.18) indicates the symmetric and square $(3N_\mathrm{e}+6)$ matrix which is a generalized moment of inertia tensor. The components of the $\underset{\sim}{\boldsymbol{\theta}}$-matrix may be easily deduced from Eq. (IV.14), with

$$\boldsymbol{\omega} \times \boldsymbol{r}_\varepsilon = \begin{pmatrix} 0 & c_\varepsilon & -b_\varepsilon \\ -c_\varepsilon & 0 & a_\varepsilon \\ b_\varepsilon & -a_\varepsilon & 0 \end{pmatrix} \cdot \begin{pmatrix} \omega_a \\ \omega_b \\ \omega_c \end{pmatrix} \tag{IV.19}$$

and similar relations for the other vector products. The resultant $\underset{\sim}{\boldsymbol{\theta}}$ matrix is given by Eq. (IV.20) (see next page) where for printing purposes only the first two electrons are included explicitly.

The Zeeman contribution [the last three terms of Eq. (IV.14)] depends linearly on the velocities and may be written as

$$\mathscr{L}_{\text{Zeeman}} = \underset{\sim}{\boldsymbol{\Omega}}^t \cdot \underset{\sim}{\boldsymbol{\Gamma}} = \underset{\sim}{\boldsymbol{\Omega}}_q^t \cdot \underset{\sim}{\boldsymbol{T}}_q^t \cdot \underset{\sim}{\boldsymbol{\Gamma}} = \underset{\sim}{\boldsymbol{\Omega}}_q^t \cdot \underset{\sim}{\boldsymbol{\Gamma}}_q, \tag{IV.21}$$

where $\underset{\sim}{\boldsymbol{\Gamma}}$ is a column matrix depending on the molecular charge distribution and on the magnetic field. It is given in Eq. (IV.22).

$$
\boldsymbol{\Theta} = \left(\begin{array}{c|c|c|c|c|c|c|c|c|c}
\begin{matrix} M & & \\ & M & \\ & & M \end{matrix} & & & & & & & & & \\
\hline
 & \begin{matrix} I^{(\mathrm{n})}_{aa} \\ + m \sum_{\varepsilon} (b_{\varepsilon}^2 + c_{\varepsilon}^2) \\ - M(b_0^2 + c_0^2) \end{matrix} & \begin{matrix} - m \sum_{\varepsilon} a_{\varepsilon} b_{\varepsilon} \\ + M a_0 b_0 \end{matrix} & \begin{matrix} - m \sum_{\varepsilon} a_{\varepsilon} c_{\varepsilon} \\ + M a_0 c_0 \end{matrix} & 0 & \begin{matrix} - m c_1 \\ + m c_0 \end{matrix} & \begin{matrix} + m b_1 \\ - m b_0 \end{matrix} & 0 & \begin{matrix} - m c_2 \\ + m c_0 \end{matrix} & \begin{matrix} + m b_2 \\ - m c_0 \end{matrix} \\
\hline
 & \begin{matrix} - m \sum_{\varepsilon} a_{\varepsilon} b_{\varepsilon} \\ + M a_0 b_0 \end{matrix} & \begin{matrix} I^{(\mathrm{n})}_{bb} \\ + m \sum_{\varepsilon} (c_{\varepsilon}^2 + a_{\varepsilon}^2) \\ - M(c_0^2 + a_0^2) \end{matrix} & \begin{matrix} - m \sum_{\varepsilon} b_{\varepsilon} c_{\varepsilon} \\ + M b_0 c_0 \end{matrix} & \begin{matrix} + m c_1 \\ - m c_0 \end{matrix} & 0 & \begin{matrix} - m a_1 \\ + m a_0 \end{matrix} & \begin{matrix} m c_2 \\ - m c_0 \end{matrix} & 0 & \begin{matrix} - m a_2 \\ + m a_0 \end{matrix} \\
\hline
 & \begin{matrix} - m \sum_{\varepsilon} a_{\varepsilon} c_{\varepsilon} \\ + M a_0 c_0 \end{matrix} & \begin{matrix} - m \sum_{\varepsilon} b_{\varepsilon} c_{\varepsilon} \\ + M b_0 c_0 \end{matrix} & \begin{matrix} I^{(\mathrm{n})}_{cc} \\ + m \sum_{\varepsilon} (a_{\varepsilon}^2 + b_{\varepsilon}^2) \\ - M(a_0^2 + b_0^2) \end{matrix} & \begin{matrix} - m b_1 \\ + m b_0 \end{matrix} & \begin{matrix} + m a_1 \\ - m a_0 \end{matrix} & 0 & \begin{matrix} - m b_2 \\ + m c_0 \end{matrix} & \begin{matrix} + m a_2 \\ - m a_0 \end{matrix} & 0 \\
\hline
 & 0 & m c_1 - m c_0 & - m b_1 + m b_0 & m\left(1 - \frac{m}{M}\right) & 0 & 0 & - m^2/M & 0 & 0 \\
\hline
 & - m c_1 + m c_0 & 0 & m a_1 - m a_0 & 0 & m\left(1 - \frac{m}{M}\right) & 0 & 0 & - m^2/M & 0 \\
\hline
 & m b_1 - m b_0 & - m a_1 + m a_0 & 0 & 0 & 0 & m\left(1 - \frac{m}{M}\right) & 0 & 0 & - m^2/M \\
\hline
 & 0 & m c_2 - m c_0 & - m b_2 + m b_0 & - m^2/M & 0 & 0 & m\left(1 - \frac{m}{M}\right) & 0 & 0 \\
\hline
 & - m c_2 + m c_0 & 0 & m a_2 - m a_0 & 0 & - m^2/M & 0 & 0 & m\left(1 - \frac{m}{M}\right) & 0 \\
\hline
 & m b_2 - m b_0 & - m a_2 + m a_0 & 0 & 0 & 0 & - m^2/M & 0 & 0 & m\left(1 - \frac{m}{M}\right)
\end{array}\right) \qquad \text{(IV.20)}
$$

$$\underset{\sim}{\Gamma} = \begin{pmatrix} \Gamma_{X0} \\ \Gamma_{Y0} \\ \Gamma_{Z0} \\ \\ \Gamma_{a} \\ \Gamma_{b} \\ \Gamma_{c} \\ \\ \Gamma_{a1} \\ \Gamma_{b1} \\ \Gamma_{c1} \\ \\ \Gamma_{a2} \\ \Gamma_{b2} \\ \Gamma_{c2} \end{pmatrix} \left| \begin{array}{l} + \frac{1}{c} \begin{pmatrix} 0 & \mu_Z & -\mu_Y \\ -\mu_Z & 0 & \mu_X \\ \mu_Y & -\mu_X & 0 \end{pmatrix} \cdot \begin{pmatrix} H_X \\ H_Y \\ H_Z \end{pmatrix} \\ \\ \left[+ \frac{|c|}{2c} \begin{pmatrix} S_{aa} - s_{aa} & S_{ab} - s_{ab} & S_{ac} - s_{ac} \\ S_{ab} - s_{ab} & S_{bb} - s_{bb} & S_{bc} - s_{bc} \\ S_{ac} - s_{ac} & S_{bc} - s_{bc} & S_{cc} - s_{cc} \end{pmatrix} - \frac{1}{c} \begin{pmatrix} (c_0\mu_c + b_0\mu_b) & -b_0\mu_a & -c_0\mu_a \\ -a_0\mu_b & (a_0\mu_a + c_0\mu_c) & -c_0\mu_b \\ -a_0\mu_c & -b_0\mu_c & (b_0\mu_b + a_0\mu_a) \end{pmatrix} \right] \cdot \begin{pmatrix} H_a \\ H_b \\ H_c \end{pmatrix} \\ \\ \left[- \frac{|e|}{2c} \begin{pmatrix} 0 & c_1 & -b_1 \\ -c_1 & 0 & a_1 \\ b_1 & -a_1 & 0 \end{pmatrix} - \frac{1}{c} \frac{m}{M} \begin{pmatrix} 0 & \mu_c & -\mu_b \\ -\mu_c & 0 & \mu_a \\ \mu_b & -\mu_a & 0 \end{pmatrix} \right] \cdot \begin{pmatrix} H_a \\ H_b \\ H_c \end{pmatrix} \\ \\ \left[- \frac{|e|}{2c} \begin{pmatrix} 0 & c_2 & -b_2 \\ -c_2 & 0 & a_2 \\ b_2 & -a_2 & 0 \end{pmatrix} - \frac{1}{c} \frac{m}{M} \begin{pmatrix} 0 & \mu_c & -\mu_b \\ -\mu_c & 0 & \mu_a \\ \mu_b & -\mu_a & 0 \end{pmatrix} \right] \cdot \begin{pmatrix} H_a \\ H_b \\ H_c \end{pmatrix} \\ \quad \vdots \qquad\qquad\qquad \vdots \qquad\qquad \vdots \end{array} \right| \tag{IV.22}$$

with $S_{aa} = \sum_{\nu} Z_\nu (b_\nu^2 + c_\nu^2)$ $\qquad S_{ab} = -\sum_{\nu} Z_\nu a_\nu b_\nu$ $\qquad S_{ac} = -\sum_{\nu} Z_\nu a_\nu c_\nu$, etc.

$s_{aa} = \sum_{\varepsilon} (b_\varepsilon^2 + c_\varepsilon^2)$ $\qquad s_{ab} = -\sum_{\varepsilon} a_\varepsilon b_\varepsilon$ $\qquad s_{ac} = -\sum_{\varepsilon} a_\varepsilon c_\varepsilon$, etc.

$\mu_X = \mu_a \cos a_X + \mu_b \cos b_X + \mu_c \cos c_X$, etc.

$\mu_a = |e| \sum_{\nu} Z_\nu a_\nu - |e| \sum_{\varepsilon} a_\varepsilon$, etc.

X- and a-component of electric dipole moment, etc.

$H_a = H_X \cos aX + H_Y \cos aY + H_Z \cos aZ$, etc. $\quad$ a-component of magnetic field.

In the third step of our derivation of the effective rotational Hamiltonian we introduce the moments π_i and π_{qi} which correspond to the i-th components of $\underset{\sim}{\Omega}$ and $\underset{\sim}{\Omega}_q$ respectively:

$$\pi_i = \frac{\partial \mathscr{L}}{\partial \Omega_i} \tag{IV.23}$$

$$\pi_{qi} = \frac{\partial \mathscr{L}}{\partial \Omega_{qi}} \tag{IV.23'}$$

Arranging these moments into column vectors $\underset{\sim}{\pi}$ and $\underset{\sim}{\pi}_q$ leads to the more compact matrix notation given by [for printing purposes, Eqs. (IV.24) and (IV.24') show the transposed vectors]:

$$\underset{\sim}{\pi}^t = \left(\frac{\partial \mathscr{L}}{\partial \dot{x}_0}, \frac{\partial \mathscr{L}}{\partial \dot{y}_0}, \frac{\partial \mathscr{L}}{\partial \dot{z}_0}, \frac{\partial \mathscr{L}}{\partial \omega_a}, \frac{\partial \mathscr{L}}{\partial \omega_b}, \frac{\partial \mathscr{L}}{\partial \omega_c}, \ldots \frac{\partial \mathscr{L}}{\partial \dot{a}_\varepsilon}, \frac{\partial \mathscr{L}}{\partial \dot{b}_\varepsilon}, \frac{\partial \mathscr{L}}{\partial \dot{c}_\varepsilon}, \ldots \frac{\partial \mathscr{L}}{\partial \dot{a}_{N_e}}, \frac{\partial \mathscr{L}}{\partial \dot{b}_{N_e}}, \frac{\partial \mathscr{L}}{\partial \dot{c}_{N_e}}\right) \tag{IV.24}$$

$$\underset{\sim}{\pi}_q^t = \left(\frac{\partial \mathscr{L}}{\partial \dot{x}_0}, \frac{\partial \mathscr{L}}{\partial \dot{y}_0}, \frac{\partial \mathscr{L}}{\partial \dot{z}_0}, \frac{\partial \mathscr{L}}{\partial \dot{\phi}}, \frac{\partial \mathscr{L}}{\partial \dot{\theta}}, \frac{\partial \mathscr{L}}{\partial \dot{\chi}}, \ldots \frac{\partial \mathscr{L}}{\partial \dot{a}_\varepsilon}, \frac{\partial \mathscr{L}}{\partial \dot{b}_\varepsilon}, \frac{\partial \mathscr{L}}{\partial \dot{c}_\varepsilon}, \ldots \frac{\partial \mathscr{L}}{\partial \dot{a}_{N_e}}, \frac{\partial \mathscr{L}}{\partial \dot{b}_{N_e}}, \frac{\partial \mathscr{L}}{\partial \dot{c}_{N_e}}\right) \tag{IV.24'}$$

$$\underset{\sim}{\pi} = \underset{\sim}{\theta} \cdot \underset{\sim}{\Omega} + \underset{\sim}{\Gamma} \tag{IV.25}$$

$$\underset{\sim}{\pi}_q = \underset{\sim}{\theta}_q \cdot \underset{\sim}{\Omega}_q + \underset{\sim}{\Gamma}_q \tag{IV.25'}$$

$$\underset{\sim}{\Omega} = \underset{\sim}{\theta}^{-1} \cdot (\underset{\sim}{\pi} - \underset{\sim}{\Gamma}) \tag{IV.26}$$

$$\underset{\sim}{\Omega}_q = \underset{\sim}{\theta}_q^{-1} \cdot (\underset{\sim}{\pi}_q - \underset{\sim}{\Gamma}_q) \tag{IV.26}$$

$\underset{\sim}{\theta}^{-1}$ stands for the inverse of the $\underset{\sim}{\theta}$-matrix, etc. We note that the fourth, fifth and sixth component of $\underset{\sim}{\pi}$ are not conjugate to any generalized coordinates since the corresponding rotational angles ϕ_a, ϕ_b and ϕ_c about the molecular a, b, and c axes do not uniquely define the orientation of the molecular coordinate system in space. However, $\underset{\sim}{\pi}$ may be expressed by the generalized moments as is shown in the following sequence of equations:

$$\begin{aligned}
\underset{\sim}{\pi}_q &= \underset{\sim}{\theta}_q \cdot \underset{\sim}{\Omega}_q + \underset{\sim}{\Gamma}_q \\
&= (\underset{\sim}{T}_q^t \cdot \underset{\sim}{\theta} \cdot \underset{\sim}{T}_q) \cdot \underset{\sim}{\Omega}_q + \underset{\sim}{T}_q^t \cdot \underset{\sim}{\Gamma} \\
&= \underset{\sim}{T}_q^t \cdot \underset{\sim}{\theta} \cdot (\underset{\sim}{T}_q \cdot \underset{\sim}{\Omega}_q) + \underset{\sim}{T}_q^t \cdot \underset{\sim}{\Gamma} \\
&= \underset{\sim}{T}_q^t \cdot (\underset{\sim}{\theta} \cdot \underset{\sim}{\Omega} + \underset{\sim}{\Gamma})
\end{aligned}$$

$$\underset{\sim}{\pi}_q = \underset{\sim}{T}_q^t \cdot \underset{\sim}{\pi} \qquad \text{(IV.27)}$$

$$\underset{\sim}{\pi} = (\underset{\sim}{T}_q^t)^{-1} \cdot \underset{\sim}{\pi}_q \qquad \text{(IV.28)}$$

We will need Eqs. (IV.27) and (IV.28) when we translate the classical Hamilton function into quantum mechanical form [the inverse of the matrix $\underset{\sim}{T}_q^t$ which enters Eq. (IV.28) is easily obtained from Eq. (IV.16) and (IV.16′)].

In the fourth step of our derivation we set up the Hamilton function according to the general relation:

$$\mathscr{H}(q_K,p_K) = \sum_k \dot{q}_K p_K - \mathscr{L}(q_K,\dot{q}_K)$$

Depending on whether the mixed moments π or the pure conjugated moments π_q are used, the following relations are obtained:

$\mathscr{H}$ with mixed moments:

$$\mathscr{H} = \underset{\sim}{\Omega}^t \cdot \underset{\sim}{\pi} - \mathscr{L} \qquad \text{(IV.29)}$$

$$\mathscr{H} = \frac{1}{2} \underset{\sim}{\Omega}^t \cdot \underset{\sim}{\theta} \cdot \underset{\sim}{\Omega} + V_{\text{Coulomb}} \qquad \text{(IV.30)}$$

$\mathscr{H}$ with pure moments:

$$\mathscr{H} = \underset{\sim}{\Omega}_q^t \cdot \underset{\sim}{\pi}_q - \mathscr{L} \qquad \text{(IV.29′)}$$

$$\mathscr{H} = \frac{1}{2} \underset{\sim}{\Omega}_q^t \cdot \underset{\sim}{\theta} \cdot \underset{\sim}{\Omega}_q + V_{\text{Coulomb}} \qquad \text{(IV.30′)}$$

Compare Eqs. (IV.18), (IV.25), and (IV.25′).

$$\mathscr{H} = \frac{1}{2} (\underset{\sim}{\pi}^t - \underset{\sim}{\Gamma}^t) \cdot \underset{\sim}{\theta}^{-1} \cdot (\underset{\sim}{\pi} - \underset{\sim}{\Gamma}) + V_{\text{Coulomb}} \qquad \text{(IV.31)}$$

$$\mathscr{H} = \frac{1}{2} (\underset{\sim}{\pi}_q^t - \underset{\sim}{\Gamma}_q^t) \cdot \underset{\sim}{\theta}_q^{-1} \cdot (\underset{\sim}{\pi}_q - \underset{\sim}{\Gamma}_q) + V_{\text{Coulomb}} \qquad \text{(IV.31′)}$$

[Compare Eqs. (IV.26) and (IV.26′)].

In proceeding from Eqs. (IV.30) and (IV.30′) to Eqs. (IV.31) and (IV.31′) the fact that the inverse of a symmetrix matric ($\underset{\sim}{\theta}$ or $\underset{\sim}{\theta}_q$ in our case) is symmetric itself has been used implicitly.

Since $\underset{\sim}{\boldsymbol{\pi}}$, $\underset{\sim}{\boldsymbol{\Gamma}}$, $\underset{\sim}{\boldsymbol{\pi}}_q$, and $\underset{\sim}{\boldsymbol{\Gamma}}_q$ are all known, the classical Hamiltonian is obtained explicitly by calculating the inverse of the $\underset{\sim}{\boldsymbol{\theta}}$-matrix, given by Eq. (IV.20). [$\underset{\sim}{\boldsymbol{\theta}}_q^{-1}$ follows from $\boldsymbol{\theta}^{-1}$ by matrix multiplication: $\underset{\sim}{\boldsymbol{\theta}}_q^{-1} = \underset{\sim}{\boldsymbol{T}}_q^{-1} \cdot \underset{\sim}{\boldsymbol{\theta}}^{-1} \cdot (\underset{\sim}{\boldsymbol{T}}_q^{t})^{-1}$.] Although this inversion may appear a formidable task, it can be performed easily by breaking it up into two steps, a prediagonalization followed by a then almost trivial inversion.

The transformation that prediagonalizes the $\underset{\sim}{\boldsymbol{\theta}}$-matrix is most easily seen if one neglects the comparatively small $\boldsymbol{r}_0$ contributions giving a $\underset{\sim}{\boldsymbol{\theta}}$ which is already almost diagonal with the only nonvanishing off-diagonal elements in the fourth, fifth and sixth columns and rows respectively. These off-diagonal elements may be removed by applying a transformation according to

$$\tilde{\underset{\sim}{\boldsymbol{\theta}}} = \underset{\sim}{\boldsymbol{T}}^{\mathrm{t}} \cdot \underset{\sim}{\boldsymbol{\theta}} \cdot \underset{\sim}{\boldsymbol{T}} \tag{IV.32}$$

$$\underset{\sim}{\boldsymbol{T}} = \left(\begin{array}{ccc|ccc|ccc|ccc}
1 & & & & & & & & & & & \\
 & 1 & & & & & & & & & & \\
 & & 1 & & & & & & & & & \\
 & & & 1 & & & & & & & & \\
 & & & & 1 & & & & & & & \\
 & & & & & 1 & & & & & & \\
\hline
 & & & 0 & -c_1 & b_1 & 1 & & & & & \\
 & & & c_1 & 0 & -a_1 & & 1 & & & & \\
 & & & -b_1 & a_1 & 0 & & & 1 & & & \\
\hline
 & & & 0 & -c_2 & b_2 & & & & 1 & & \\
 & & & c_2 & 0 & -a_2 & & & & & 1 & \\
 & & & -b_2 & a_2 & 0 & & & & & & 1
\end{array}\right) \tag{IV.33}$$

If $\boldsymbol{r}_0$ contributions are neglected, $\tilde{\underset{\sim}{\boldsymbol{\theta}}}$ is diagonal and the $\tilde{\underset{\sim}{\boldsymbol{\theta}}}^{-1}$ matrix is obtained simply by replacing the diagonal elements of $\tilde{\underset{\sim}{\boldsymbol{\theta}}}$ by their reciprocals. Remembering Eq. (IV.32), the inverse of $\underset{\sim}{\boldsymbol{\theta}}$ itself then follows from $\tilde{\underset{\sim}{\boldsymbol{\theta}}}^{-1}$ by lefthand and righthand multiplication with $\underset{\sim}{\boldsymbol{T}}$ and $\underset{\sim}{\boldsymbol{T}}^t$ respectively:

$$\underset{\sim}{\boldsymbol{\theta}} = (\underset{\sim}{\boldsymbol{T}}^t)^{-1} \cdot \tilde{\underset{\sim}{\boldsymbol{\theta}}} \cdot \underset{\sim}{\boldsymbol{T}}^{-1} \tag{IV.34}$$

$$\underset{\sim}{\boldsymbol{\theta}}^{-1} = \underset{\sim}{\boldsymbol{T}} \cdot \tilde{\underset{\sim}{\boldsymbol{\theta}}}^{-1} \cdot \underset{\sim}{\boldsymbol{T}}^{\mathrm{t}} \tag{IV.35}$$

In the more general case, that is if the $\boldsymbol{r}_0$ contributions are not neglected, the same transformation $\underset{\sim}{\boldsymbol{T}}$ and the same line of thinking may still be used. Although not yet diagonal, $\tilde{\underset{\sim}{\boldsymbol{\theta}}}$ as defined by Eq. (IV.32) with $\underset{\sim}{\boldsymbol{T}}$ from Eq. (IV.33) and $\underset{\sim}{\boldsymbol{\theta}}$ from Eq. (IV.20) has a very simple structure as given in Eq. (IV.36), and its inverse may be obtained from the "Ansatz" shown in Eq. (IV.37) with the unknowns a and b yet to be determined.

$$
\tilde{\underset{\sim}{\theta}} = \left(\begin{array}{ccc|ccc|ccc|ccc|ccc}
M & 0 & 0 & & & & & & & & & & & & \\
0 & M & 0 & & & & & & & & & & & & \\
0 & 0 & M & & & & & & & & & & & & \\
\hline
 & & & I_{aa}^{(\mathrm{n})} & 0 & 0 & & & & & & & & & \\
 & & & 0 & I_{bb}^{(\mathrm{n})} & 0 & & & & & & & & & \\
 & & & 0 & 0 & I_{cc}^{(\mathrm{n})} & & & & & & & & & \\
\hline
 & & & & & & m-\frac{m^2}{M} & 0 & 0 & -\frac{m^2}{M} & 0 & 0 & -\frac{m^2}{M} & 0 & 0 \\
 & & & & & & 0 & m-\frac{m^2}{M} & 0 & 0 & -\frac{m^2}{M} & 0 & 0 & -\frac{m^2}{M} & 0 \\
 & & & & & & 0 & 0 & m-\frac{m^2}{M} & 0 & 0 & -\frac{m^2}{M} & 0 & 0 & -\frac{m^2}{M} \\
\hline
 & & & & & & -\frac{m^2}{M} & 0 & 0 & m-\frac{m^2}{M} & 0 & 0 & -\frac{m^2}{M} & 0 & 0 \\
 & & & & & & 0 & -\frac{m^2}{M} & 0 & 0 & m-\frac{m^2}{M} & 0 & 0 & -\frac{m^2}{M} & 0 \\
 & & & & & & 0 & 0 & -\frac{m^2}{M} & 0 & 0 & m-\frac{m^2}{M} & 0 & 0 & -\frac{m^2}{M} \\
\hline
 & & & & & & -\frac{m^2}{M} & 0 & 0 & -\frac{m^2}{M} & 0 & 0 & m-\frac{m^2}{M} & 0 & 0 \\
 & & & & & & 0 & -\frac{m^2}{M} & 0 & 0 & -\frac{m^2}{M} & 0 & 0 & m-\frac{m^2}{M} & 0 \\
 & & & & & & 0 & 0 & -\frac{m^2}{M} & 0 & 0 & -\frac{m^2}{M} & 0 & 0 & m-\frac{m^2}{M}
\end{array}\right) \tag{IV.36}
$$

$$\tilde{\underset{\sim}{\theta}}^{-1} = \begin{pmatrix} \frac{1}{M} & 0 & 0 & & & & & & & & & & \vdots \\ 0 & \frac{1}{M} & 0 & & & & & & & & & & \vdots \\ 0 & 0 & \frac{1}{M} & & & & & & & & & & \vdots \\ & & & \frac{1}{I_{aa}^{(n)}} & 0 & 0 & & & & & & & \vdots \\ & & & 0 & \frac{1}{I_{bb}^{(n)}} & 0 & & & & & & & \vdots \\ & & & 0 & 0 & \frac{1}{I_{cc}^{(n)}} & & & & & & & \vdots \\ & & & & & & a & 0 & 0 & b & 0 & 0 & \vdots \\ & & & & & & 0 & a & 0 & 0 & b & 0 & \vdots \\ & & & & & & 0 & 0 & a & 0 & 0 & b & \vdots \\ & & & & & & b & 0 & 0 & a & 0 & 0 & \vdots \\ & & & & & & 0 & b & 0 & 0 & a & 0 & \vdots \\ & & & & & & 0 & 0 & b & 0 & 0 & a & \vdots \\ & \cdots & & & \cdots & & & \cdots & & & \cdots & & \end{pmatrix} \qquad \text{(IV.37)}$$

The defining equation for $\tilde{\underset{\sim}{\theta}}^{-1}$:

$$\boldsymbol{E} = \tilde{\underset{\sim}{\theta}} \cdot \tilde{\underset{\sim}{\theta}}^{-1} \qquad (\underset{\sim}{\boldsymbol{E}} = \text{unit matrix}) \qquad \text{(IV.38)}$$

together with Eqs. (IV.36) and (IV.37) then lead to a system of two linear equations [Eqs. (IV.39) and (IV.40)] from which the unknowns a and b may be calculated;

$$\left(1 - \frac{m}{M}\right) ma - (N_{\text{e}} - 1)\, \frac{m^2}{M}\, b \overset{!}{=} 1 \qquad \text{(IV.39)}$$

from the diagonal elements of Eq. (IV.38), and

$$-\frac{m^2}{M}\, a + \left(1 - (N_{\text{e}} - 1)\, \frac{m}{M}\right) m\, b \overset{!}{=} 0 \qquad \text{(IV.40)}$$

from the off-diagonal elements of Eq. (IV.38).

$$a = \frac{1}{m}\left[1 + \frac{m}{M\left(1 - N_{\text{e}}\, \frac{m}{M}\right)}\right] \qquad \text{(IV.41)}$$

$$b = \frac{1}{M\left(1 - N_{\text{e}}\, \frac{m}{M}\right)} \qquad \text{(IV.42)}$$

Neglecting higher order terms in $(m/M)^n$ with $n \geq 2$, these values become:

$$a = \frac{1}{m} + \frac{1}{M} \tag{IV.41'}$$

$$b = \frac{1}{M} \tag{IV.42'}$$

From Eq. (IV.31) together with Eq. (IV.35) the final Hamiltonian function is given by

$$\mathscr{H} = \frac{1}{2} (\underset{\sim}{\pi}^t - \underset{\sim}{\Gamma}^t) \cdot \{\underset{\sim}{T} \cdot \tilde{\underset{\sim}{\theta}}^{-1} \cdot \underset{\sim}{T}^t\} \cdot (\underset{\sim}{\pi} - \underset{\sim}{\Gamma}) + V_{\mathrm{Coulomb}} \tag{IV.43}$$

Looking at this equation it is tempting to change the bracketing and to define a new set of mixed moments as shown in the following equations:

$$\tilde{\underset{\sim}{\pi}} = \underset{\sim}{T}^t \cdot \underset{\sim}{\pi} = \underset{\sim}{T}^t \cdot (\underset{\sim}{T}_q^t)^{-1} \cdot \underset{\sim}{\pi}_q \tag{IV.44}$$

$$\underset{\sim}{\pi}_q = \underset{\sim}{T}_q^t \cdot (\underset{\sim}{T}^t)^{-1} \cdot \tilde{\underset{\sim}{\pi}} \tag{IV.45}$$

$$\tilde{\underset{\sim}{\Gamma}} = \underset{\sim}{T}^t \cdot \underset{\sim}{\Gamma} \tag{IV.46}$$

Written as a function of $\tilde{\underset{\sim}{\pi}}$ and $\tilde{\underset{\sim}{\Gamma}}$ the Hamiltonian then takes the simple form given by

$$\begin{aligned} \mathscr{H} &= \frac{1}{2} \{(\underset{\sim}{\pi}^t - \underset{\sim}{\Gamma}^t) \cdot \underset{\sim}{T}\} \cdot \tilde{\underset{\sim}{\theta}}^{-1} \cdot \{\underset{\sim}{T}^t \cdot (\underset{\sim}{\pi} - \underset{\sim}{\Gamma})\} + V_{\mathrm{Coulomb}} \\ \mathscr{H} &= \frac{1}{2} (\tilde{\underset{\sim}{\pi}}^t - \tilde{\underset{\sim}{\Gamma}}^t) \cdot \tilde{\underset{\sim}{\theta}}^{-1} \cdot (\tilde{\underset{\sim}{\pi}} - \tilde{\underset{\sim}{\Gamma}}) + V_{\mathrm{Coulomb}} \end{aligned} \tag{IV.47}$$

The main advantage of this expression lies in the fact that the matrix elements of $\tilde{\underset{\sim}{\theta}}^{-1}$ are constants. This will be very helpful at the state when the translation into quantum mechanics is performed.

Except for the fourth, fifth and sixth components, $\tilde{\underset{\sim}{\pi}}$ and $\underset{\sim}{\pi}$ are identical with the set of pure conjugate moments, $\underset{\sim}{\pi}_q$, and it may be interesting to note that in Eq. (IV.47) the combinations $(\tilde{\pi}_4 - \tilde{\Gamma}_4)$, $(\tilde{\pi}_5 - \tilde{\Gamma}_5)$, and $(\tilde{\pi}_6 - \tilde{\Gamma}_6)$ have a simple physical meaning. From the definitions:

$$(\tilde{\underset{\sim}{\pi}} - \tilde{\underset{\sim}{\Gamma}}) = \underset{\sim}{T}^t \cdot (\underset{\sim}{\pi} - \underset{\sim}{\Gamma}) = \underset{\sim}{T}^t \cdot \underset{\sim}{\theta} \cdot \underset{\sim}{\Omega}$$

and after multiplications of the matrices $\underset{\sim}{T}^t$ and $\underset{\sim}{\theta}$ [compare Eqs. (IV.33) and (IV.20)], it immediately follows that the following relations hold:

$$\begin{aligned} (\tilde{\pi}_4 - \tilde{\Gamma}_4) &= I_{aa}^{(\mathrm{n})}\, \omega_a \\ (\tilde{\pi}_5 - \tilde{\Gamma}_5) &= I_{bb}^{(\mathrm{n})}\, \omega_b \\ (\tilde{\pi}_6 - \tilde{\Gamma}_6) &= I_{cc}^{(\mathrm{n})}\, \omega_c \end{aligned}$$

Thus, these three components are nothing else than the contributions of the nuclear frame to the angular momentum components about the molecular axes.

From Eq. (IV.47) the final Hamiltonian takes the following form:

$$\mathscr{H} = \frac{1}{2m} \sum_{\varepsilon}^{\text{electrons}} (p_{a\varepsilon}^2 + p_{b\varepsilon}^2 + p_{c\varepsilon}^2) + V_{\text{Coulomb}} \quad \text{(a)}$$

$$+ \frac{1}{2M} \left\{ (\sum_{\varepsilon} p_{a\varepsilon})^2 + (\sum_{\varepsilon} p_{b\varepsilon})^2 + (\sum_{\varepsilon} p_{c\varepsilon})^2 \right\} \quad \text{(b)}$$

$$+ \frac{1}{2} \left\{ \frac{P_a^2}{I_{aa}^{(n)}} + \frac{P_b^2}{I_{bb}^{(n)}} + \frac{P_c^2}{I_{cc}^{(n)}} \right\} \quad \text{(c)}$$

$$+ \frac{1}{2} \left\{ \frac{L_a^2}{I_{aa}^{(n)}} + \frac{L_b^2}{I_{bb}^{(n)}} + \frac{L_c^2}{I_{cc}^{(n)}} \right\} \quad \text{(d)}$$

$$- \frac{1}{2} \left\{ \frac{L_a P_a + P_a L_a}{I_{aa}^{(n)}} + \frac{L_b P_b + P_b L_b}{I_{bb}^{(n)}} + \frac{L_c P_c + P_c L_c}{I_{cc}^{(n)}} \right\} \quad \text{(e)}$$

$$+ \frac{1}{2M} \left\{ P_{X_0}^2 + P_{Y_0}^2 + P_{Z_0}^2 \right\} \quad \text{(f)}$$

$$- \frac{1}{Mc} \boldsymbol{\mu}_{\text{el}} \cdot (\boldsymbol{P}_0 \times \boldsymbol{H}) \quad \text{(g)}$$

$$- \frac{|e|}{4c} \left\{ (P_a, P_b, P_c) \cdot \begin{pmatrix} \frac{1}{I_{aa}^{(n)}} & & \\ & \frac{1}{I_{bb}^{(n)}} & \\ & & \frac{1}{I_{cc}^{(n)}} \end{pmatrix} \cdot \begin{pmatrix} S_{aa} & S_{ab} & S_{ac} \\ S_{ab} & S_{bb} & S_{bc} \\ S_{ac} & S_{bc} & S_{cc} \end{pmatrix} \cdot \begin{pmatrix} H_a \\ H_b \\ H_c \end{pmatrix} \right. \quad \text{(h)}$$

$$+ (H_a, H_b, H_c,) \cdot \begin{pmatrix} S_{aa} & S_{ab} & S_{ac} \\ S_{ab} & S_{bb} & S_{bc} \\ S_{ac} & S_{bc} & S_{cc} \end{pmatrix} \cdot \begin{pmatrix} \frac{1}{I_{aa}^{(n)}} & & \\ & \frac{1}{I_{bb}^{(n)}} & \\ & & \frac{1}{I_{cc}^{(n)}} \end{pmatrix} \cdot \left. \begin{pmatrix} P_a \\ P_b \\ P_c \end{pmatrix} \right\} \quad \text{(i)}$$

$$+ \frac{|e|}{4c} \left\{ (L_a, L_b, L_c) \cdot \begin{pmatrix} \frac{1}{I_{aa}^{(n)}} & & \\ & \frac{1}{I_{bb}^{(n)}} & \\ & & \frac{1}{I_{cc}^{(n)}} \end{pmatrix} \cdot \begin{pmatrix} S_{aa} & S_{ab} & S_{ac} \\ S_{ab} & S_{bb} & S_{bc} \\ S_{ac} & S_{bc} & S_{cc} \end{pmatrix} \cdot \begin{pmatrix} H_a \\ H_b \\ H_c \end{pmatrix} \right. \quad \text{(j)}$$

$$+ (H_a, H_b, H_c) \cdot \begin{pmatrix} S_{aa} & S_{ab} & S_{ac} \\ S_{ab} & S_{bb} & S_{bc} \\ S_{ac} & S_{bc} & S_{cc} \end{pmatrix} \cdot \begin{pmatrix} \frac{1}{I_{aa}^{(n)}} & & \\ & \frac{1}{I_{bb}^{(n)}} & \\ & & \frac{1}{I_{cc}^{(n)}} \end{pmatrix} \cdot \left. \begin{pmatrix} L_a \\ L_b \\ L_c \end{pmatrix} \right\} \quad \text{(k)}$$

$$+\frac{|e|}{2mc}\left\{H_aL_a + H_bL_b + H_cL_c\right\} \tag{l}$$

$$\begin{aligned}+\frac{|e|}{2Mc}\Big\{&[(\sum_\varepsilon b_\varepsilon)(\sum_\varepsilon p_{c\varepsilon}) - (\sum_\varepsilon c_\varepsilon)(\sum_\varepsilon p_{b\varepsilon})]\,H_a\\ &+[(\sum_\varepsilon c_\varepsilon)(\sum_\varepsilon p_{a\varepsilon}) - (\sum_\varepsilon a_\varepsilon)(\sum_\varepsilon p_{c\varepsilon})]\,H_b\\ &+[(\sum_\varepsilon a_\varepsilon)(\sum_\varepsilon p_{b\varepsilon}) - (\sum_\varepsilon b_\varepsilon)(\sum_\varepsilon p_{a\varepsilon})]\,H_c\Big\}\end{aligned} \tag{m}$$

$$\begin{aligned}+\frac{1}{Mc}\Big\{&[\mu_b(\sum_\varepsilon p_{c\varepsilon}) - \mu_c(\sum_\varepsilon p_{b\varepsilon})]\,H_a + [\mu_c(\sum_\varepsilon p_{a\varepsilon}) - \mu_a(\sum_\varepsilon p_{c\varepsilon})]\,H_b\\ &+[\mu_a(\sum_\varepsilon p_{b\varepsilon}) - \mu_b(\sum_\varepsilon p_{a\varepsilon})]\,H_c\Big\}\end{aligned} \tag{n}$$

$$+\frac{e^2}{8mc^2}(H_a,H_b,H_c)\cdot\begin{pmatrix} s_{aa} & s_{ab} & s_{ac}\\ s_{ab} & s_{bb} & s_{bc}\\ s_{ac} & s_{bc} & s_{cc}\end{pmatrix}\cdot\begin{pmatrix}H_a\\ H_b\\ H_c\end{pmatrix} \tag{o}$$

$$+\frac{e^2}{8c^2}(H_a,H_b,H_c)\cdot\begin{bmatrix} S_{aa} & S_{ab} & S_{ac}\\ S_{ab} & S_{bb} & S_{bc}\\ S_{ac} & S_{bc} & S_{cc}\end{bmatrix}\cdot\begin{bmatrix}\frac{1}{I_{aa}^{(\mathrm{n})}} & & \\ & \frac{1}{I_{bb}^{(\mathrm{n})}} & \\ & & \frac{1}{I_{cc}^{(\mathrm{n})}}\end{bmatrix}\cdot\begin{bmatrix} S_{aa} & S_{ab} & S_{ac}\\ S_{ab} & S_{bb} & S_{bc}\\ S_{ac} & S_{bc} & S_{ac}\end{bmatrix}\cdot\begin{bmatrix}H_a\\ H_b\\ H_c\end{bmatrix} \tag{p}$$

$$+\frac{1}{2Mc^2}(H_a,H_b,H_c)\cdot\begin{bmatrix}(\mu_b^2+\mu_c^2) & -\mu_a\mu_b & -\mu_a\mu_c\\ -\mu_a\mu_b & (\mu_c^2+\mu_a^2) & -\mu_b\mu_c\\ -\mu_a\mu_c & -\mu_b\mu_c & (\mu_a^2+\mu_b^2)\end{bmatrix}\cdot\begin{bmatrix}H_a\\ H_b\\ H_c\end{bmatrix} \tag{q}$$

$$\begin{aligned}+\frac{e^2}{8Mc^2}(H_a,H_b,H_c)\cdot&\begin{bmatrix}[(\sum_\varepsilon b_\varepsilon)^2+(\sum_\varepsilon c_\varepsilon)^2] & -(\sum_\varepsilon a_\varepsilon)(\sum_\varepsilon b_\varepsilon) & -(\sum_\varepsilon a_\varepsilon)(\sum_\varepsilon c_\varepsilon)\\ -(\sum_\varepsilon a_\varepsilon)(\sum_\varepsilon b_\varepsilon) & [(\sum_\varepsilon c_\varepsilon)^2+(\sum_\varepsilon a_\varepsilon)^2] & -(\sum_\varepsilon b_\varepsilon)(\sum_\varepsilon c_\varepsilon)\\ -(\sum_\varepsilon a_\varepsilon)(\sum_\varepsilon c_\varepsilon) & -(\sum_\varepsilon b_\varepsilon)(\sum_\varepsilon c_\varepsilon) & [(\sum_\varepsilon a_\varepsilon)^2+(\sum_\varepsilon b_\varepsilon)^2]\end{bmatrix}\\ &\cdot\begin{bmatrix}H_a\\ H_b\\ H_c\end{bmatrix}\end{aligned} \tag{r}$$

$$\begin{aligned}&+\frac{|e|}{4Mc^2}(H_a,H_b,H_c)\,\cdot\\ &\begin{bmatrix}[2(\sum_\varepsilon b_\varepsilon)\mu_b+2(\sum_\varepsilon c_\varepsilon)\mu_c] & -[(\sum_\varepsilon a_\varepsilon)\mu_b+(\sum_\varepsilon b_\varepsilon)\mu_a] & -[(\sum_\varepsilon a_\varepsilon)\mu_c+(\sum_\varepsilon c_\varepsilon)\mu_a]\\ -[(\sum_\varepsilon a_\varepsilon)\mu_b+(\sum_\varepsilon b_\varepsilon)\mu_a] & 2[(\sum_\varepsilon c_\varepsilon)\mu_c+(\sum_\varepsilon a_\varepsilon)\mu_a] & -[(\sum_\varepsilon b_\varepsilon)\mu_c+(\sum_\varepsilon c_\varepsilon)\mu_b]\\ -[(\sum_\varepsilon a_\varepsilon)\mu_c+(\sum_\varepsilon c_\varepsilon)\mu_a] & -[(\sum_\varepsilon b_\varepsilon)\mu_c+(\sum_\varepsilon c_\varepsilon)\mu_b] & 2[(\sum_\varepsilon a_\varepsilon)\mu_a+(\sum_\varepsilon b_\varepsilon)\mu_b]\end{bmatrix}\\ &\cdot\begin{bmatrix}H_a\\ H_b\\ H_c\end{bmatrix}\end{aligned} \tag{s}$$

Only the $\boldsymbol{r}_0$ contributions which are of first order in m/M are explicitly included in this Hamiltonian and the following definitions are used (ε indicates electrons and ν indicates nuclei):

$$P_a = \frac{\partial \mathscr{L}}{\partial \omega_a},\; P_b = \frac{\partial \mathscr{L}}{\partial \omega_b},\; P_c = \frac{\partial \mathscr{L}}{\partial \omega_c};\; p_{a\varepsilon} = \frac{\partial \mathscr{L}}{\partial \dot{a}_\varepsilon},\; p_{b\varepsilon} = \frac{\partial \mathscr{L}}{\partial \dot{b}_\varepsilon},\; p_{c\varepsilon}\, \frac{\partial \mathscr{L}}{\partial \dot{c}_a}$$

$$L_a = \sum_\varepsilon (b_\varepsilon p_{c\varepsilon} - c_\varepsilon p_{b\varepsilon}),\; L_b = \sum_\varepsilon (c_\varepsilon p_{a\varepsilon} - a_\varepsilon p_{c\varepsilon}),\; L_c = \sum_\varepsilon (a_\varepsilon p_{b\varepsilon} - b_\varepsilon p_{a\varepsilon})$$

$$S_{aa} = \sum_\nu Z_\nu (b_\nu^2 + c_\nu^2),\; S_{ab} = - \sum_\nu Z_\nu a_\nu b_\nu,\; \text{and cyclic permutations}$$

$$s_{aa} = \sum_\varepsilon (b_\varepsilon^2 + c_\varepsilon^2),\; s_{ab} = - \sum_\varepsilon a_\varepsilon b_\varepsilon,\; \text{and cyclic permutations}$$

$$\mu_a = |e| \left(\sum_\nu Z_\nu a_\nu - \sum_\varepsilon a_\varepsilon\right)$$

$$\boldsymbol{P}_0 = P_{X_0} \boldsymbol{e}_X + P_{Y_0} \boldsymbol{e}_Y + P_{Z_0} \boldsymbol{e}_Z \text{ with } P_{X_0} = \frac{\partial \mathscr{L}}{\partial x_0},\; \text{etc.}$$

The Hamilton function given in Eq. (IV.48) differs slightly from those reported previously [1, 73, 74)] mainly as far as the contributions in Eqs. (IV.48b, m, n, q, r, and s) are concerned. However, at present these differences are merely of academical interest, since in view of order of magnitude considerations, these $\boldsymbol{r}_0$ contributions have been neglected in all practical calculations. As an example, Eq. (IV.48r), being weighted with $1/M$ rather than $1/m$ is neglected as compared to the similar term in Eq. (IV.48o). Apart from the heavy difference in the weighting factor, mutual cancellation in the sums $(\sum a_\varepsilon)$, etc., will even further decrease the absolute values of the diagonal elements of Eq. (IV.48r) as compared to those of Eq. (IV.48o). Similar arguments lead to the neglect of the other $\boldsymbol{r}_0$ contributions mentioned above.

Since in most practical cases (m/M) is in the order of 10^{-5}, one may expect that the neglected terms would lead to corrections of the theoretical expressions for the g- and χ-values which fall into the 10 to 100 ppm range of the experimental values. With experimental uncertainties typically on the order of 0.1 to 1% of the observed g-values and susceptibility anisotropies, the above neglections therefore appear to be justified at least as far as microwave spectroscopy is concerned. The neglect of vibrations in the theoretical treatment is certainly more serious.

B. Translation into Quantum Mechanics and Derivation of the Effective Rotational Hamiltonian Operator

In the preceeding Section we have derived two equations for the classical Hamiltonian:

$$\mathscr{H} = \frac{1}{2} (\underset{\sim}{\pi}_q - \underset{\sim}{\Gamma}_q) \cdot \underset{\sim}{\theta}_q^{-1} \cdot (\underset{\sim}{\pi}_q - \underset{\sim}{\Gamma}_q) + V_{\text{Coulomb}} \tag{IV.31'}$$

and

$$\mathscr{H} = \frac{1}{2}(\tilde{\pi}^t - \tilde{\Gamma}^t) \cdot \tilde{\theta}^{-1} \cdot (\tilde{\pi} - \tilde{\Gamma}) + V_{\text{Coulomb}} \tag{IV.47}$$

$\tilde{\theta}^{-1}$ is given by Eq. (IV.37) and θ_q^{-1} was given by [compare Eqs. (IV.18) and (IV.35)]:

$$\theta_q^{-1} = T_q^{-1} \cdot \theta^{-1} \cdot (T_q^t)^{-1} = T_q^{-1} \cdot T \cdot \tilde{\theta}^{-1} \cdot T^t \cdot (T_q^t)^{-1} \tag{IV.49}$$

with T from Eq. (IV.33) and T_q from Eq. (IV.17). In Eq. (IV.31′) the Hamiltonian is expressed as a function of the generalized coordinates:

$$(X_0, Y_0, Z_0, \phi, \theta, \chi, a_1, b_1, c_1 \ldots a_\varepsilon, b_\varepsilon, c_\varepsilon, \ldots a_{Ne}, b_{Ne}, c_{Ne})$$

and the conjugate momenta

$$\pi_q^t = \left(\frac{\partial\mathscr{L}}{\partial\dot{X}_0}, \frac{\partial\mathscr{L}}{\partial\dot{Y}_0}, \frac{\partial\mathscr{L}}{\partial\dot{Z}_0}, \frac{\partial\mathscr{L}}{\partial\dot{\phi}}, \frac{\partial\mathscr{L}}{\partial\dot{\theta}}, \frac{\partial\mathscr{L}}{\partial\dot{\chi}}, \frac{\partial\mathscr{L}}{\partial\dot{a}_1}, \frac{\partial\mathscr{L}}{\partial\dot{b}_1}, \frac{\partial\mathscr{L}}{\partial\dot{c}_1}, \ldots, \frac{\partial\mathscr{L}}{\partial\dot{a}_\varepsilon}, \frac{\partial\mathscr{L}}{\partial\dot{b}_\varepsilon}, \frac{\partial\mathscr{L}}{\partial\dot{c}_\varepsilon}, \ldots, \frac{\partial\mathscr{L}}{\partial\dot{a}_{Ne}}, \frac{\partial\mathscr{L}}{\partial\dot{b}_{Ne}}, \frac{\partial\mathscr{L}}{\partial\dot{c}_{Ne}}\right)$$

with the Lagrangian $\mathscr{L}$ as given by Eq. (IV.18). In Eq. (IV.47) the Hamiltonian is expressed with the same generalized coordinates, however, instead of the conjugate momenta π_q, a set of mixed momenta $\tilde{\pi}$ are used which differs from the set of conjugate momenta in the fourth, fifth, and sixth components [compare Eq. (IV.44)]:

$$\tilde{\pi}_4 = P_a - L_a;\ P_a = \frac{\partial\mathscr{L}}{\partial\omega_a} = -\frac{\cos\chi}{\sin\theta} P_\phi + \sin\chi\, P_\theta + \frac{\cos\theta}{\sin\theta}\cos\chi\, P_\chi \tag{IV.50}$$

P_a is the a component of the overall angular momentum and

$$L_a = \sum_\varepsilon^{\text{electrons}} (b_\varepsilon p_{c\varepsilon} - c_\varepsilon p_{b\varepsilon})\,.$$

Continuing, we write

$$\tilde{\pi}_5 = P_b - L_b;\ P_b = \frac{\partial\mathscr{L}}{\partial\omega_b} = \frac{\sin\chi}{\sin\theta} P_\phi + \cos\chi\, P_\theta - \frac{\cos\theta}{\sin\theta}\sin\chi\, P_\chi$$

$$L_b = \sum_\varepsilon^{\text{electrons}} (c_\varepsilon p_{a\varepsilon} - a_\varepsilon p_{c\varepsilon})$$

$$\tilde{\pi}_6 = P_c - L_c;\ P_c = \frac{\partial\mathscr{L}}{\partial\omega_c} = P_\chi$$

$$L_c = \sum_\varepsilon^{\text{electrons}} (a_\varepsilon p_{b\varepsilon} - b_\varepsilon p_{a\varepsilon})$$

In order to obtain the quantum mechanical Hamiltonian operator from the classical Eqs. (IV.31′) or (IV.47) respectively, we will follow closely the treatment of Wilson *et al.* [72] and we will use their symbols g^{ij} for the elements of the $\underset{\sim}{\theta}_q^{-1}$ matrix and G^{mn} for the elements of the $\underset{\sim}{\tilde{\theta}}^{-1}$ matrix. With this notation, the compact matrix Eqs. (IV.31) and (IV.47) take the more explicit form:

$$\mathscr{H} = \frac{1}{2} \sum_{ij} (\pi_{qi} - \Gamma_{qi}) \cdot g^{ij} \cdot (\pi_{qj} - \Gamma_{qj}) + V_{\text{Coulomb}} \tag{IV.31''}$$

and

$$\mathscr{H} = \frac{1}{2} \sum_{mn} (\tilde{\pi}_m - \tilde{\Gamma}_m) \cdot G^{mn} \cdot (\tilde{\pi}_n - \tilde{\Gamma}_n) + V_{\text{Coulomb}} \tag{IV.47'}$$

We now introduce the quantum operators, which correspond to the conjugate momenta:

$$\begin{aligned}
\underline{\pi}_{q_1} &= \underline{P}_{X_0} = \frac{\hbar}{\mathrm{i}} \frac{\partial}{\partial_{X0}} \\
\underline{\pi}_{q_2} &= \underline{P}_{Y_0} = \frac{\hbar}{\mathrm{i}} \frac{\partial}{\partial_{Y0}} \\
\underline{\pi}_{q_3} &= \underline{P}_{Z_0} = \frac{\hbar}{\mathrm{i}} \frac{\partial}{\partial_{Z0}} \\
\underline{\pi}_{q_4} &= \underline{P}_{\phi} = \frac{\hbar}{\mathrm{i}} \frac{\partial}{\partial\phi} \\
\underline{\pi}_{q_5} &= \underline{P}_{\theta} = \frac{\hbar}{\mathrm{i}} \frac{\partial}{\partial\theta} \\
\underline{\pi}_{q_6} &= \underline{P}_{\chi} = \frac{\hbar}{\mathrm{i}} \frac{\partial}{\partial\chi} \\
\underline{\pi}_{q_7} &= \underline{p}_{a_1} = \frac{\hbar}{\mathrm{i}} \frac{\partial}{\partial a_1} \\
\underline{\pi}_{q_8} &= \underline{p}_{b_1} = \frac{\hbar}{\mathrm{i}} \frac{\partial}{\partial b_1} \\
\underline{\pi}_{q_9} &= \underline{p}_{c_1} = \frac{\hbar}{\mathrm{i}} \frac{\partial}{\partial c_1} \\
&\text{etc.}
\end{aligned} \tag{IV.52}$$

Using Eq. (IV.31″) and the above notation, the customary method [79] of obtaining the correct wave mechanical operator leads to:

$$\underline{\mathscr{H}} = \frac{1}{2} g^{1/4} \sum_{ij} (\underline{\pi}_{qi} - \underline{\Gamma}_{qi}) \frac{g^{ij}}{\sqrt{g}} (\underline{\pi}_{qj} - \underline{\Gamma}_{qj}) g^{1/4} + V_{\text{Coulomb}} \tag{IV.31'''}$$

where the symbol g is used for the determinant of the $\underset{\sim}{\theta}_q^{-1}$ matrix and where it is assumed that the quantum mechanical state functions Ψ are normalized. Although the Hamiltonian given in Eq. (IV.31‴) is correct, we will not use it further but turn to the simpler Eq. (IV.47′), where all matrix elements G^{mn} are constants. In analogy to Eq. (IV.31‴) one might expect that the quantum mechanical Hamiltonian operator should have the form

$$\mathscr{H} = \frac{1}{2} G^{1/4} \sum_{mn} (\tilde{\underline{\pi}}_m - \tilde{\underline{\Gamma}}_m) \frac{G^{mn}}{\sqrt{G}} (\tilde{\underline{\pi}}_n - \tilde{\underline{\Gamma}}_n) G^{1/4} + V_{\text{Coulomb}} \qquad \text{(IV.47‴)}$$

where G is the determinant of the (G^{mn}) matrix, or

$$\mathscr{H} = \frac{1}{2} \sum_{mn} (\tilde{\underline{\pi}}_m - \tilde{\underline{\Gamma}}_m) G^{mn} (\tilde{\underline{\pi}}_n - \tilde{\underline{\Gamma}}_n) + V_{\text{Coulomb}} \qquad \text{(IV.47‴)}$$

Equations (IV.47‴) would immediately follow from Eq. (IV.47″) due to the fact that the constants G^{mn} commute with any operator. In Eqs. (IV.47″) and (IV.47‴) the operators $\underline{\pi}_m$ are defined as

$$\begin{aligned}
&\tilde{\underline{\pi}}_1 = \underline{P}_{x_0};\ \tilde{\underline{\pi}}_2 = \underline{P}_{y_0};\ \tilde{\underline{\pi}}_3 = \underline{P}_{z_0} \\
&\tilde{\underline{\pi}}_4 = \underline{P}_a - \underline{L}_a = \underline{P}_a - \frac{\hbar}{i} \sum_{\varepsilon} \left(b_\varepsilon \frac{\partial}{\partial c_\varepsilon} - c_\varepsilon \frac{\partial}{\partial b_\varepsilon} \right) \\
&\tilde{\underline{\pi}}_5 = \underline{P}_b - \underline{L}_b = \underline{P}_b - \frac{\hbar}{i} \sum_{\varepsilon} \left(c_\varepsilon \frac{\partial}{\partial a_\varepsilon} - a_\varepsilon \frac{\partial}{\partial c_\varepsilon} \right) \qquad \text{(IV.53)} \\
&\tilde{\underline{\pi}}_6 = \underline{P}_c - \underline{L}_c = \underline{P}_c - \frac{\hbar}{i} \sum_{\varepsilon} \left(a_\varepsilon \frac{\partial}{\partial b_\varepsilon} - b_\varepsilon \frac{\partial}{\partial a_\varepsilon} \right) \\
&\tilde{\underline{\pi}}_7 = \underline{p}_{a_1};\ \tilde{\underline{\pi}}_8 = \underline{p}_{b_1};\ \tilde{\underline{\pi}}_9 = \underline{p}_{c_1};\ \text{etc.}
\end{aligned}$$

with $\underline{P}_a$, $\underline{P}_b$, and $\underline{P}_c$ operators for the *a*-, *b*-, and *c*-component of the overall angular momentum. Because of difficulties which might arise from non-commuting operators, it might well be that Eq. (IV.47‴) is not equivalent to the correct Hamiltonian operator in Eq. (IV.31‴). However, Wilson *et al.* [72] have shown that Eq. (IV.47‴) indeed follows from Eq. (IV.31″) provided that the transformation which leads from the conjugated moments $\boldsymbol{\pi}_q$ to the mixed momenta $\tilde{\boldsymbol{\pi}}$ complies to a certain condition. In our case this transformation is given by Eq. (IV.44) and in analogy to Ref. [72], we will use the symbol s^{mi} for the matrix elements of the $\boldsymbol{T}^t \cdot (\boldsymbol{T}_q^t)^{-1}$ matrix and s_{im} for the elements of the inverse [Eq. (IV.45)]:

$$\tilde{\underline{\pi}}_m = \sum_i s^{mi} \underline{\pi}_{qi} \qquad \text{(IV.44′)}$$

$$\underline{\pi}_{qi} = \sum_m s_{im} \tilde{\underline{\pi}}_m \,. \qquad \text{(IV.45′)}$$

Now the condition which would justify the use of Eq. (IV.47‴) instead of Eq. (IV.31‴) consists of two parts:

a) The operator equivalence given in Eq. (IV.54) must hold

$$s^{1/2} \sum_{im} s^{mi}\, \underline{\pi}_{qi}\, \frac{s_{km}}{s} = s^{-1/2}\, \underline{\pi}_{qk} \tag{IV.54}$$

where s is the determinant of the (s_{im}) matrix, and

b) the state functions Ψ which are used in connection with the Hamiltonian in Eq. (IV.47‴) have to be normalized using the absolute value of s as a weighting factor. In our case, $s = -\sin(\theta)$ so that the normalization condition simply corresponds to the use of the volume element normally used in connection with the Eulerian angles: $\sin(\theta)\, d\phi\, d\theta\, d\chi$ rather than $d\phi\, d\theta\, d\chi$ that would have to be used in connection with Eq. (IV.31‴).

It is fairly easy to show that $\underline{T}^t \cdot (\underline{T}_q^t)^{-1}$ indeed complies to the condition given in Eq. (IV.54). As far as the Eulerian angles are concerned, the proof is identical to the treatment given elsewhere [72] and will not be repeated here. As far as the electronic moments are concerned the proof is even simpler, since in each case where an electronic operator (for instance, $\underline{p}_{a\varepsilon}$) does not commute with a matrix element s_{km} (for instance, a_ε), the corresponding matrix element s^{mi} is zero.

As a consequence of the above considerations the classical Hamiltonian given in Eq. (IV.47) already has the correct form for direct translation into quantum mechanics and after the neglect of the $\boldsymbol{r}_0$ contributions which has been discussed already, the Hamiltonian operator is given by:

$$\mathscr{H} = \frac{1}{2m} \sum_{\varepsilon}^{\text{electrons}} (\underline{p}_{a\varepsilon}^2 + \underline{p}_{b\varepsilon}^2 + \underline{p}_{c\varepsilon}^2) + V_{\text{Coulomb}} \qquad \text{(a) (IV.55)}$$

$$+ \frac{1}{2M} (\underline{P}_{X_0}^2 + \underline{P}_{Y_0}^2 + \underline{P}_{Z_0}^2) \qquad \text{(b)}$$

$$+ \frac{1}{2} \left\{ \frac{\underline{L}_a^2}{I_{aa}^{(n)}} + \frac{\underline{L}_b^2}{I_{bb}^{(n)}} + \frac{\underline{L}_c^2}{I_{cc}^{(n)}} \right\} \qquad \text{(c)}$$

$$- \left\{ \frac{\underline{L}_a \underline{P}_a}{I_{aa}^{(n)}} + \frac{\underline{L}_b \underline{P}_b}{I_{bb}^{(n)}} + \frac{\underline{L}_c \underline{P}_c}{I_{cc}^{(n)}} \right\} \qquad \text{(d)}$$

$$+ \frac{1}{2} \left\{ \frac{\underline{P}_a^2}{I_{aa}^{(n)}} + \frac{\underline{P}_b^2}{I_{bb}^{(n)}} + \frac{\underline{P}_c^2}{I_{cc}^{(n)}} \right\} \qquad \text{(e)}$$

$$- \frac{1}{Mc}\, \boldsymbol{\mu}_{\text{el}} \cdot (\underline{\boldsymbol{P}}_0 \times \boldsymbol{H}) \qquad \text{(f)}$$

$$- \frac{|e|}{4c} H_Z \sum_{\gamma\gamma'} \left[\underline{P}_\gamma \frac{S_{\gamma\gamma'}}{I_{\gamma\gamma}^{(n)}} \underline{\cos \gamma' Z} + \underline{\cos \gamma Z} \frac{S_{\gamma\gamma'}}{I_{\gamma'\gamma'}^{(n)}} \underline{P}_{\gamma'} \right] \qquad \text{(g)}$$

$$+ \frac{|e|}{4c} H_Z \sum_{\gamma\gamma'} \left[\underline{L}_\gamma \frac{S_{\gamma\gamma'}}{I_{\gamma\gamma}^{(n)}} \underline{\cos \gamma' Z} + \underline{\cos \gamma Z} \frac{S_{\gamma\gamma'}}{I_{\gamma'\gamma'}^{(n)}} \underline{L}_{\gamma'} \right] \qquad \text{(h)}$$

$$+ \frac{|e|}{2mc} H_Z^2 \sum_{\gamma} \underline{\cos \gamma Z L_\gamma} \qquad \text{(i)}$$

$$+ \frac{e^2}{8c^2} H_Z^2 \sum_{\gamma\gamma'\gamma''} \underline{\cos\ \gamma Z} \frac{S_{\gamma\gamma''} S_{\gamma''\gamma'}}{I^{(n)}_{\gamma''\gamma''}} \underline{\cos\ \gamma' Z} \qquad \text{(j)}$$

$$+ \frac{e^2}{8mc^2} H_Z^2 \sum_{\gamma\gamma'} \underline{\cos\ \gamma Z}\ \underline{s_{\gamma\gamma'}}\ \underline{\cos\ \gamma' Z} \qquad \text{(k)}$$

$$+ \frac{1}{2\,Mc^2} (\underline{\mu}_{el} \times \boldsymbol{H})^2 \qquad \text{(l)}$$

where the sums over γ are over a, b, and c and where we have specialized the magnetic field to point into the space fixed Z-direction, *i.e.*:

$$H_a = H_Z \underline{\cos aZ}\,, \quad H_b = H_Z \underline{\cos bZ}\,, \quad H_c = H_Z \underline{\cos cZ}\,.$$

In Eq. (IV.55), part (a) is identical with the Hamiltonian operator of the electrons if moving in the Coulomb field of the non-rotating nuclear frame. Part (b) corresponds to the translational energy connected with the center of mass motion. Part (e) corresponds to the rotational energy of the nuclear frame. Part (d) leads to a coupling of electronic motion and rotation which corresponds to the Coriolis forces and it leads to an electronic contribution to the effective moments of inertia. Part (f) corresponds to the Lorentz forces which, in the presence of an exterior magnetic field, act on the molecular charge distribution because of its translational motion relative to the field. Part (g) will lead to a nuclear contribution to the g-values, while part (i) will contribute to the g-values as well as to the susceptibilities. Part (k) gives a contribution to the susceptibilities alone.

In our further treatment we will neglect part (h) as compared to part (i) and part (j) as compared to part (k). The latter would lead to a nuclear contribution to the susceptibilities. In both cases the contributions of the neglected terms to the Zeeman energy may be estimated to be roughly a factor of m/M_p smaller than the contributions of the terms retained. By a similar argument, part (l) may be neglected as compared to part (k) due to the weighting with $1/M$ as compared to $1/m$.

As the next step we further simplify the problem by treating the translational motion classically, *i.e.*, we will replace $\boldsymbol{P}_0/M$ by the molecular center of mass velocity $\boldsymbol{V}_0$ and we will assume that between collisions $\boldsymbol{V}_0$ is a constant of the motion. In standard microwave spectroscopy, a Maxwell-Boltzmann distribution may be assumed for the molecular velocities $\boldsymbol{V}_0$ (compare the detailed discussion in Section III). Under these assumptions, Eq. (IV.55f) may be rewritten as

$$-\underline{\mu}_{el} \cdot \boldsymbol{E}_{TS} = -\frac{1}{c}\,\underline{\mu}_{el} \cdot (\boldsymbol{V}_0 \times \boldsymbol{H}) \qquad \text{(IV.55f)}$$

where we have introduced a virtual exterior field,

$$\boldsymbol{E}_{TS} = \frac{1}{c} (\boldsymbol{V}_0 \times \boldsymbol{H})\,.$$

Accordingly the final operator is given by:

$$\begin{aligned}
\mathscr{H} = {} & \frac{1}{2m}\sum_{\varepsilon}(p_{\underline{a}\varepsilon}{}^2 + p_{\underline{b}\varepsilon}{}^2 + p_{\underline{c}\varepsilon}{}^2) + V_{\text{Coulomb}} & \text{(a)}\\
& + \frac{1}{2}\left\{\frac{\underline{L}_a^2}{I_{aa}^{(n)}} + \frac{\underline{L}_b^2}{I_{bb}^{(n)}} + \frac{\underline{L}_c^2}{I_{cc}^{(n)}}\right\} & \text{(b)}\\
& - \left\{\frac{\underline{L}_a\underline{P}_a}{I_{aa}^{(n)}} + \frac{\underline{L}_b\underline{P}_b}{I_{bb}^{(n)}} + \frac{\underline{L}_c\underline{P}_c}{I_{cc}^{(n)}}\right\} & \text{(c)}\\
& + \frac{1}{2}\left\{\frac{\underline{P}_a^2}{I_{aa}^{(n)}} + \frac{\underline{P}_b^2}{I_{bb}^{(n)}} + \frac{\underline{P}_c^2}{I_{cc}^{(n)}}\right\} & \text{(d) (IV.56)}\\
& - \frac{|e|}{4c}H_z\sum_{\gamma\gamma'}\left[\underline{P}_\gamma \frac{S_{\gamma\gamma'}}{I_{\gamma\gamma}^{(n)}}\underline{\cos\gamma' Z} + \underline{\cos\gamma Z}\frac{S_{\gamma\gamma'}}{I_{\gamma'\gamma'}^{(n)}}\underline{P}_{\gamma'}\right] & \text{(e)}\\
& + \frac{|e|}{2mc}H_z\sum_{\gamma}\underline{\cos\gamma Z\, \mathrm{L}_\gamma} & \text{(f)}\\
& + \frac{e^2}{8mc^2}H_z{}^2\sum_{\gamma\gamma'}\underline{\cos\gamma Z}\, \underline{s}_{\gamma\gamma'}\, \underline{\cos\gamma' Z} & \text{(g)}\\
& - |E_{\text{TS}}|\sum_{\gamma}\mu_\gamma\cos\gamma Y & \text{(h)}
\end{aligned}$$

In Eq. (IV.56h) we have chosen the space fixed Y-axis to point in the direction of the virtual electric field $\boldsymbol{E}_{\text{TS}}$. The eigenvalue problem which corresponds to the above simplified Hamiltonian may be solved by a perturbation treatment. For this purpose the corresponding Hamiltonian matrix is calculated in a basis of functions built from products between electronic functions $\phi_{\text{el}}(a_1,b_1,c_1,a_2,b_2,c_2,\ldots,a_{N_e},b_{N_e},c_{N_e})$ and rotational functions $\psi_{\text{rot}}(\phi,\theta,\chi)$. As electronic functions we will use the eigenfunctions of Eq. (IV.56a), *i.e.*,

$$\phi_{\text{el}}(a_1,b_1,c_1,\ldots) = \phi_n(a_1,b_1,c_1,\ldots) \quad \text{with}$$

$$\underline{\mathscr{H}}_0\phi_n = \frac{1}{2m}\sum_{\varepsilon}(\underline{p}_{a\varepsilon}^2 + \underline{p}_{b\varepsilon}^2 + \underline{p}_{c\varepsilon}^2)\phi_n + V_{\text{Coulomb}}\phi_n = E_n\phi_n \qquad \text{(IV.56')}$$

In Eq. (IV.56′) n stands for all electronic quantum numbers and according to our model, the geometry of the nuclear frame which enters into the expression for the Coulomb energy, V_{Coulomb}, is assumed to be frozen in the equilibrium structure corresponding to the electronic ground state. As rotational functions we will use the eigenfunctions of the operator in Eq. (IV.56d), *i.e.*,

$$\psi_{\text{rot}}(\phi,\theta,\chi) = \psi_j(\phi,\theta,\chi)$$

$$\frac{1}{2}\left[\frac{\underline{P}_a^2}{I_{aa}^{(n)}} + \frac{\underline{P}_b^2}{I_{bb}^{(n)}} + \frac{\underline{P}_c^2}{I_{cc}^{(n)}}\right]\psi_j(\phi,\theta,\chi) = E_j\psi_j(\phi,\theta,\chi) \qquad \text{(IV.56d')}$$

In Eq. (IV.56d') j stands for all rotational quantum numbers.

In view of the fact that the energy difference between the electronic ground state and the electronically excited states is usually large compared to the rotational energy differences, we will use second order perturbation theory for degenerate systems, *i.e.*, a Van Vleck transformation [75] aiming at the electronic ground state with Eq. (IV.56a) as the zero order Hamiltonian. All other contributions including Eq. (IV.56d) are regarded as perturbations and are summarized by the symbol $\mathscr{H}'$. Retaining only the leading terms of the Van Vleck expansion, the effective Hamiltonian matrix takes a form given by

$$\langle j,n|\underline{\mathscr{H}}_{\text{eff}}|j',n\rangle = E_n\delta_{jj'} + \langle j,n|\underline{\mathscr{H}}'|j',n\rangle + \sum_{\substack{n'' \\ n''\neq n}}\sum_{j''} \frac{\langle j,n|\underline{\mathscr{H}}'|j''n''\rangle\langle j''n''|\underline{\mathscr{H}}'|j'n\rangle}{E_n - E_{n'}} + \ldots \tag{IV.57}$$

Projecting the Hamiltonian given in Eq. (IV.56) into the electronic ground state according to Eq. (IV.57) leads to a rather lengthy expression from which only those terms which contribute most to the effective rotational Hamiltonian will be given explicitly. A compact notation will be used for the perturbation sums:

$$\left(\frac{AB}{\Delta}\right) = \sum_{\substack{n \\ n\neq 0}} \frac{\langle 0|\underline{A}|n\rangle\langle n|\underline{B}|0\rangle}{E_0 - E_n} \tag{IV.58}$$

where $\underline{A}$ and $\underline{B}$ are operators acting on the electronic variables. With this notation the leading terms of the effective rotational Hamiltonian which follow from Eqs. (IV.56) and (IV.57) are given by

$$\text{(a)}\quad \langle j|\underline{\mathscr{H}}_{\text{eff}}|j'\rangle = \sum_{\gamma} \frac{\langle j|\underline{P}_\gamma^2|j'\rangle}{2I_{\gamma\gamma}^{(\text{n})}} \tag{IV.59}$$

arising in first order from Eq. (IV.56d),

$$\text{(b)}\quad \sum_{\gamma\gamma'} \frac{1}{I_{\gamma\gamma}^{(\text{n})} I_{\gamma'\gamma'}^{(\text{n})}} \left\{ \left(\frac{L_\gamma L_{\gamma'}}{\Delta}\right) \langle j|\underline{P}_\gamma \underline{P}_{\gamma'}|j'\rangle + \left(\frac{L_{\gamma'} L_\gamma}{\Delta}\right) \langle j|\underline{P}_{\gamma'}\underline{P}_\gamma|j'\rangle \right\}$$

arising in second order from Eq. (IV.56c),

$$\text{(c)}\quad -\frac{|e|}{4c} H_z \sum_{\gamma\gamma'} \frac{S_{\gamma\gamma'}}{I_{\gamma\gamma}^{(\text{n})}} \left\{ \langle j|\underline{P}_\gamma \,\underline{\cos\gamma' Z}|j'\rangle + \langle j|\underline{\cos\gamma' Z}\,\underline{P}_\gamma|j'\rangle \right\}$$

arising in first order from Eq. (IV.56e), and

$$\text{(d)}\quad -\frac{|e|}{2mc} \sum_{\gamma\gamma'} \frac{1}{I_{\gamma\gamma}^{(\text{n})}} \left\{ \left(\frac{L_\gamma L_{\gamma'}}{\Delta}\right) \langle j|\underline{P}_\gamma \,\underline{\cos\gamma' Z}|j'\rangle + \left(\frac{L_{\gamma'} L_\gamma}{\Delta}\right) \langle j|\underline{\cos\gamma' Z}\,\underline{P}_\gamma|j'\rangle \right\}$$

arising in second order from Eqs. (IV.56c) and (IV.56f). Included also are

$$\text{(e)} \quad + \frac{e^2}{8mc^2} H_z^2 \sum_{\gamma\gamma'} \langle 0 | \underline{s}_{\gamma\gamma'} | 0 \rangle \langle j | \underline{\cos \gamma Z} \; \underline{\cos \gamma' Z} | j' \rangle$$

$$\text{(f)} \quad + \frac{e^2}{4m^2c^2} H_z^2 \sum_{\gamma\gamma'} \left[\left(\frac{\underline{L}_\gamma \underline{L}_{\gamma'}}{\Delta} \right) \langle j | \underline{\cos \gamma Z} \; \underline{\cos \gamma' Z} | j' \rangle \right.$$

$$\left. + \left(1 - \delta_{\gamma\gamma'}\right) \left(\frac{\underline{L}_{\gamma'} \underline{L}_\gamma}{\Delta} \right) \langle j | \underline{\cos \gamma' Z} \; \underline{\cos \gamma Z} | j' \rangle \right]$$

$$\text{(g)} \quad - E_{\mathrm{TS}} \sum_{\gamma} \langle 0 | \mu_\gamma | 0 \rangle \langle j | \cos \gamma Y | j' \rangle$$

$\delta_{\gamma\gamma'}$ = Kronecker symbol defined as $\delta_{\gamma\gamma'} = 1$ for $\gamma = \gamma'$
$\delta_{\gamma\gamma'} = 0$ for $\gamma \neq \gamma'$

The following relations have been used implicitly when proceeding from Eqs. (IV.56) to (IV.58) via Eq. (IV.57):

$$\langle j | \underline{\mathrm{Rot}} | j' \rangle \langle n | \underline{E}_{\mathrm{e}} | n' \rangle = \langle jn | \underline{\mathrm{Rot}} \; \underline{E}_{\mathrm{e}} | j'n' \rangle \qquad \text{(IV.60)}$$

where Rot and E_{e} are operators acting only on the rotational or only on the electronic variables respectively.

$$\langle j | \underline{\mathrm{Rot}}_1 \; \underline{\mathrm{Rot}}_2 | j' \rangle = \sum_{j''} \langle j | \underline{\mathrm{Rot}}_1 | j'' \rangle \langle j'' | \underline{\mathrm{Rot}}_2 | j' \rangle \qquad \text{(IV.61)}$$

where $\underline{\mathrm{Rot}}_1$ and $\underline{\mathrm{Rot}}_2$ are operators acting on the rotational variables (direction cosines and angular momenta).

Contributions which do not depend on the rotational states are not included in Eq. (IV.59) since they would cancel in pure rotational transitions and are thus of no immediate interest as far as rotational spectroscopy is concerned. Although we will not discuss in detail all neglected terms, we will examine some specific cases. For this purpose we arbitrarily pick a contribution which stems from Eqs. (IV.56b) and (IV.56g) and we will specialize to the case where $\gamma = \gamma' = \gamma'' = a$. Using the symbol N_{c} for this neglected contribution, we get:

$$N_{\mathrm{c}} = \sum_{\substack{n \\ n'' \neq 0}}^{\text{electronic states}} \; \sum_{j''}^{\text{rotational states}} \frac{\langle j,0 | \frac{\underline{L}_a^2}{2 I_{aa}^{(\mathrm{n})}} | j''n \rangle \langle j''n | \frac{e^2}{8mc^2} H_Z^2 \, \underline{\cos^2 aZ} \; \underline{s_{aa}} | j'0 \rangle}{E_0 - E_n}$$

Since the first operator $\frac{\underline{L}_a^2}{2 I_{aa}^{(\mathrm{n})}}$ does not depend on rotational variables, it is diagonal in the rotational quantum numbers j and we obtain:

$$N_{\mathrm{c}} = \frac{e^2}{16 mc^2 I_{aa}^{(\mathrm{n})}} H_Z^2 \langle j | \underline{\cos^2 aZ} | j' \rangle \left(\frac{\underline{L}_a^2 \underline{s}_{aa}}{\Delta} \right)$$

To get a rough estimate for the perturbation sum, we replace the individual energy differences in the denominator, *i.e.*, $E_0 - E_1$, $E_0 - E_2$, etc. by an average value ΔE_{av} to give

$$\left(\frac{\underline{L}_a{}^2 \underline{s}_{aa}}{\Delta}\right) \approx \frac{1}{\Delta E_{av}} \left(\langle 0 | \underline{L}_a^2 \underline{s_{aa}} | 0 \rangle - \langle 0 | \underline{L}_a^2 | 0 \rangle \langle 0 | \underline{s_{aa}} | 0 \rangle\right)$$

With the assumption that the two expectation values in the bracket have the same order of magnitude, we obtain as an order of magnitude equivalence:

$$|N_c| \approx \left| \frac{e^2}{16 mc^2 I_{aa}^{(n)}} H_Z{}^2 \langle j | \underline{\cos^2 aZ} | j' \rangle \frac{\langle 0 | \underline{L}_a{}^2 | 0 \rangle \langle 0 | \underline{s_{aa}} | 0 \rangle}{\Delta E_{av}} \right| \tag{IV.62}$$

We now compare this value with a similar contribution included in Eq. (IV.59e) which we will call R_c for retained contribution.

$$R_c = \frac{e^2}{8 mc^2} H_Z^2 \langle j | \cos^2 aZ | j' \rangle \langle 0 | s_{aa} | 0 \rangle \tag{IV.63}$$

From Eqs. (IV.62) and (IV.63) the ratio between the neglected and the retained contribution is approximately given by:

$$\left| \frac{N_c}{R_c} \right| \approx \left| \frac{\langle 0 | \underline{L}_a^2 | 0 \rangle / 2 I_{aa}^{(n)}}{\Delta E_{av}} \right| . \tag{IV.64}$$

If we now assume that $\langle 0 | \underline{L}_a^2 | 0 \rangle$ will roughly have a value up to $N_e \hbar^2$ where N_e is the number of electrons (see below), then the ratio of $|N_c/R_c|$ should have the same order of magnitude as the ratio between rotational energies and the average electronic energy difference and this indeed justifies the neglect of N_c as compared to R_c. In order to justify the rough equivalence, $\langle 0 | \underline{L}_a^2 | 0 \rangle \approx N_e \hbar^2$, we give some typical experimental values. In orthofluoropyridine,[48] for example, the following values were obtained for the perturbation sums:

$$\left(\frac{\underline{L}_a \underline{L}_a}{\Delta}\right) = 2.60 \times 10^{-42} \text{ g cm}^2$$

$$\left(\frac{\underline{L}_b \underline{L}_b}{\Delta}\right) = 4.8 \times 10^{-42} \text{ g cm}^2$$

$$\left(\frac{\underline{L}_c \underline{L}_c}{\Delta}\right) = 6.5 \times 10^{-42} \text{ g cm}^2$$

Replacing the sums by approximative values as above:

$$\left(\frac{\underline{L}_a \underline{L}_a}{\Delta}\right) \approx \frac{1}{\Delta E_{av}} \left\{ \langle 0 | \underline{L}_a^2 | 0 \rangle - \langle 0 | \underline{L}_a | 0 \rangle \langle 0 | \underline{L}_a | 0 \rangle \right\}, \text{ etc.}$$

and taking into account that in general $\langle 0|\underset{\sim}{L}_a|0\rangle = 0$ for a closed shell molecule, we get the rough correspondence:

$$\Delta E_{\mathrm{av}} \left(\frac{\underset{\sim}{L}_a \underset{\sim}{L}_a}{\Delta}\right) \approx \langle 0|\underset{\sim}{L}_{\mathrm{a}}^2|0\rangle$$

With $\Delta E_{\mathrm{av}} = 1$ eV as a reasonable choice together with the fluoropyridine value for $\left(\frac{\underset{\sim}{L}_a \underset{\sim}{L}_a}{\Delta}\right)$ we get:

$$\Delta E_{\mathrm{av}} \left(\frac{\underset{\sim}{L}_a \underset{\sim}{L}_a}{\Delta}\right) \approx 3.7\, \hbar^2$$

as compared to 50 $\hbar^2$ obtained by the approximation used above. This shows that $\left(\frac{\underset{\sim}{L}_a \underset{\sim}{L}_a}{\Delta}\right) \approx \frac{N_e \hbar^2}{\Delta E_{\mathrm{av}}}$ with $\Delta E_{\mathrm{av}} \approx 1\, eV$ leads to a rather conservative estimate of the perturbation sum and confirms the order of magnitude estimations carried out above.

In addition to the order of magnitude considerations, group theoretical arguments may be applied in a manner similar to those carried out in Chapter III to show that certain contributions must vanish provided the molecule contains some symmetry elements. However, we will not elaborate on this point further.

C. Theoretical Expressions for the $\underset{\sim}{g}$- and $\underset{\sim}{\chi}$- Tensor Elements

In order to obtain the theoretical expressions for the $\underset{\sim}{g}$- and $\underset{\sim}{\chi}$-tensor elements, we have to compare the theoretical Hamiltonian given in Eq. (IV.59) with the phenomenological Hamiltonian in Eq. (I.4). For this comparison we have to keep in mind that $\underset{\sim}{J}_\gamma$ in the phenomenological expression corresponds to $P_\gamma/\hbar$ in Eq. (IV.59) and that in order to make the phenomenological Hamiltonian Hermitian, products such as $\underline{\cos \gamma Z}\, J_{\gamma'}$ must be symmetrisized to $(\underline{\cos \gamma Z} J_{\gamma'} + J_{\gamma'}\, \underline{\cos \gamma Z})/2$, etc.

Comparing the terms which are of first order in H_z and those which are of second order in H_Z respectively, the following relations are obtained for the diagonal elements of the $\underset{\sim}{g}$- and $\underset{\sim}{\chi}$-tensors respectively.

$$g_{aa} = \frac{M_{\mathrm{p}}}{I_{aa}^{(\mathrm{n})}} \sum_{\nu}^{\text{nuclei}} Z_\nu (b_\nu^2 + c_\nu^2) + \frac{M_{\mathrm{p}}}{I_{aa}^{(\mathrm{n})}} \frac{2}{m} \sum_{\substack{n \\ n \neq 0}}^{\substack{\text{excited} \\ \text{states}}} \frac{|\langle 0|\underset{\sim}{L}_a|n\rangle|^2}{E_0 - E_n} + \ldots \qquad \text{(IV.65)}$$

$$\chi_{aa} = -\frac{e^2}{4mc^2} \left\{ \langle 0| \sum_{\varepsilon} \underline{(b_\varepsilon^2 + c_\varepsilon^2)}|0\rangle + \frac{2}{m} \sum_{\mathrm{n}}^{\substack{\text{excited} \\ \text{states}}} \frac{|\langle 0|\underset{\sim}{L}_a|n\rangle|^2}{E_0 - E_n} + \ldots \right\} \qquad \text{(IV.66)}$$

$$= \chi_{aa}^{(\mathrm{dia})} + \chi_{aa}^{(\mathrm{para})} .$$

These are the equations which have been used in Section II in order to extract ground state properties such as the second moments of the electron charge distribution, the molecular electric quadrupole moments and the sign of the electric dipole moment from the Zeeman data.

A comparison of the contributions which do not depend explicitly on the magnetic field leads to effective values for the inverse momenta of inertia:

$$\frac{1}{I_{aa}} = \frac{1}{I_{aa}^{(\mathrm{n})}} \left\{ 1 + \frac{2}{I_{aa}^{(\mathrm{n})}} \sum_{n}^{\substack{\text{excited}\\ \text{states}}} \frac{|\langle 0|L_a|n\rangle|^2}{E_0 - E_n} \right\} + \dots \tag{IV.67}$$

Since the perturbation sum is necessarily negative ($E_0 < E_n$), the effective value of $1/I_{aa}$ is in general slightly smaller than $1/I_{aa}^{(\mathrm{n})}$. This, of course, is quite reasonable in view of electron contributions to the moment of inertia. In Eq. (IV.67), the off-diagonal contributions of the electrons have been neglected. Actually I_{aa} in Eq. (I.4) has the meaning of the a- component of the *molecular* moment of inertia tensor whose principal axis system may differ slightly from the principal axis system of the nuclear moment of inertia tensor. Due to the small mass of the electrons, however, inclusion of the electronic off-diagonal elements (off-diagonal in the nuclear principal axis system), will cause corrections of the relation given in Eq. (IV.67) which fall into the ppm range and therefore have been neglected.

Acknowledgement. We gratefully acknowledge the support of the Deutsche Forschungsgemeinschaft and the National Science Foundation.

D. H. S. would also like to thank his friends and colleagues H. Dreizler, A. Guarnieri, E. Hamer, L. Engelbrecht, W. Czieslik, and J. Wiese for many helpful discussions on the rotational Zeeman effect.

Appendix AI. Molecular Zeeman Effect Data on Several Molecules

In this Appendix we tabulate the molecular Zeeman data collected since the review of Flygare and Benson [1]. The format follows the tables published by Flygare and Benson [1].

Table A1. Linear and symmetric top molecules

Molecule	$g_\perp$		$\chi_\perp - \chi_\parallel$ [10^{-6} erg/(G^2 mole)]	Q [esu cm^{-26}]	Refs.
$^{12}C^{32}S$	—0.2702(4)	$v = 0$	+24.2(12)	+0.8(14)	1)
$^{12}C^{34}S$	—0.2659(7)	$v = 0$			
$^{13}C^{32}S$	—0.2529(5)	$v = 0$			
$^{12}C^{80}Se$	—0.2431(16)	$v = 0$	27.8(14)	—2.6(16)	
$^{28}Si^{16}O$	—0.1546(3)	$v = 0$	11.1(9)	—4.56(110)	2)
$^{28}Si^{32}S$	—0.090974(65) $-0.000296(24)[v+1/2]$		+20.4(1) $+0.33(5)[v+1/2]$		3)
$^{46}Ge^{16}O$	—0.14104(11) $-0.00066(60)[v+1/2]$		+14.9(2) $+0.29(4)[v+1/2]$		
$^{120}Sn^{16}O$	—0.14631(38)		22.1(2)		4)
$^{208}Pb^{16}O$	—0.16233(39)		31.1(2)		5)
^{208}Pb Te	—0.01800(18)				
$^{27}Al^{19}F$	—0.08051(8)		5.2(5)	—5.91(48)	6)
$^{69}Ga^{19}F$	—0.06012(12)		6.2(2)	—6.44(24)	7)
$^{63}Cu^{19}F$	—0.0628(2)		6.5(7)	—6.05(82)	7)
$^{205}Tl^{19}F$	0.05370(15)		7.2(24)		8)
$^{133}Cs^{19}F$	0.06413(18)	$v = 0$	9.6(7)		8)
	0.06444(30)	$v = 1$	10.5(18)		
	0.06384(22)	$v = 2$	10.3(24)		
	0.06355(21)	$v = 3$	8.8(18)		
	0.06421(88)	$v = 4$	9.0(3.6)		
$^{133}Cs^{35}Cl$	—0.02803(7)	$v = 0$	14.4(24)		8)
	—0.02756(17)	$v = 1$			
	—0.02699(25)	$v = 2$			
	—0.02699(23)	$v = 3$			
	—0.02699(26)	$v = 4$			
	—0.02648(28)	$v = 5$			
	—0.02630(23)	$v = 6$			
	$g_\perp(v) = -0.02815(7) + 0.00031(8)[v+1/2]$				
$^{133}Cs^{79}Br$	—0.0099(10)	$v = 0$			8)
$^{133}Cs^{127}I$	$\lvert g_\perp \rvert < 0.0036$	$v = 0$			8)

Table A1 (continued)

Molecule	$g_\perp$	$\chi_\perp - \chi_\parallel$ [10^{-6} erg/(G^2 mole)]	Q [esu cm^{-26}]	Refs.
$H—^{11}B{=}^{32}S$	—0.0414(2) $v=0$	+ 7.2(5) $v=0$		9)
$D^{11}B^{32}S$	—0.0356(2) $v=0$	+ 9.8(21) $v=0$		10)
$H—^{12}C{\equiv}^{15}N$	—0.0904(3) $v=0$	+ 7.2(4) $v=0$	3.1(6)	11)
$^{79}Br^{12}C^{15}N$	—0.03165(50) $v=0$	11.37(150) $v=0$	—6.46(175) $v=0$	12)
$^{81}Br^{12}C^{15}N$	—0.02981(75) $v=0$	10.37(300) $v=0$	—6.48(400) $v=0$	13)
$H_3C—F$	—0.062(2) $=g_\perp$	+ 8.5(6)	—0.4(10)*	14)
	+0.265(8) $=g_\parallel$			15)

1) McGurk, J., Tigelaar, H. L., Rock, S. L., Flygare, W. H.: J. Chem. Phys. *58*, 1420 (1973).
2) Honerjäger, R., Tischer, R.: Z. Naturforsch. *29a*, 1695 (1974).
3) Honerjäger, R., Tischer, R.: Z. Naturforsch. *28a*, 1374 (1973).
4) Honerjäger, R., Tischer, R.: Z. Naturforsch. *28a*, 1372 (1973).
5) Honerjäger, R., Tischer, R.: Z. Naturforsch. *29a*, 1695 (1974).
6) Honerjäger, R., Tischer, R.: Z. Naturforsch. *29a*, 342, (1974).
7) Honerjäger, R., Tischer, R.: Z. Naturforsch. *29a*, 1919 (1974).
8) Honerjäger, R., Tischer, R.: Z. Naturforsch. *28a*, 458 (1973).
9) Pearson, E. F., Norris, C. L., Flygare, W. H.: J. Chem. Phys. *60*, 1761 (1974).
10) $+H—B{=}S^-$.
11) Hartford, S. L., Allen, W. C., Norris, C. L., Pearson, E. F., Flygare, W. H.: Chem. Phys. Letters *18*, 153 (1973).
12) Blackman, G. L., Brown, R. D., Burden, F. R.: J. Chem. Phys. *59*, 3760, (1973).
13) Compare too: Ewing, J. J., Tigelaar, H. L., Flygare, W. H.: J. Chem. Phys. *56*, 1957 (1972).
14) Norris, C. L., Pearson, E. F., Flygare, W. H.: J. Chem. Phys. *60*, 1758 (1974).
15) * $g_\parallel$ not corrected for translational Lorentz effect. The corrected value is close to +0.245.

Table A2. Asymmetric top molecules

Molecule 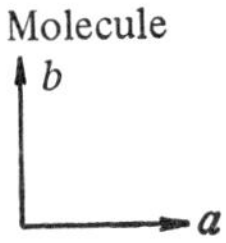	g_{aa} / g_{bb} / g_{cc}	$2\chi_{aa}-\chi_{bb}-\chi_{cc}$ / $2\chi_{bb}-\chi_{cc}-\chi_{aa}$ [10^{-6} erg/(G^2 mole)]	Q_{aa} / Q_{bb} / Q_{cc} [10^{-26} esu cm^2]	Refs.
$H\diagdown{}^{16}O—F$	+0.642(1)	—19.6(6)	— 0.2(4)	1)
	—0.119(1)	+12.8(12)	— 1.9(8)	
	—0.061(1)		— 2.1(1.1)	
$H\diagdown{}^{16}O—{}^{35}Cl$	+0.6390(55)	—24.8(3)	+ 0.50(35)	2)
	—0.0752(9)	+15.8(7)	+ 0.74(67)	
	—0.0616(10)		+ 1.24(102)	
$H_3C\diagdown{}^{16}O\diagup{}^{35}Cl$	—0.0595(7)	—24.7(11)	+ 4.6(10)	3)
	—0.0249(8)	+ 1.8(24)	— 0.9(19)	
	—0.0378(5)		— 3.5	

Table A2 (continued)

Molecule 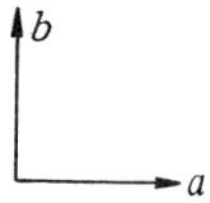	g_{aa} g_{bb} g_{cc}	$2\chi_{aa}-\chi_{bb}-\chi_{cc}$ $2\chi_{bb}-\chi_{cc}-\chi_{aa}$ [10^{-6} erg/(G^2 mole)]	Q_{aa} Q_{bb} [10^{-26} esu cm^2] Q_{cc}	Refs.
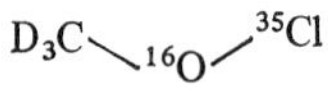	—0.0479(9) —0.0220(9) —0.0333(8)	—23.2(11) + 0.4(20)	+ 4.0(10) — 0.8(17) — 3.2	
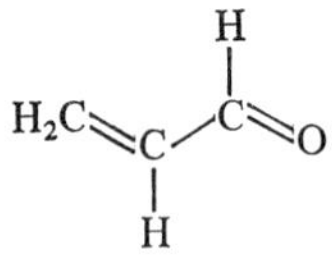	—0.5512(19) —0.0567(10) —0.0080(10)	24.1(9) 17.1(15)	— 2.5(11) + 3.3(17) — 0.8(22)	4)
	—0.09692(12) 0.01848(15) 0.03361(21)	18.46(21) — 0.05(30)	2.60(15) — 3.70(21) + 1.10(33)	5)
	—0.07794(27) +0.01436(51) +0.02683(33)	18.45(48) 0.24(66)	2.70(39) — 4.20(51) + 1.50(78)	
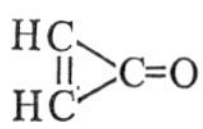	—0.2900(13) —0.0963(4) —0.0121(4)	13.6(11) 22.0(8)	— 3.0(9) + 4.0(7) — 1.0(13)	6)
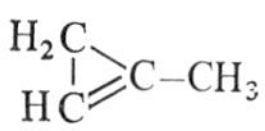	—0.0813(7) —0.0261(4) 0.0166(3)	13.9(3) 16.4(6)	+ 0.6(4) — 0.3(6) — 0.3(8)	7)
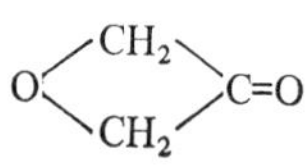	—0.1059(8) —0.0581(4) —0.0437(4)	9.6(5) — 7.8(6)	—12.8(8) + 7.9(8) + 4.9(8)	8)
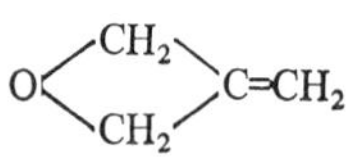	—0.0510(18) —0.0435(10) —0.0313(10)	—10.9(7) + 2.3(9)	— 5.4(10) + 5.1(12) + 0.2(15)	8)
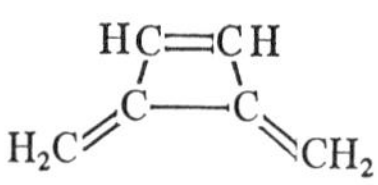	—0.0703(7) —0.0532(7) +0.0023(7)	22.1(7) 21.2(6)	4.0(12) 4.0(12) — 8.0(10)	4)
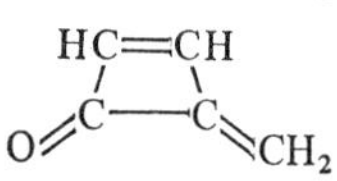	—0.0925(3) —0.0729(3) —0.0086(3)	23.0(5) 25.5(5)	1.0(6) 2.8(6) — 3.8(10)	9)
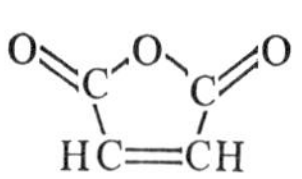	—0.1131(10) —0.0499(14) —0.0150(12)	30.3(20) 22.9(15)	—11.6 + 5.9 + 5.7	4)
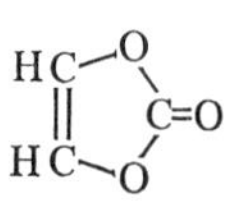	—0.0856(16) —0.0502(9) —0.0112(10)	7.2(12) 21.7(14)	— 1.1(18) + 0.9(19) + 0.2(28)	4)
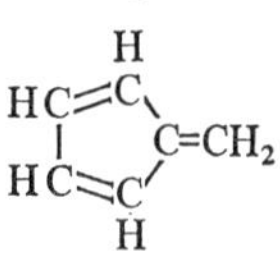	—0.1059(14) —0.0482(7) +0.0219(7)	35.9(7) 38.1(11)	5.8(14) 3.6(16) — 9.4(21)	4)

Table A2 (continued)

Molecule 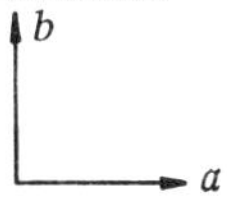	g_{aa} g_{bb} g_{cc}	$2\chi_{aa}-\chi_{bb}-\chi_{cc}$ $2\chi_{bb}-\chi_{cc}-\chi_{aa}$ $[10^{-6}$ erg/(G^2 mole)]	Q_{aa} Q_{bb} $[10^{-26}$ esu cm$^2]$ Q_{cc}	Refs.
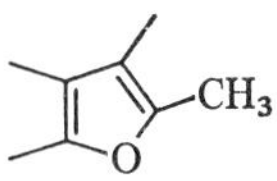	—0.0704(5) —0.0335(4) +0.0253(5)	23.6(9) 44.2(7)	6.4(10) 0.0(10) — 6.4(16)	[10]
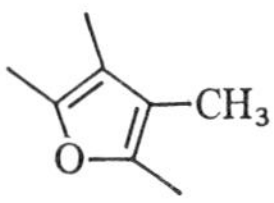	—0.0729(5) —0.0369(4) +0.0220(3)	25.3(6) 43.4(6)	+ 2.6(8) + 2.8(8) — 5.5(12)	[11]
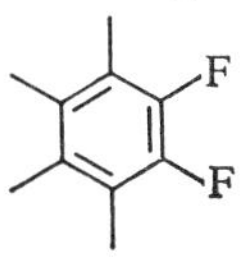	—0.0412(12) —0.0371(8) +0.0163(7)	50.8(15) 57.9(10)	— 3.6(26) + 7.0(24) — 3.4(33)	[12]
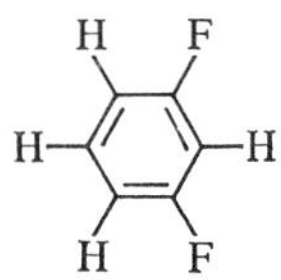	—0.0486(2) —0.0316(3) +0.0116(3)	46.6(4) 55.7(4)	— 5.0(9) + 7.6(10) — 2.6(13)	[13]
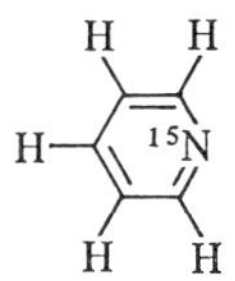	—0.08086(19) —0.09974(16) +0.04101(17)	54.12(40) 62.02(40)	— 2.6(4) + 8.0(4) — 5.4(7)	[14]
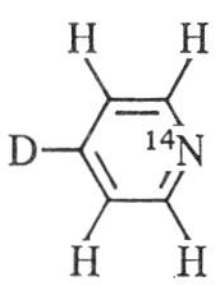	—0.07935(70) —0.09438(60) +0.04063(55)	53.3(10) 60.1(16)	— 2.3(12) 9.3(15) — 6.9(22)	

[1] Rock, S. L., Pearson, E. F., Appleman, E. H., Norris, C. L., Flygare, W. H.: J. Chem. Phys. *59*, 3940 (1973).
[2] Suzuki, M., Guarnieri, A.: Z. Naturforsch. *30a*, 497 (1975). Nuclear g value of ^{35}Cl also determined $g_I = 0.5490(14)$.
[3] Suzuki, M., Guarnieri, A.: private communication.
[4] Benson, R. C., Flygare, W. H.: J. Chem. Phys. *58*, 2366 (1973).
[5] Hamer, E.: Thesis, Universität Kiel, Germany (1974). 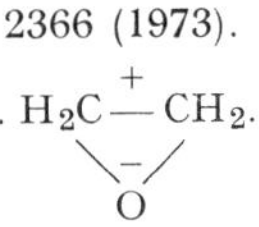
[6] Benson, R. C., Flygare, W. H., Oda, M., Breslow, R.: J. Am. Chem. Soc. *95*, 2772 (1973).
[7] Benson, R. C., Flygare, W. H.: J. Chem. Phys. *58*, 2651 (1973).
[8] Norris, C. L., Tigelaar, H. L., Flygare, W. H.: Chem. Phys. *1*, 1 (1973).
[9] Czieslik, W., Sutter, D. H.: Z. Naturforsch. *29a*, 1820 (1974).
[10] Czieslik, W., Andresen, U., Dreizler, H.: Z. Naturforsch. *28a*, 1906 (1973).
[11] Czieslik, W., Sutter, D. H.: private communication.
[12] Sutter, D. H.: Z. Naturforsch. *29a*, 786 (1974).
[13] Wiese, A.: Diplomarbeit, Universität Kiel, Germany (1975).
[14] Hamer, E.: Thesis, Universität Kiel, Germany (1974).

Table A.3. Molecules with low barrier internal rotation

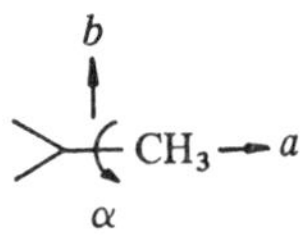	g_{aa} g_{bb} g_{cc} g_{aa} frame g_{α}	$2\chi_{aa}-\chi_{bb}-\chi_{cc}$ $2\chi_{bb}-\chi_{cc}-\chi_{aa}$ $\left[10^{-6}\ \frac{\text{erg}}{\text{G}^2\ \text{mole}}\right]$	Q_{aa} Q_{bb} Q_{cc} $[10^{-26}$ esu cm$^2]$	Ref.
^{16}O, ^{16}O >$^{14}N—^{12}CH_3$	—0.1095(5) —0.12128(9) —0.04219(8) —0.14880(10) +0.3463(8)	— 6.04(19) +22,79(26)	2.4(2) —4.6(2) 2.2(4)	1)
^{16}O, ^{16}O >$^{15}N—^{12}CH_3$	—0.1099(5) —0.12130(7) —0.04223(6) —0.14918(8) +0.3463(4)	— 5.88(15) +22.85(18)	2.4(2) —4.7(2) 2.3(3)	
^{16}O, ^{16}O >$^{14}N—^{12}CD_3$	—0.1018(5) —0.10146(13) —0.03886(8) —0.14903(12) +0.1749(5)	— 5.64(23) +21.69(34)	1.8(3) —3.9(3) 2.1(4)	
F, F >$^{11}B—CH_3$	—0.0178(9) —0.01730(12) —0.00995(11) —0.04238(14) +0.3415(5)	—1.24(19) 5.96(22)	—0.1(4) —3.1(4) 3.2(4)	

1) Engelbrecht, L.: Ph. D. Thesis, Christian Albrechts Universität, Kiel (1975).

AII. Matrix Elements of the Direction Cosines and Angular Momentum Operators in the Symmetric Top Basis

In the following the commutation relations and the nonvanishing matrix elements of the direction cosines and angular momentum operators are summarized. For an excellent discussion of the theory of angular momentum operators the reader is referred to Ref. [76]. Derivations of the matrix elements may also be found in many textbooks on quantum mechanics.

Commutation relations for the angular momentum operators:

X, Y, and Z are the space fixed axes with running index F and a, b, and c are the rotating molecular axes with running index γ. The angular momentum is given in units of $\hbar$.

$$\underline{J}_X\underline{J}_Y - \underline{J}_Y\underline{J}_X = \mathrm{i}\underline{J}_Z\,,\ \text{etc.} \tag{1}$$

$$\underline{J}^2\underline{J}_F - \underline{J}_F\underline{J}^2 = 0 \qquad F = X,Y,Z \quad \underline{J}^2 = \underline{J}_X^2 + \underline{J}_Y^2 + \underline{J}_Z^2 \tag{2}$$

$$\underline{J}_a\underline{J}_b - \underline{J}_b\underline{J}_a = -\,\mathrm{i}\underline{J}_c\,,\ \text{etc.} \tag{3}$$

$$\underline{J}^2\underline{J}_\gamma - \underline{J}_\gamma\underline{J}^2 = 0 \qquad \gamma = a,b,c \tag{4}$$

$$\underline{J}_F \underline{J}_\gamma - \underline{J}_\gamma \underline{J}_F = 0 \qquad \begin{array}{l} F = X \text{ or } Y \text{ or } Z \\ \gamma = a \text{ or } b \text{ or } c \end{array} \tag{5}$$

Commutation relations involving the direction cosines:

$$\underline{J}_F \, \underline{\cos \gamma F} - \underline{\cos \gamma F} \, \underline{J}_F = 0 \tag{6}$$

$$\underline{J}_\gamma \, \underline{\cos \gamma F} - \underline{\cos \gamma F} \, \underline{J}_\gamma = 0 \tag{7}$$

But:

$$\underline{J}_a \, \underline{\cos bF} - \underline{\cos bF} \, \underline{J}_a = -\,\mathrm{i}\, \underline{\cos cF} \tag{8}$$

$$\underline{J}_X \, \underline{\cos \gamma Y} - \underline{\cos \gamma Y} \, \underline{J}_X = +\,\mathrm{i}\, \underline{\cos \gamma Z} \tag{9}$$

$$\sum_F \underline{\cos^2 \gamma F} = 1\,, \qquad \sum_\gamma \underline{\cos^2 \gamma F} = 1\,. \tag{10}$$

Nonvanishing matrix elements of the angular momentum operators:

Because $\underline{J}^2$ commutes for instance with $\underline{J}_Z$ [Eq. (2))] and with $\underline{J}_a$ [Eq. (4)], there exists a basis $\phi_{JKM}(\phi,\theta,\chi)$ which simultaneously diagonalizes all three operators. Such a basis is frequently called a symmetric top basis since it also diagonalizes the rotational Hamiltonian of a symmetric top molecule:

$$\mathscr{H}_{\text{eff}} = A\underline{J}_a^2 + C(\underline{J}^2 - \underline{J}_a^2)\,.$$

Operator	Nonvanishing matrix elements
$\underline{J}^2$	$\langle J,K,M \vert \underline{J}^2 \vert J,K,M \rangle = J(J+1)$
$\underline{J}_X$	$\langle J,K,M \vert \underline{J}_X \vert J,K,M \pm 1 \rangle = \mp \frac{\mathrm{i}}{2} \sqrt{J(J+1) - M(M \pm 1)}$
$\underline{J}_Y$	$\langle J,K,M \vert \underline{J}_Y \vert J,K,M \pm 1 \rangle = \frac{1}{2} \sqrt{J(J+1) - M(M \pm 1)}$
$\underline{J}_Z$	$\langle J,K,M \vert \underline{J}_Z \vert J,K,M \rangle = M$
$\underline{J}_a$	$\langle J,K,M \vert \underline{J}_a \vert J,K \pm 1,M \rangle = \pm \frac{\mathrm{i}}{2} \sqrt{J(J+1) - K(K \pm 1)}$
$\underline{J}_b$	$\langle J,K,M \vert \underline{J}_b \vert J,K \pm 1,M \rangle = \frac{1}{2} \sqrt{J(J+1) - K(K \pm 1)}$
$\underline{J}_c$	$\langle J,K,M \vert \underline{J}_c \vert J,K,M \rangle = K$

Nonvanishing matrix elements of the direction cosines:

The nonvanishing matrixelements of the direction cosines may be factored into a product of a J-, a (J,K)-, and a (J,M)-dependent term, called reduced matrix elements:

$$\langle J,K,M \,|\underline{\cos \gamma F}|\, J',K',M' \rangle = \langle J \,\|\underline{\cos \gamma F}\|\, J' \rangle \langle J,K \,\|\underline{\cos \gamma F}\|\, J',K' \rangle \cdot \langle J,M \,\|\underline{\cos \gamma F}\|\, J',M' \rangle\,.$$

Reduced matrix element	Nonvanishing matrixelement for $J' =$ $J-1$	J	$J+1$
$\langle J \Vert \underline{\cos \gamma F} \Vert J' \rangle$	$\{4J(4J^2-1)^{\frac{1}{2}}\}^{-1}$	$\{4J(J+1)\}^{-1}$	$\{4(J+1)((2J+1)(2J+3))^{\frac{1}{2}}\}^{-1}$
$\langle J,K \Vert \underline{\cos cF} \Vert J',K' \rangle$	$2(J^2-K^2)^{\frac{1}{2}}$	$2K$	$2((J+1)^2-K^2)^{\frac{1}{2}}$
$\langle J,K \Vert \underline{\cos bF} \Vert J'K \pm 1 \rangle$ $= \mp i \langle J,K \Vert \underline{\cos aF} \Vert J',K \pm 1 \rangle$	$\pm [(J \mp K)(J \mp K-1)]^{\frac{1}{2}}$	$[J(J+1)-K(K \pm 1)]^{\frac{1}{2}}$	$\mp [(J \pm K+1)(J \pm K+2)]^{\frac{1}{2}}$
$\langle J,M \Vert \underline{\cos \gamma Z} \Vert J',K \pm 1 \rangle$	$2[J^2-M^2]^{\frac{1}{2}}$	$2M$	$2[(J+1)^2-M^2]^{\frac{1}{2}}$
$\langle J,M \Vert \underline{\cos \gamma Y} \Vert J',M \pm 1 \rangle$ $= \pm i \langle J,M \Vert \underline{\cos \gamma X} \Vert J',M \pm 1 \rangle$	$\pm [(J \mp M)(J \mp M-1)]^{\frac{1}{2}}$	$[J(J+1)-M(M \pm 1)]^{\frac{1}{2}}$	$\mp [(J \pm M+1)(J \pm M+2)]^{\frac{1}{2}}$

As a demonstration for the use of this table the matrix elements of $\underline{\cos cY}$ are calculated for $J=J'$:

$$\langle J,K,M|\underline{\cos cY}|J,K',M'\rangle = \langle J||\underline{\cos cY}||J\rangle\langle J,K||\underline{\cos cY}||JK'\rangle\langle J,M||\underline{\cos cY}||J,M'\rangle$$

$$\langle J,K,M\,|\underline{\cos cY}|\,J,K,M\pm 1\rangle = \frac{2K}{4J(J+1)}\sqrt{J(J+1)-M(M\pm 1)}\,.$$

For $J=2$ and $K=1$ the following submatrix is obtained (with M ranging from -2 to $+2$).

$$(\langle 2,1,M\,|\cos cY|\,2,1,M'\rangle) = \begin{bmatrix} 0 & 1/6 & 0 & 0 & 0 \\ 1/6 & 0 & \sqrt{6}/12 & 0 & 0 \\ 0 & \sqrt{6}/12 & 0 & \sqrt{6}/12 & 0 \\ 0 & 0 & \sqrt{6}/12 & 0 & 1/6 \\ 0 & 0 & 0 & 1/6 & 0 \end{bmatrix}$$

This submatrix is used in the discussion of the translational Zeeman effect of symmetric top molecules (see Chapter II).

AIII. The Zeeman Effect of a Rotationally Restricted Diatomic Vibrator

The basic effects of nondegenerate vibrational motions on the rotational constants, molecular g-values, and susceptibility anisotropies may be demonstrated by the model of a diatomic molecule with its center of mass fixed in space and with the nuclear motion restricted to a plane perpendicular to the exterior magnetic field.

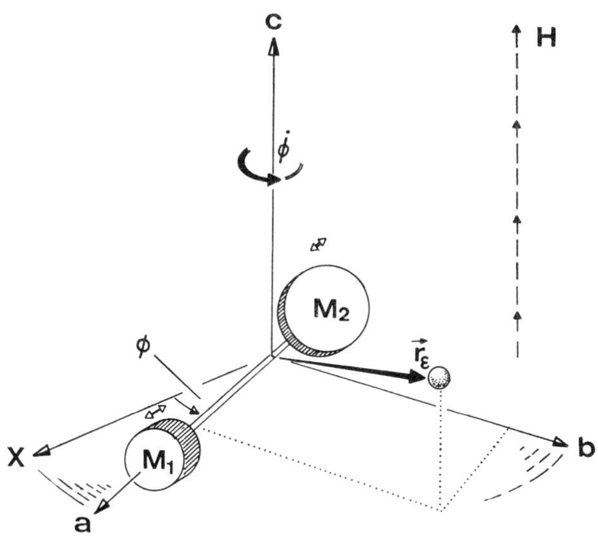

Fig. AIII.1. Coordinates used for the model of a rotationally restricted diatomic vibrator. For simplicity the center of mass of the nuclei is assumed to be fixed in space and the motion of the nuclei is assumed to be restricted to a plane perpendicular to the c-axis. No restrictions are imposed on the motions of the electrons. The positions of M_1 and M_2 should be exchanged in order to agree with the text.

Choosing the molecular coordinate system as shown in Fig. AIII.1 the remaining two coordinates needed to describe the nuclear motion are the internuclear distance, R, and the polar angle, ϕ. The time derivatives of the basis vectors of the molecular coordinate system which is assumed to have its c-axis parallel to the exterior field and its a-axis pointing in the direction of the internuclear axis, are given by:

$$\frac{d\boldsymbol{e}_a}{dt} = \dot{\phi}\boldsymbol{e}_c \times \boldsymbol{e}_a = \dot{\phi}\boldsymbol{e}_b\,, \qquad \frac{d\boldsymbol{e}_b}{dt} = \dot{\phi}\boldsymbol{e}_c \times \boldsymbol{e}_b = -\,\dot{\phi}\boldsymbol{e}_a\,, \qquad \frac{d\boldsymbol{e}_c}{dt} = 0\,. \tag{1}$$

Then, by referring the position vectors and velocities of the electrons and nuclei to the rotating molecular coordinate system, $\boldsymbol{e}_a(t)$, $\boldsymbol{e}_b(t)$, $\boldsymbol{e}_c$, the following expressions are obtained:

electrons:

$$\boldsymbol{r}_\varepsilon = a_\varepsilon \boldsymbol{e}_a + b_\varepsilon \boldsymbol{e}_b + c_\varepsilon \boldsymbol{e}_c$$

$$\boldsymbol{v}_\varepsilon = \frac{d\boldsymbol{r}_\varepsilon}{dt} = (\dot{a}_\varepsilon - b_\varepsilon\dot{\phi})\boldsymbol{e}_a + (\dot{b}_\varepsilon + a_\varepsilon\dot{\phi})\boldsymbol{e}_b + \dot{c}_\varepsilon \boldsymbol{e}_c$$

nuclei:

$$\boldsymbol{R}_1 = -\frac{M_2}{M_1+M_2} R\boldsymbol{e}_a; \qquad \boldsymbol{V}_1 = \frac{d\boldsymbol{R}_1}{dt} = -\frac{M_2}{M_1+M_2}\dot{R}\boldsymbol{e}_a - \frac{M_2}{M_1+M_2} R\dot{\phi}\boldsymbol{e}_b$$

$$\boldsymbol{R}_2 = \frac{M_1}{M_1+M_2} R\boldsymbol{e}_a; \qquad \boldsymbol{V}_2 = \frac{d\boldsymbol{R}_2}{dt} = \frac{M_1}{M_1+M_2}\dot{R}\boldsymbol{e}_a + \frac{M_1}{M_1+M_2} R\dot{\phi}\boldsymbol{e}_b\,.$$

Introducing these expressions into the general Lagrangian and neglecting all spin contributions gives

$$\mathscr{L} = \frac{M_1}{2}\boldsymbol{V}_1^2 + \frac{M_2}{2}\boldsymbol{V}_2^2 + \sum_{\varepsilon}^{\text{electrons}} \frac{m}{2}\boldsymbol{v}_\varepsilon^2 + \sum_{\nu=1}^{2} \frac{Z_\nu|e|}{2c}\boldsymbol{V}_\nu \cdot (\boldsymbol{H} \times \boldsymbol{R}_\nu) \tag{2}$$
$$- \frac{|e|}{2c}\sum_{\varepsilon}^{\text{electrons}} \boldsymbol{v}_\varepsilon \cdot (\boldsymbol{H} \times \boldsymbol{r}_\varepsilon) - V_{\text{coulomb}}$$

Following the standard procedure of the Lagrange-Hamilton formalism, the classical Hamiltonian function is obtained:

$$\mathscr{H} = \frac{1}{2m}\sum_{\varepsilon}^{\text{electrons}} (p_{a_\varepsilon}^2 + p_{b_\varepsilon}^2 + p_{c_\varepsilon}^2) + V_{\text{Coulomb}} \tag{a}$$

$$+ \frac{P_R^2}{2M} \tag{b}$$

$$+ \frac{P_\phi^2}{2MR^2} \tag{c}$$

$$- \frac{P_\phi L}{MR^2} \tag{d}$$

$$
\begin{aligned}
&+\frac{L^2}{2MR^2} && \text{(e)} \quad (3)\\
&-\frac{|e|Z}{2cM} H_c P_\phi && \text{(f)}\\
&+\frac{|e|Z}{2cM} H_c L && \text{(g)}\\
&+\frac{|e|}{2cm} H_c L && \text{(h)}\\
&+\frac{e^2 Z^2}{8Mc^2} H_c^2 R^2 && \text{(i)}\\
&+\frac{e^2}{8mc^2} H_c^2 \sum_\varepsilon (a_\varepsilon^2 + b_\varepsilon^2) && \text{(j)}
\end{aligned}
$$

In Eq. (3) the symbols have the following meanings:
conjugated momenta:

$$P_R = \frac{\partial \mathscr{L}}{\partial \dot{R}}, \quad P_\phi = \frac{\partial \mathscr{L}}{\partial \dot{\phi}}, \quad \ldots, \quad p_{a_\varepsilon} = \frac{\partial \mathscr{L}}{\partial \dot{a}_\varepsilon}, \quad p_{b_\varepsilon} = \frac{\partial \mathscr{L}}{\partial \dot{b}_\varepsilon}, \quad p_{c_\varepsilon} = \frac{\partial \mathscr{L}}{\partial \dot{c}_\varepsilon}, \quad \ldots$$

electronic angular momentum about the axis of the magnetic field:

$$L = \sum_\varepsilon (a_\varepsilon p_{b_\varepsilon} - b_\varepsilon p_{a_\varepsilon})$$

reduced nuclear mass:

$$M = \frac{M_1 M_2}{M_1 + M_2}$$

reduced nuclear charge:

$$Z = Z_1 \left(\frac{M_2}{M_1 + M_2}\right)^2 + Z_2 \left(\frac{M_1}{M_1 + M_2}\right)^2.$$

Except for the neglect of the center of mass motions and the additional term accounting for the vibrational kinetic energy, Eq. (3.b), this Hamiltonian closely corresponds to the Hamiltonian given in Eq. (IV.48). This Hamiltonian is written in a form appropriate for direct translation into quantum mechanics by replacing the conjugate momenta by the corresponding differential operators:

$$p_R \to \frac{\hbar}{\mathrm{i}} \frac{\partial}{\partial R} \; P_\phi \to \frac{\hbar}{\mathrm{i}} \frac{\partial}{\partial \phi}, \; \ldots \; p_{a_\varepsilon} \to \frac{\hbar}{\mathrm{i}} \frac{\partial}{\partial a_\varepsilon}, \; p_{b_\varepsilon} \to \frac{\hbar}{\mathrm{i}} \frac{\partial}{\partial b_\varepsilon}, \; p_{c_\varepsilon} \to \frac{\hbar}{\mathrm{i}} \frac{\partial}{\partial c_\varepsilon}, \; \ldots$$

The resulting Schrödinger equation may be solved by a perturbation treatment. As zeroth order basis functions we use products of rotational $\left(\frac{1}{\sqrt{2\pi}} \mathrm{e}^{\mathrm{i}J\phi}\right)$, vibrational $[\omega_{vn}(R)]$, and electronic $[\eta_n(R, a_1, b_1, c_1, a_2, b_2, c_2, \ldots)]$ wave functions. We see immediately that the Hamiltonian is diagonal with respect to the rotational quantum number J (J = integer). This is a consequence of the simplifying restriction to only one rotational degree of freedom.

We define the electronic wavefunctions as the eigenfunctions of the fixed-center Schrödinger equation, where R is a constant:

$$-\frac{\hbar^2}{2m}\sum_{\varepsilon}\Delta_{\varepsilon}\eta_n(R,\ldots,a_{\varepsilon},b_{\varepsilon},c_{\varepsilon},\ldots)+V_{\text{coulomb}}\,\eta_n(R,\ldots,a_{\varepsilon},b_{\varepsilon},c_{\varepsilon},\ldots) = E_n(R)\,\eta_n(R,\ldots,a_{\varepsilon},b_{\varepsilon},c_{\varepsilon},\ldots)\,. \tag{4}$$

$E_n(R)$, the eigenvalues of the fixed center Hamiltonian of the electrons depend parametrically on the internuclear distance. They define the potential surface for the vibrational Schrödinger equation:

$$-\frac{\hbar^2}{2M}\frac{\partial^2\omega_{vn}(R)}{\partial R^2}+E_n(R)\,\omega_{vn}(R)=E_{vn}\,\omega_{vn}(R)\,. \tag{5}$$

With these definitions and under the neglect of the first and second derivatives of the electronic wavefunctions, $\eta_n(R,\ldots,a_{\varepsilon},b_{\varepsilon},c_{\varepsilon},\ldots)$, with respect to the internuclear distance R, the product wavefunctions $\omega_{vn}(R)\,\eta_n(R,\ldots,a_{\varepsilon},b_{\varepsilon},c_{\varepsilon},\ldots)$ are the eigenfunctions of the vibronic part of the Hamiltonian [Eqs. (3.a) and (3.b)]. Taking the rest of the Hamiltonian as perturbation operator and applying standard second order perturbation theory then leads to the electronic ground state energy expression given by

$$\begin{aligned}
E'_{v0J} &= E_{v0}\\
&+\frac{\hbar^2J^2}{2M}\left\langle \omega_{v0}\eta_0\left|\frac{1}{R^2}\right|\omega_{v0}\eta_0\right\rangle\\
&+\frac{\hbar^2J^2}{M^2}\sum_{v'n'}\frac{\left|\left\langle\omega_{v0}\eta_0\left|\frac{L}{R^2}\right|\omega_{v'n'}\eta_{n'}\right\rangle\right|^2}{E_{v0}-E_{v'n'}}-\frac{|e|\hbar}{2Mc}\,Z\,H_cJ\\
&-\frac{|e|\hbar J}{2cmM}\,H_c\sum_{v'n'}\left\{\frac{\left\langle\omega_{v0}\eta_0\left|\frac{L}{R^2}\right|\omega_{v'n'}\eta_{n'}\right\rangle\left\langle\omega_{v'n'}\eta_{n'}\,|\underline{L}|\,\omega_{v0}\eta_0\right\rangle}{E_{v0}-E_{v'n'}}+\text{c.c.}\right\}\\
&+\frac{e^2}{8mc^2}\,H_c^2\,\langle\omega_{v0}\eta_0\,|\textstyle\sum_{\varepsilon}(\underline{a_{\varepsilon}^2+b_{\varepsilon}^2})|\,\omega_{v0}\eta_0\rangle\\
&+\frac{e^2}{4m^2c^2}\,H_c^2\sum\frac{|\langle\omega_{v0}\eta_0\,|\underline{L}|\,\omega_{v'n'}\eta_{n'}\rangle|^2}{E_{v0}-E_{v'n'}}\\
&+\ldots
\end{aligned} \tag{6}$$

In Eq. (6) only those perturbation contributions are given explicitly which are most important for the present discussion of vibrational effects on rotational constants, g-values, and magnetic susceptibilities.

Comparing with the standard rotational Hamiltonian Eq. (I.7) which for the restricted rotor reduces to

$$E_{\text{rot},J}=hCJ^2-\mu_{\text{N}}g_{\perp}H_cJ-\frac{1}{2}\chi_{\perp}H_c^2\,, \tag{7}$$

gives the following expressions:

Rotational constant:

$$C = \frac{h}{8\pi^2 M}\left[\left\langle \omega_{v0}\eta_0 \left| \frac{1}{R^2} \right| \omega_{v0}\eta_0 \right\rangle + \frac{2}{M}\sum_{v'n'} \frac{\left|\left\langle \omega_{v0}\eta_0 \left| \frac{L}{R^2} \right| \omega_{v'n'}\eta_{n'} \right\rangle\right|^2}{E_{v0} - E_{v'n'}}\right]. \tag{8}$$

Rotational g-factor:

$$g_\perp = \frac{M_p}{M}\left(Z + \frac{1}{m}\sum_{v'n'}\left\{\frac{\left\langle \omega_{v0}\eta_0 \left| \frac{L}{R^2} \right| \omega_{v'n'}\eta_{n'} \right\rangle \left\langle \omega_{v'n'}\eta_{n'} | \underline{L} | \omega_{v0}\eta_0 \right\rangle}{E_{v0} - E_{v'n'}} + \text{c.c.}\right\}\right) \tag{9}$$

(c.c. = complex conjugate)

Magnetic susceptibility:

$$\chi_\perp = -\frac{e^2}{4mc^2}\left(\langle \omega_{v0}\eta_0 | \sum_\varepsilon (a_\varepsilon^2 + b_\varepsilon^2) | \omega_{v0}\eta_0 \rangle - \frac{2}{m}\sum_{v'n'} \frac{|\langle \omega_{v0}\eta_0 | \underline{L} | \omega_{v'n'}\eta_{n'} \rangle|^2}{E_{v0} - E_{v'n'}}\right). \tag{10}$$

Although Franck-Condon type matrix elements and summations over the different excited vibrational states enter into the above expressions, it may be shown that they are closely related to the vibrational averages of the corresponding rigid rotor expressions. As an example, we will demonstrate this for the perturbation sum entering into the expression for the rotational constant. We further specify $v = 0$ to treat the vibrational ground state. Then as a first approximation, we replace the actual energy differences $E_{v0} - E_{v'n'}$ by the corresponding vertical excitation energy $E_0(Re) - E_{n'}(Re)$ (compare Fig. AIII.2). Carrying out first the integrations over the electronic coordinates, the following sequence of equations is obtained:

$$\begin{aligned}
&\sum_{v'n'} \frac{\left\langle \omega_{00}\eta_0 \left| \frac{L}{R^2} \right| \omega_{v'n'}\eta_{n'} \right\rangle \left\langle \omega_{v'n'}\eta_{n'} \left| \frac{L}{R^2} \right| \omega_{00}\eta_0 \right\rangle}{E_{00} - E_{v'n'}} \\
&\approx \sum_{n'} \frac{1}{E_0(Re) - E_{n'}(Re)} \sum_{v'} \left\langle \omega_{00}\eta_0 \left| \frac{L}{R^2} \right| \omega_{v'n'}\eta_{n'} \right\rangle \left\langle \omega_{v'n'}\eta_{n'} \left| \frac{L}{R^2} \right| \omega_{00}\eta_0 \right\rangle \\
&= \sum_{n'} \frac{1}{E_0(Re) - E_{n'}(Re)} \sum_{v'} \left\langle \omega_{00} \left| \left\langle \eta_0 \left| \frac{L}{R^2} \right| \eta_{n'} \right\rangle \right| \omega_{v'n'} \right\rangle \qquad (11) \\
&\qquad \cdot \left\langle \omega_{v'n'} \left| \left\langle \eta_{n'} \left| \frac{L}{R^2} \right| \eta_0 \right\rangle \right| \omega_{00} \right\rangle
\end{aligned}$$

Using the fact that the vibrational functions of each electronic energy surface form a complete orthonormal set (the problem of the continuum may be by-

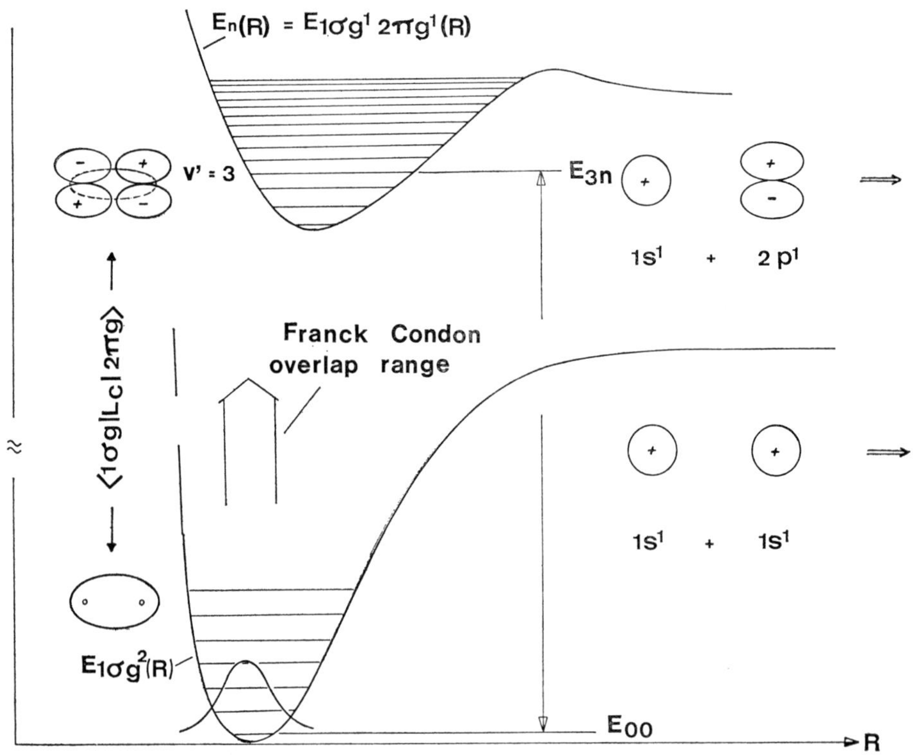

Fig. AIII.2. Potential energy curves, electronic orbitals, and vibrational levels are schematically depicted for the electronic ground state and an excited electronic state of a homonuclear diatomic molecule such as H_2. The molecular c-axis is assumed to be perpendicular to the plane. The electronic angular momentum operator, $L_c = \frac{\hbar}{i} \sum_{\varepsilon} \frac{\partial}{\partial \phi_{\varepsilon c}}$, leads to nonzero matrix elements between the bonding σ ground state orbital and the antibonding π-state which contributes to the paramagnetic susceptibility as well as to the electronic contribution to the molecular g-factor

passed by introducing a sufficiently large periodicity box) $|\omega_{v'n'}\rangle$ and $\langle\omega_{v'n'}|$ may be expanded in terms of the electronic ground state vibrational functions:

$$|\omega_{v'n'}\rangle = \sum_{v''} \langle\omega_{v'n'}\,|\omega_{v''0}\rangle|\,\omega_{v''0}\rangle$$
$$\langle\omega_{v'n'}| = \sum_{v'''} \langle\omega_{v'''0}\,|\,\omega_{v'n'}\rangle\,\langle\omega_{v'''0}|\,. \tag{12}$$

Together with the orthonormality relation

$$\sum_{v'} \langle\omega_{v'''0}\,|\,\omega_{v'n'}\rangle\,\langle\omega_{v'n'}\,|\,\omega_{v''0}\rangle = \langle\omega_{v'''0}\,|\,\omega_{v''0}\rangle = \delta_{v'''v''} \tag{13}$$

this yields:

$$\sum_{n'} \frac{1}{E_0(Re) - E_{n'}(Re)} \sum_{v''} \left\langle \omega_{00} \middle| \left\langle \eta_0 \middle| \frac{L}{R^2} \middle| \eta_{n'} \right\rangle \middle| \omega_{v''0} \right\rangle \left\langle \omega_{v''0} \middle| \left\langle \eta_{n'} \middle| \frac{L}{R^2} \middle| \eta_0 \right\rangle \middle| \omega_{00} \right\rangle$$
$$= \sum_{n'} \frac{1}{E_0(Re) - E_{n'}(Re)} \left\langle \omega_{00} \middle| \left\langle \eta_0 \middle| \frac{L}{R^2} \middle| \eta_{n'} \right\rangle \left\langle \eta_{n'} \middle| \frac{L}{R^2} \middle| \eta_0 \right\rangle \middle| \omega_{00} \right\rangle . \tag{14}$$

Exactly the same expression is obtained if the vertical energy approximation is used in the vibrational ground state average of the corresponding rigid rotor expression [compare Eq. (3) with $P_R = 0$]:

$$\left\langle \omega_{00} \middle| \sum_{n'} \frac{\left\langle \eta_0 \middle| \frac{L}{R^2} \middle| \eta_{n'} \right\rangle \left\langle \eta_{n'} \middle| \frac{L}{R^2} \middle| \eta_0 \right\rangle}{E_0(R) - E_{n'}(R)} \middle| \omega_{00} \right\rangle$$
$$\approx \sum_{n'} \frac{1}{E_0(Re) - E_{n'}(Re)} \left\langle \omega_{00} \middle| \left\langle \eta_0 \middle| \frac{L}{R^2} \middle| \eta_{n'} \right\rangle \left\langle \eta_{n'} \middle| \frac{L}{R^2} \middle| \eta_0 \right\rangle \middle| \omega_{00} \right\rangle . \tag{15}$$

We therefore conclude that replacing the rigid rotor expressions Eqs. (I.2), (I.4) and (I.8) by the corresponding ground state average values should be a good approximation to the theoretical expressions for the observed rotational constants, molecular g-values, and magnetic susceptibilities.

AIV. The Gauge Transformation of the Classical Lagrangian

In order to remove the explicit dependence on the position of the molecular coordinate system, $(\boldsymbol{R}_0 - \boldsymbol{r}_0)$, the Lagrangian given in Eq. (IV.13) may be modified by subtracting the total differential of an appropriate scalar function with respect to time:

$$\frac{dF}{dt} = \underbrace{\frac{\partial F}{\partial x_0}\dot{x}_0 + \frac{\partial F}{\partial y_0}\dot{y}_0 + \frac{\partial F}{\partial z_0}\dot{z}_0}_{a} + \underbrace{\sum_{\varepsilon}^{\text{electrons}} \left(\frac{\partial F}{\partial a_\varepsilon}\dot{a}_\varepsilon + \frac{\partial F}{\partial b_\varepsilon}\dot{b}_\varepsilon + \frac{\partial F}{\partial c_\varepsilon}\dot{c}_\varepsilon \right)}_{b} + \underbrace{\frac{\partial F}{\partial \phi_a}\dot{\phi}_a + \frac{\partial F}{\partial \phi_b}\dot{\phi}_b + \frac{\partial F}{\partial \phi_c}\dot{\phi}_c}_{c} \tag{1}$$

Since the angular velocities about the molecular coordinate axes, ω_a, ω_b, and ω_c have been used in the Lagrangian, Eq. (IV.13), we also use ϕ_a, ϕ_b, and ϕ_c in Eq. (1) in order to express the orientational dependence of F. This is permissible as long as only infinitesimal rotations are considered.

In Eq. (1) the partial derivatives of F with respect to ϕ_a, ϕ_b, and ϕ_c may be most conveniently expressed as:

$$\frac{\partial F}{\partial \phi_a} = \sum_i (\boldsymbol{e}_a \times \boldsymbol{r}_i) \cdot \boldsymbol{\nabla}_i F; \quad \frac{\partial F}{\partial \phi_b} = \sum_i (\boldsymbol{e}_b \times \boldsymbol{r}_i) \cdot \boldsymbol{\nabla}_i F; \quad \frac{\partial F}{\partial \phi_c} = \sum_i (\boldsymbol{e}_c \times \boldsymbol{r}_i) \cdot \boldsymbol{\nabla}_i F. \tag{2}$$

Eq. (2) follows from the fact that under an infinitesimal rotation $\delta\phi_a$, $\delta\phi_b$, and $\delta\phi_c$ about the *a*-, *b*-, and *c*-molecular axes, respectively, the change of the position vector of the *i*-th particle, $\delta\boldsymbol{r}_i$, is given by:

$$\delta\boldsymbol{r}_i = \delta\phi_a(\boldsymbol{e}_a \times \boldsymbol{r}_i) + \delta\phi_b(\boldsymbol{e}_b \times \boldsymbol{r}_i) + \delta\phi_c(\boldsymbol{e}_c \times \boldsymbol{r}_i)\,, \tag{3}$$

and from the comparison of

$$\delta F = \frac{\partial F}{\partial \phi_a}\,\delta\phi_a + \frac{\partial F}{\partial \phi_b}\,\delta\phi_b + \frac{\partial F}{\partial \phi_c}\,\delta\phi_c\,, \tag{4}$$

and

$$\delta F = \sum_i \boldsymbol{\nabla}_i F \cdot \delta\boldsymbol{r}_i\,, \tag{5}$$

with $\delta\boldsymbol{r}_i$ given by Eq. (3).

The appropriate gauge function, F, is most easily guessed, if one concentrates for a moment on the dominant contribution to be compensated in the Lagrangian, namely

$$-\frac{|e|}{2c} \sum_{\varepsilon}^{\text{electrons}} \boldsymbol{v}_\varepsilon \cdot (\boldsymbol{H} \times \boldsymbol{R}_0)\,,$$

comparing with Eq. (1b) immediately leads to the conclusion that F should include $-\frac{1}{2c}|e| \sum_\varepsilon \boldsymbol{r}_\varepsilon \cdot (\boldsymbol{H} \times \boldsymbol{R}_0)$. Generalizing this result, F is obtained as:

$$F = \frac{1}{2c}\,\boldsymbol{\mu}_{\text{el}} \cdot [\boldsymbol{H} \times (\boldsymbol{R}_0 - \boldsymbol{r}_0)] = \frac{1}{2c}\,(\boldsymbol{R}_0 - \boldsymbol{r}_0) \cdot (\boldsymbol{\mu}_{\text{el}} \times \boldsymbol{H}) \tag{6}$$

with

$$\boldsymbol{\mu}_{\text{el}} = |e|\,\Big(\sum_\nu^{\text{nuclei}} Z_\nu \boldsymbol{r}_\nu - \sum_\varepsilon^{\text{electrons}} \boldsymbol{r}_\varepsilon\Big)\,,$$

the molecular electric dipole moment. Keeping in mind that the electronic coordinates a_ε, b_ε, c_ε also enter into $\boldsymbol{r}_0 = \frac{M}{m} \sum_\varepsilon \boldsymbol{r}_\varepsilon$ and using Eq. (2), we obtain:

$$\frac{dF}{dt} = -\frac{1}{2c}\,\boldsymbol{\mu}_{\text{el}} \cdot (\boldsymbol{V}_0 \times \boldsymbol{H}) \qquad \text{from (1.a)}$$

$$+\frac{1}{2c}\,\boldsymbol{\mu}_{\mathrm{el}}\cdot(\boldsymbol{v}_0\times\boldsymbol{H})-\frac{|e|}{2c}\,[\boldsymbol{v}_\varepsilon\cdot(\boldsymbol{H}\times(\boldsymbol{R}_0-\boldsymbol{r}_0))] \qquad \text{from (1.b)}$$

$$+\frac{1}{2c}\,(\boldsymbol{\omega}\times\boldsymbol{\mu}_{\mathrm{el}})\cdot[\boldsymbol{H}\times(\boldsymbol{R}_0-\boldsymbol{r}_0)]-\frac{1}{2c}\,(\boldsymbol{\omega}\times\boldsymbol{r}_0)\cdot(\boldsymbol{\mu}_{\mathrm{el}}\times\boldsymbol{H})\,. \qquad \text{from (1.c)} \tag{7}$$

Subtracting this result from the Lagrangian in Eq. (IV.13) shows that the $(\boldsymbol{R}_0-\boldsymbol{r}_0)$-dependence is indeed removed and the field independent contributions which also depend on the velocity of the origin of the molecular coordinate system, $\boldsymbol{V}_0-[\boldsymbol{v}_0+(\boldsymbol{\omega}\times\boldsymbol{r}_0)]$, add up to the potential energy of the molecular electric dipole moment within a virtual electric Stark-field
$\boldsymbol{E}_{\mathrm{TS}}=\frac{1}{2c}\,[(\boldsymbol{V}_0-(\boldsymbol{v}_0+(\boldsymbol{\omega}\times\boldsymbol{r}_0)))\times\boldsymbol{H}]$.

V. References

[1] Flygare, W. H., Benson, R. C.: Mol. Phys. *20*, 225 (1971).
[2] Landolt-Börnstein: New Series, Group II: Atomic and molecular physics, Vol. 6.
[3] Wick, G. C.: Z. Physik *85*, 25 (1933). — Eshbach, J. R., Strandberg, M. W. P.: Phys. Rev. *85*, 24 (1952).
[4] Van Vleck, J. H.: The theory of electric magnetic susceptibilities. London: Oxford University Press 1932.
[5] Townes, C. H., Schawlow, A. L.: Microwave spectroscopy. New York: McGraw-Hill, Inc. 1956.
[6] Hüttner, W., Lo, M.-K., Flygare, W. H.: J. Chem. Phys. *48*, 1206 (1968).
[7] McGurk, J., Tigelaar, H. L., Rock, S. L., Norris, C. L., Flygare, W. H.: J. Chem. Phys. *58*, 1420 (1973).
[8] Buckingham, A. D., Utting, B. D.: Ann. Rev. Phys. Chem. *21*, 287 (1970).
[9] Gierke, T. D., Tigelaar, H. L., Flygare, W. H.: J. Am. Chem. Soc. *94*, 330 (1972).
[10] Hamer, E., Engelbrecht, L., Sutter, D. H.: Z. Naturforsch. *29a*, 924 (1974).
[11] Townes, C. H., Dousmanis, G. C., White, A. D., Schwarz, R. F.: Discussions Faraday Soc. *19*, 56 (1955).
[12] Benson, R. C.: Ph. D. thesis, University of Illinois, Urbana (1973).
[13] A discussion of molecular magnetic moments due to vibrational angular momentum in molecules with degenerate vibrational states has been given by Moss, R. E., Perry, A. J.: Mol. Phys. *25*, 1121 (1973).
[14] The torsional magnetic moment due to methyl top internal rotations has been treated extensively by Engelbrecht, L.: Ph. D. thesis, Kiel University (1975).
[15] Ramsey, N. F.: Phys. Rev. *87*, 1075 (1952).
[16] While the dependence of the rotational constants on the vibrational states has been investigated for many molecules (see Ref. [2]), the vibrational dependence of the g-values and susceptibility anisotropies has been studied only for few diatomic molecules (compare Honerjäger, R., Tischer, R.: Z. Naturforsch. *28a*, 458 (1973); *28a*, 1374 (1973)).
[17] For a provocative discussion on the concept of aromaticity compare Labarre, J., Crasnier, F.: Fortschr. Chem. Forsch. *24*, 33 (1971). See also Jones, A. J.: Rev. Pure Appl. Chem. *18*, 253 (1968).
[18] For experimental methods to determine the bulk susceptibility compare, for instance: Sellwood, P. W.: Magnetochemistry. New York: Interscience Publ. 1956. — Weiss, A., Witte, H.: Magnetochemie, Grundlagen und Anwendungen. Weinheim: Verlag Chemie 1973.
[19] Schmalz, T. G., Norris, C. L., Flygare, W. H.: J. Am. Chem. Soc. *94*, 7961 (1973).
[20] Pascal, P.: Ann. Chim. Phys. *19*, 5 (1910).
[21] Pacault, A., Hooran, J., Marchand, A.: Advan. Chem. Phys. *VIII*, 171 (1961).
[22] Wiese, A.: Diplomarbeit, Universität Kiel (1975).
[23] $\chi_{aa} - \chi = +0.15 \times 10^{-6}$ erg/(G mole), $\chi_{bb} - \chi = -0.07 \times 10^{-6}$ erg/(G^2 mole) and $\chi_{cc} - \chi = -0.09 \times 10^{-6}$ erg/(G^2 mole) from Verhoeven, J., Dymanus, A.: J. Chem. Phys. *52*, 3222 (1970) were used together with a bulk value of $\chi = -13 \cdot 10^{-6}$ erg/(G^2 mole).
[24] Bak, B., Christensen, D., Dixon, W. B., Hansen-Nygaard, L., Rastrup-Andersen, J., Schonlander, M.: J. Mol. Spectry. *9*, 124 (1962).
[25] Sutter, D. H., Flygare, W. H.: J. Am. Chem. Soc. *91*, 4063 (1969). — Bak, B., Hamer, E., Sutter, D. H., Dreizler, H.: Z. Naturforsch. *27a*, 705 (1972).
[26] See also Schmalz, T. G., Gierke, T. D., Beak, P., Flygare, W. H.: Tetrahedron Letters *1974*, 2885.
[27] Sutter, D. H., Flygare, W. H.: J. Am. Chem. Soc. *91*, 4063 (1969). — Norris, C. L., Benson, R. C., Beak, P., Flygare, W. H.: J. Am. Chem. Soc. *93*, 5591 (1971). — Benson, R. C., Flygare, W. H.: J. Chem. Phys. *58*, 2366 (1973).
[28] Hückel, E.: Z. Physik *70*, 204 (1931); *72*, 310 (1931); *76*, 628 (1932). — Elvidge, J. A., Jackman, L. M.: J. Chem. Soc. *1961*, 859. — Abraham, R. J., Sheppard, R. C., Thomas, W. A., Turner, S.: Chem. Commun. *1965*, 43. — Elvidge, J. A.: Chem. Commun. *1965*, 160. — Davis, D. W.: Chem. Commun. *1965*, 258. — Black, P. J., Brown, R. D., Heffer-man, M. L.: Australian J. Chem. *20*, 1305 (1967).

[29] Salem, L.: The molecular orbital theory of conjugated systems. New York: W. A. Benjamin, Inc. 1966. — Murrell, J. N., Harget, A. J.: Semiempirical SCF-MO theory of molecules. New York: Wiley-Interscience, 1972.
[30] Benassi, R., Lazzeretti, P., Taddei, F.: J. Phys. Chem. *79*, 848 (1975).
[31] Norris, C. L., Tigelaar, H. L., Flygare, W. H.: Chem. Phys. *1*, 1 (1973). — Czieslik, W., Sutter, D. H.: Z. Naturforsch. *29a*, 1820 (1974).
[32] Benson, R. C., Flygare, W. H., Oda, M., Breslow, R.: J. Am. Chem. Soc. *95*, 2772 (1973).
[33] Norris, C. L., Benson, R. C., Beak, P., Flygare, W. H.: J. Am. Chem. Soc. *93*, 5592 (1971).
[34] Sutter, D. H.: Z. Naturforsch. *29a*, 786 (1974).
[35] Bogaard, M. P., Buckingham, A. D., Corfield, M. G., Dunmur, D. A., White, A. H.: Chem. Phys. Letters *12*, 558 (1972).
[36] The difference between the Cotton-Mouton curves and the curves connecting the rotational Zeeman results may be partly due to the neglect of a possible field dependence of the electric polarizabilities in the analysis of the Cotton-Mouton data.
[37] Jijima, T.: Bull. Chem. Soc. Japan *45*, 3526 (1972).
[38] Yen, C. K.: Phys. Rev. *74*, 1396 (1948); *76*, 1494 (1949). — Yen, C. K., Barghausen, J. W. B. Stanley, R. W.: Phys. Rev. *85*, 717 (1952).
[39] Eshbach, J. R., Strandberg, M. W. P.: Phys. Rev. *85*, 24 (1952).
[40] Flygare, W. H.: J. Chem. Phys. *42*, 1563 (1965).
[41] Burrus, C. A.: J. Chem. Phys. *30*, 976 (1959).
[42] Hughes, R. H., Wilson, E. B., Jr.: Phys. Rev. *72*, 1265 (1946).
[43] Flygare, W. H., Hüttner, W., Shoemaker, R. L., Foster, P. D.: J. Chem. Phys. *50*, 1714 (1969).
[44] Honerjäger, R., Tischer, R.: Z. Naturforsch. *28a*, 458 (1973).
[45] Bhattachayya, P. K., Taft, H., Smith, N., Daily, B. P.: Rev. Sci. Instr. *46*, 608 (1975)
[46] Rudolph, H. D.: Z. Angew. Phys. *13*, 401 (1961).
[47] Andresen, U., Dreizler, H.: Z. Angew. Phys. *30*, 207 (1970).
[48] Sutter, D. H.: Z. Naturforsch. *26a*, 1644 (1971).
[49] Hamer, E.: Ph. D. Thesis, Christian Albrechts Universität, Kiel (1973).
[50] Sperner, E.: Einführung in die Analytische Geometrie und Algebra, Vol. II, p. 140. Göttingen: Vandenhoek u. Ruprecht 1951.
[51] Mueller, D. J.: Num. Math. *8*, 72 (1966).
[52] Kivelson, D., Wilson, E. B., Jr.: J. Chem. Phys. *20*, 1575 (1952).
[53] Compare Landau, L., Lifschitz, E. M.: Lehrbuch der theoretischen Physik, Band III, Quantenmechanik, Chapt. II, Sect. 11, Exercize.
[54] See Ref. [5], p. 92.
[55] van Vleck, J. H.: Rev. Mod. Phys. *23*, 213 (1951). — Edmonds, A. R.: Angular Moment in Quantum Mechanics, Chap. V, and Tables 1 and 2 of Appendix II. (Princeton University Press).
[56] Hüttner, W., Flygare, W. H.: J. Chem. Phys. *47*, 4137 (1967).
[57] Engelbrecht, L.: Ph. D. Thesis, Christian Albrechts Universität, Kiel (1975).
[58] Norris, C. L., Pearson, E. F., Flygare, W. H.: J. Chem. Phys. *60*, 1758 (1974).
[59] Hüttner, W., Morgenstern, K., Z. Naturforsch. *25a*, 547 (1970).
[60] Moss, R. E., Perry, A. J.: Mol. Phys. *25*, 1121 (1973).
[61] Hüttner, W., Flygare, W. H.: Trans. Faraday Soc. *65*, 1953 (1969).
[62] Tannenbaum, E., Myers, R. J., Gwinn, W. D.: J. Chem. Phys. *25*, 42 (1956).
[63] Naylor, R. E., Wilson, E. B., Jr.: J. Chem. Phys. *26*, 1057 (1957).
[64] Sutter, D. H., Guarnieri, A.: Z. Naturforsch. *25a*, 1036 (1970).
[65] The Nuclear quadrupole definition used here is consistent with Eq. (II.1). Normally nuclear quadrupole moments are listed as $2Q/|e|$ where Q is given by Eq. (II.1).
[66] Rose, M. E.: Elementary theory of angular momentum. London: J. Wiley and Sons, Inc. 1957.
[67] Kemp, M. K., Flygare, W. H.: J. Am. Chem. Soc. *90*, 6267 (1968).
[68] VanderHart, D., Flygare, W. H.: Mol. Phys. *18*, 77 (1970).
[69] Ewing, J. J., Tigelaar, H. C., Flygare, W. H.: J. Chem. Phys. *56*, 1957 (1972).
[70] Suzuki, M., Guarnieri, A.: Z. Naturforsch. *30a*, 497 (1975).
[71] Weizel, W.: Lehrbuch der Theoretischen Physik, Band 1.

[72] Wilson, Jr., E. B., Decius, J. C., Cross, P. C.: Molecular Vibrations, p. 281. New York-Toronto-London: McGraw-Hill 1955.
[73] Sutter, D. H.: Guarnieri, A., Dreizler, H.: Z. Naturforsch. *25a*, 222 (1970); *25a*, 2005 (1970).
[74] Howard, B. J., Moss, R. E.: Z. Naturforsch. *25a*, 2004 (1970); Mol. Phys. *19*, 433 (1970).
[75] Kemble, E. C.: The fundamental principles of quantum mechanics, Chap. VII, Sect. 35. New York: Dover Publ. 1958.
[76] Slater, J. C.: Quantum theory of atomic structure, Vol. II, Appendix 31. New York: McGraw-Hill Book Co. 1960.
[77] Weast, R. C.: Handbook of chemistry and physics. The Chemical Rubber Co., 53 Ed. (1972—1973).
[78] Fuller, G. H. and Cohen, V. W.: Appendix to Nuclear Data Sheets, (May 1965).
[79] See Ref. [75] Chap. XI Sec. 48c. Compare however: Jensen, H. and Koppe, H.: Quantum mechanics with constraints, in: Anals of Physics *63*, 586 (1970).

Received August 25, 1975

Author Index Volumes 26—66

The volume numbers are printed in italics

Topics in Current Chemistry
Fortschritte der chemischen Forschung

Vol. 17
W. Demtröder, Laser Spectroscopy.
2nd., enlarged edition. 16 figures.
3 tables. III, 106 pages. 1973

Vol. 22
W. Kutzelnigg and G. Berthier, σ and π Electrons in Organic Compounds.
11 figures. IV, 122 pages. 1971

Vol. 23
Molecular Orbitals.
40 figures. 5 tables. III, 123 pages. 1971

Vol. 24
Electronic Structure of Organic Compounds. 12 figures. III, 54 pages

Vol. 28
π Complexes of Transition Metals.
11 figures. III, 181 pages. 1972

Vol. 30
Nuclear Quadrupole Resonance.
23 figures. III, 173 pages. 1972

Vol. 37
K. L. Kompa, Chemical Lasers.
31 figures. III, 92 pages. 1973

Vol. 39
Computers in Chemistry.
13 figures. III, 195 pages. 1973

Vol. 49
C. A. Mead: Symmetry and Chirality.
23 figures. IV, 88 pages. 1974

Vol. 53
Gas-Phase Electron Diffraction.
38 figures. IV, 119 pages. 1975

Vol. 56
Theoretical Inorganic Chemistry.
22 figures. 18 tables. IV, 159 pages.
1975

Vol. 58
New Theoretical Aspects.
60 figures. 7 tables. IV, 186 pages. 1975

Springer-Verlag Berlin Heidelberg New York